Universitext

To *Eberhard Zeidler,*
for building up our Institute with so
much enthusiasm and human skill

Preface

What is the title of this book intended to signify, what connotations is the adjective "Postmodern" meant to carry? A potential reader will surely pose this question. To answer it, I should describe what distinguishes the approach to analysis presented here from what has by its protagonists been called "Modern Analysis". "Modern Analysis" as represented in the works of the Bourbaki group or in the textbooks by Jean Dieudonné is characterized by its systematic and axiomatic treatment and by its drive towards a high level of abstraction. Given the tendency of many prior treatises on analysis to degenerate into a collection of rather unconnected tricks to solve special problems, this definitely represented a healthy achievement. In any case, for the development of a consistent and powerful mathematical theory, it seems to be necessary to concentrate solely on the internal problems and structures and to neglect the relations to other fields of scientific, even of mathematical study for a certain while. Almost complete isolation may be required to reach the level of intellectual elegance and perfection that only a good mathematical theory can acquire. However, once this level has been reached, it can be useful to open one's eyes again to the inspiration coming from concrete external problems. The axiomatic approach started by Hilbert and taken up and perfected by the Bourbaki group has led to some of the most important mathematical contributions of our century, most notably in the area of algebraic geometry. This development was definitely beneficial for many areas of mathematics, but for other fields this was not true to the same extent. In geometry, the powerful tool of visual imagination was somewhat neglected, and global nonlinear phenomena connected with curvature could not always be addressed adequately. In analysis, likewise, perhaps too much emphasis was laid on the linear theory, while the genuinely nonlinear problems were found to be too diverse to be subjected to a systematic and encompassing theory. This effect was particularly noticable in the field of partial differential equations. This branch of mathematics is one of those that have experienced the most active and mutually stimulating interaction with the sciences, and those equations that arise in scientific applications typically exhibit some genuinely nonlinear structure because of self-interactions and other effects.

   Thus, modern mathematics has been concerned with its own internal structure, and it has achieved great successes there, but perhaps it has lost a little of the stimulation that a closer interaction with the sciences can offer. This trend has been reversed somewhat in more recent years, and in particular rather close ties have been formed again between certain areas of mathematics and theoretical physics. Also, in mathematical research, the emphasis has perhaps shifted a bit from general theories back to more concrete problems that require more individual methods.

I therefore felt that it would be appropriate to present an introduction to advanced analysis that preserves the definite achievements of the theory that calls itself "modern", but at the same time transcends the latter's limitations.

For centuries, "modern" in the arts and the sciences has always meant "new", "different from the ancient", some times even "revolutionary", and so it was an epithet that was constantly shifting from one school to its successor, and it never stuck with any artistic style or paradigm of research. That only changed in our century, when abstract functionality was carried to its extreme in architecture and other arts. Consequently, in a certain sense, any new theory or direction could not advance any further in that direction, but had to take some steps back and take up some of the achievements of "premodern" theories. Thus, the denomination "modern" became attached to a particular style and the next generation had to call itself "postmodern". As argued above, the situation in mathematics is in certain regards comparable to that, and it thus seems logical to call "postmodern" an approach that tries to build upon the insights of the modern theory, but at the same time wishes to take back the latter's exaggerations.

Of course, the word "postmodern" does not at present have an altogether positive meaning as it carries some connotations of an arbitrary and unprincipled mixture of styles. Let me assure the potential reader that this is not intended by the title of the present book. I wish rather to give a coherent introduction to advanced analysis without abstractions for their own sake that builds a solid basis for the areas of partial differential equations, the calculus of variations, functional analysis and other fields of analysis, as well as for their applications to analytical problems in the sciences, in particular the ones involving nonlinear effects.

Of course, calculus is basic for all of analysis, but more to the point, there seem to be three key theories that mathematical analysis has developed in our century, namely the concept of Banach space, the Lebesgue integral, and the notion of abstract differentiable manifold. Of those three, the first two are treated in the present book, while the third one, although closest to the author's own research interests, has to wait for another book (this is not quite true, as I did treat that topic in more advanced books, in particular in *"Riemannian Geometry and Geometric Analysis"*, Springer, 1995).

The Lebesgue integration theory joins forces with the concept of Banach spaces when the L^p and Sobolev spaces are introduced, and these spaces are basic tools for the theory of partial differential equations and the calculus of variations. (In fact, this is the decisive advantage of the Lebesgue integral

over the older notion, the so-called Riemann integral, that it allows the construction of <u>complete</u> normed function spaces, i.e. Hilbert or Banach spaces,
namely the L^p and Sobolev spaces.) This is the topic that the book will lead
the reader to.

The organization of the book is guided by pedagogical principles. After
all, it originated in a course I taught to students in Bochum at the beginning
level of a specialized mathematics education. Thus, after carefully collect-
ing the prerequisites about the properties of the real numbers, we start with
continuous functions and calculus for functions of one variable. The introduction of Banach spaces is motivated by questions about the convergence
of sequences of continuous or differentiable functions. We then develop some
notions about metric spaces, and the concept of compactness receives particular attention. Also, after the discussion of the one-dimensional theory, we
hope that the reader is sufficiently prepared and motivated to be exposed
to the more general treatment of calculus in Banach spaces. After present-
ing some rather abstract results, the discussion becomes more concrete again
with calculus in Euclidean spaces. The implicit function theorem and the
Picard-Lindelöf theorem on the existence and uniqueness of solutions of ordinary differential equations (ODEs) are both derived from the Banach fixed
point theorem.

In the second part, we develop the theory of the Lebesgue integral in Euclidean spaces. As already mentioned, we then introduce L^p and Sobolev
spaces and give an introduction to elliptic partial differential equations
(PDEs) and the calculus of variations. Along the way, we shall see several
examples arising from physics.

In the Table of Contents, I have described the key notions and results of
each section, and so the interested reader can find more detailed information
about the contents of the book there.

This book presents an intermediate analysis course. Thus, its level is somewhat higher than the typical introductory courses in the German university
system. Nevertheless, in particular in the beginning, the choice and presentation of material are influenced by the requirement of such courses, and
I have utilized some corresponding German textbooks, namely the analysis
courses of 0. Forster (*Analysis I - III*, Vieweg 1976ff.) and H. Heuser (*Analysis I, II*, Teubner, 1980ff.). Although the style and contents of the present
book are dictated much more by pedagogical principles than is the case in
J. Dieudonné's treatise *Modern Analysis*, Academic Press, 1960ff., there is
some overlap of content. Although typically the perspective of my book is
different, the undeniable elegance of reasoning that can be found in the treatise of Dieudonné nevertheless induced me sometimes to adapt some of his
arguments, in line with my general principle of preserving the achievements
of the theory that called itself modern.

For the treatment of Sobolev spaces and the regularity of solutions of elliptic partial differential equations, I have used D. Gilbarg, N. Trudinger, *Elliptic*

Partial Differential Equations of Second Order, Springer, [2]1983, although of course the presentation here is more elementary than in that monograph.

I received the competent and dedicated help – checking the manuscript for this book, suggesting corrections and improvements, and proofreading – of Felicia Bernatzki, Christian Gawron, Lutz Habermann, Xiaowei Peng, Monika Reimpell, Wilderich Tuschmann, and Tilmann Wurzbacher. My original German text was translated into English by Hassan Azad. The typing and retyping of several versions of my manuscript was performed with patience and skill by Isolde Gottschlich. The figures were created by Micaela Krieger with the aid of Harald Wenk and Ralf Muno. I thank them all for their help without which this book would not have become possible.

Preface to the 2nd edition

For this edition, I have added some material on the qualitative behavior of solutions of ordinary differential equations, some further details on L^p and Sobolev functions, partitions of unity, and a brief introduction to abstract measure theory. I have also used this opportunity to correct some misprints and two errors from the first edition. I am grateful to Horst Lange, C. G. Simader, and Matthias Stark for pertinent comments. I also should like to thank Antje Vandenberg for her excellent TEXwork.

Preface to the 3rd edition

This edition corrects some misprints and minor inconsistencies that were kindly pointed out to me by several readers, in particular Bruce Gould, as well as an error in the proof of theorem 19.16 that was brought to my attention by Matthias Stark. I have also used this opportunity to introduce another important tool in analysis, namely covering theorems. Useful references for such results and further properties of various classes of weakly differentiable functions are W.Ziemer, *Weakly differentiable functions*, Springer, 1989, and L.C.Evans, R.Gariepy, *Measure theory and fine properties of functions*, CRC Press, 1992, as well as the fundamental H.Federer, *Geometric measure theory*, Springer, 1969.

Leipzig, January 2005   *Jürgen Jost*

Contents

Chapter II. Topological Concepts

Chapter III. Calculus in Euclidean and Banach Spaces

11. Curves in $\mathbb{R}^d$. Systems of ODEs

Chapter IV. The Lebesgue Integral

12. Preparations. Semicontinuous Functions

13. The Lebesgue Integral for Semicontinuous Functions. The Volume of Compact Sets

14. Lebesgue Integrable Functions and Sets

15. Null Functions and Null Sets. The Theorem of Fubini

16. The Convergence Theorems of Lebesgue Integration Theory

17. Measurable Functions and Sets. Jensen's Inequality. The Theorem of Egorov

24. The Maximum Principle

25. The Eigenvalue Problem for the Laplace Operator

Chapter I.

Calculus for Functions of One Variable

0. Prerequisites

We review some basic material, in particular the convergence of sequences of real numbers, and also properties of the exponential function and the logarithm.

The integers are $\mathbb{Z} = \{\ldots, -2, -1, 0, 1, 2, \ldots\}$, and the positive integers are $\mathbb{N} = \{1, 2, 3, \ldots\}$.

We shall assume the standard arithmetic operations on the real numbers $\mathbb{R}$, and occasionally also on the complex numbers $\mathbb{C}$, although the latter will not play an essential rôle. We shall also assume the ordering of the real numbers (symbols $<, \leq, >, \geq$) and the notion of absolute value $|\cdot|$ in $\mathbb{R}$, and occasionally in $\mathbb{C}$ as well.

For $a, b \in \mathbb{R}$, subsets of $\mathbb{R}$ of the form

$$(a, b) := \{x \in \mathbb{R} : a < x < b\}, (a, \infty)$$
$$:= \{x \in \mathbb{R} : a < x\}, (-\infty, b) := \{x \in \mathbb{R} : x < b\},$$

and $\mathbb{R}$ itself are called open intervals, those of the form

$$[a, b] := \{x \subset \mathbb{R} : a \leq x \leq b\}, [a, \infty)$$
$$:= \{x \in \mathbb{R} : a \leq x\}, (-\infty, b] := \{x \in \mathbb{R} : x \leq b\},$$

and $\mathbb{R}$ itself are called closed intervals.

We shall also employ the standard set theoretic symbols like $\subset$Idxaaak@$\subset$ ("subset of"), as well as the quantifiers $\forall$ ("for all") and $\exists$ ("there exists some").

We recall that a sequence $(x_n)_{n \in \mathbb{N}} \subset \mathbb{R}$ of real numbers is called a Cauchy sequence if

$$\forall \varepsilon > 0 \; \exists N \in \mathbb{N} \; \forall n, m \geq N : |x_n - x_m| < \varepsilon. \tag{1}$$

(For the reader who is not familiar with the logical quantifiers employed here, let us state the content of (1) in words:

For every $\varepsilon > 0$, there exists some $N \in \mathbb{N}$, with the property that for all $n, m \geq N$ we have

$$|x_n - x_m| < \varepsilon.)$$

A similar notion applies in $\mathbb{C}$. $\mathbb{R}$ is complete in the sense that every Cauchy sequence has a limit point, i.e. if $(x_n)_{n\in\mathbb{N}} \subset \mathbb{R}$ is a Cauchy sequence then there exists $x \in \mathbb{R}$ with

$$\forall \varepsilon > 0 \ \exists N \in \mathbb{N} \ \forall n \geq N : |x - x_n| < \varepsilon. \tag{2}$$

$\mathbb{C}$ enjoys the same completeness property.

If (2) is satisfied, we write

$$x = \lim_{n\to\infty} x_n, \tag{3}$$

and we say that $(x_n)_{n\in\mathbb{N}}$ converges to x.

Conversely, every convergent sequence is a Cauchy sequence. Also, the limit of a Cauchy sequence in (2) is unique.

We emphasize that the completeness of $\mathbb{R}$, i.e. the existence of a limit point for every Cauchy sequence in $\mathbb{R}$, is an axiom, whereas the Cauchy property of a convergent sequence and the uniqueness of the limit are theorems that hold not only in $\mathbb{R}$ but also for example in the rational numbers $\mathbb{Q}$. In order to recall the required technique – that will be used frequently in the sequel – we shall now provide proofs for those results.

Thus, assume that $(x_n)_{n\in\mathbb{N}} \subset \mathbb{R}$ converges to x. Given $\varepsilon > 0$, we may choose $N \in \mathbb{N}$ so large that for all $n, m \in \mathbb{N}$ with $n, m \geq N$

$$|x_n - x| < \varepsilon/2, |x_m - x| < \varepsilon/2. \tag{4}$$

This follows from (2) with $\varepsilon/2$ in place of ε.

With the help of the triangle inequality, (4) implies that we have for all $n, m \geq N$

$$|x_n - x_m| \leq |x_n - x| + |x_m - x| < \varepsilon/2 + \varepsilon/2 = \varepsilon$$

which verifies the Cauchy property.

To see the uniqueness of the limit of a convergent sequence, let $(x_n)_{n\in\mathbb{N}} \subset \mathbb{R}$ have x and x' as limit points. Put

$$\eta := |x - x'|.$$

We need to show $\eta = 0$, because we then get $x = x'$, hence uniqueness of the limits. We choose $\varepsilon = \eta/2$, in case we had $\eta > 0$. From (2), applied to the limit point x, we find N such that for $n \geq N$,

$$|x_n - x| < \varepsilon = \eta/2.$$

Likewise, applying (2) to the limit point x', we find $N' \in \mathbb{N}$ such that for $n \geq N'$

$$|x_n - x'| < \varepsilon = \eta/2.$$

Thus, for $n \geq \max(N, N')$, we get

$$\eta = |x - x'| \leq |x_n - x| + |x_n - x'|,$$

using the triangle inequality

$$< \eta/2 + \eta/2 = \eta.$$

Thus, the assumption $\eta > 0$ leads to a contradiction. We must therefore have $\eta = 0$, i.e. uniqueness.

We say that for a sequence $(x_n)_{n \in \mathbb{N}} \subset \mathbb{R}$,

$$\lim_{n \to \infty} x_n = \infty$$

if

$$\forall M \in \mathbb{R} \, \exists N \in \mathbb{N} \, \forall n \geq N : x_n > M,$$

and similarly

$$\lim_{n \to \infty} x_n = -\infty$$

if

$$\forall M \in \mathbb{R} \, \exists N \in \mathbb{N} \, \forall n \geq N : x_n < M.$$

We also recall the theorem of Bolzano-Weierstraß, saying that every bounded sequence $(x_n)_{n \in \mathbb{N}} \subset \mathbb{R}$ (boundedness means that there exists some $M \in \mathbb{R}$ that is independent of $n \in \mathbb{N}$, with the property that for all $n \in \mathbb{N}$

$$|x_n| \leq M)$$

has a convergent subsequence. This theorem will be used in §1. In §7, however, it will be given a proof (see Cor. 7.41) that does not depend on the results of the preceding §§. This is an instance where the pedagogical order of this textbook does not coincide with the logical one. Of course, every convergent sequence is bounded (if $x = \lim_{n \to \infty} x_n$, choose N such that for $n \geq N$

$$|x_n - x| < 1,$$

hence

$$|x_n| \leq |x_n - x| + |x| < |x| + 1,$$

and put

$$M := \max(|x_1|, \ldots, |x_{N-1}|, |x| + 1)$$

to get

$$|x_n| \leq M \quad \text{for all } n \in \mathbb{N}),$$

but bounded sequences need not converge themselves.

(Example: $x_n = 1$ for odd n, $x_n := 0$ for even n defines a bounded sequence that does not converge.)

Therefore, the selection of a subsequence in the Bolzano-Weierstraß theorem is necessary. A limit point of a subsequence of a sequence $(x_n)_{n \in \mathbb{N}}$ is called an accumulation point Bolzano-Weierstraß theorem then says that every bounded sequence in $\mathbb{R}$ has at least one accumulation point.

The standard arithmetic operations carry over to limits of convergent sequences. Thus, suppose that

$$x = \lim_{n \to \infty} x_n \tag{5}$$

$$y = \lim_{n \to \infty} y_n \tag{6}$$

for some sequences $(x_n)_{n \in \mathbb{N}}, (y_n)_{n \in \mathbb{N}}$ in $\mathbb{R}$ (or $\mathbb{C}$).

Then the sequences $(x_n + y_n)_{n \in \mathbb{N}}, (x_n y_n)_{n \in \mathbb{N}}, (\lambda x_n)_{n \in \mathbb{N}}$ for $\lambda \in \mathbb{R}$ are likewise convergent, with

$$\lim_{n \to \infty} (x_n + y_n) = x + y \tag{7}$$

$$\lim_{n \to \infty} (x_n y_n) = xy \tag{8}$$

$$\lim_{n \to \infty} (\lambda x_n) = \lambda x. \tag{9}$$

Finally, if $y \neq 0$, we may find $N \in \mathbb{N}$ such that for $n \geq N, y_n \neq 0$, as well, and the sequence $\left(\frac{x_n}{y_n} \right)_{n \geq N}$ then converges to $\frac{x}{y}$. As an example, we provide the proof of (8). First of all, as convergent sequences are bounded, we may find $M \in \mathbb{R}$ with

$$|x_n| < M.$$

Let $M' := \max(M, |y|)$. From (2), applied to (x_n) and (y_n) with $\frac{\varepsilon}{2M'}$ in place of ε, we obtain $N_1 \in \mathbb{N}$ such that for all $n \geq N_1$

$$|x_n - x| < \frac{\varepsilon}{2M'},$$

and $N_2 \in \mathbb{N}$ such that for all $n \geq N_2$

$$|y_n - y| < \frac{\varepsilon}{2M'}.$$

For $n \geq N := \max(N_1, N_2)$, we then obtain

$$\begin{aligned}
|x_n y_n - xy| &= |x_n(y_n - y) + (x_n - x)y| \\
&\leq |x_n||y_n - y| + |x_n - x||y| \\
&< M' \frac{\varepsilon}{2M'} + \frac{\varepsilon}{2M'} M' = \varepsilon.
\end{aligned}$$

Thus, the criterion (2) holds for the sequence $(x_n y_n)$ and $xy \in \mathbb{R}$. This shows the convergence of $(x_n y_n)$ towards xy.

If (5) and (6) hold for sequences $(x_n), (y_n) \subset \mathbb{R}$ and $x_n \leq y_n$ for all $n \in \mathbb{N}$, then also

$$x \leq y.$$

Similar constructions and results apply to infinite series, i.e. sequences $(x_n)_{n \in \mathbb{N}}$ of the form

$$x_n = \sum_{\nu=1}^{n} y_\nu.$$

If such a sequence converges to some $x \in \mathbb{R}$, we write

$$x = \sum_{\nu=1}^{\infty} y_\nu.$$

We say that the above series converges absolutely if the sequence $(\xi_n)_{n \in \mathbb{N}}$ given by

$$\xi_n := \sum_{\nu=1}^{n} |y_\nu|$$

converges.

Applying the above Cauchy criterion for the convergence of sequences to series yields that a series $\sum_{\nu=1}^{\infty} y_\nu$ converges precisely if

$$\forall \varepsilon > 0 \ \exists N \in \mathbb{N} \ \forall n \geq m \geq N : \left| \sum_{\nu=m}^{n} y_\nu \right| < \varepsilon.$$

Similarly, it converges absolutely if we can achieve $\sum_{\nu=m}^{n} |y_\nu| < \varepsilon$ for $n \geq m \geq N$.

The most important series is the geometric series: Let $0 < |q| < 1$. Then

$$\sum_{\nu=1}^{\infty} q^\nu = \frac{q}{1-q}.$$

This series can be used to derive the ratio test (quotient criterion) for the absolute convergence of a series $\sum_{\nu=1}^{\infty} y_\nu$ with $y_\nu \neq 0$ and

$$\left| \frac{y_{\nu+1}}{y_\nu} \right| \leq q \quad \text{for all } \nu \geq n_0,$$

where $0 < q < 1$.

Namely, from the assumption one derives that

$$|y_\nu| \leq |y_{n_0}| q^{\nu - n_0}.$$

Hence

$$\sum_{\nu=m}^{n} |y_\nu| \leq |y_{n_0}| \frac{1}{q^{n_0}} \sum_{\nu=m}^{n} q^\nu.$$

The right hand side can be made arbitrarily small for sufficiently large n, m since the geometric series converges and hence satisfies the Cauchy criterion. Therefore, the left hand side also becomes arbitrarily small, and the series $\sum_{\nu=1}^{\infty} |y_\nu|$ consequently satisfies the Cauchy criterion and converges, as claimed.

In the course of this textbook, we shall assume that the reader knows the elementary transcendental functions, namely the exponential function $\exp x = e^x$ for $x \in \mathbb{R}$ and the logarithm $\log x$ for $x > 0$. At certain points, we shall also mention the functions $\sin x, \cos x, \tan x, \cot x$, but these will not play an essential rôle. We now summarize some results about the exponential function and the logarithm. The exponential function is given by

$$e^x = \lim_{n \to \infty} (1 + \frac{x}{n})^n. \tag{10}$$

It is strictly monotonically increasing, i.e.

$$e^x < e^y \quad \text{for } x < y. \tag{11}$$

(This follows from the fact that for each n

$$(1 + \frac{x}{n})^n = 1 + x + \text{ other positive terms}$$

that are increasing in x, and consequently $(1 + \frac{y}{n})^n - (1 + \frac{x}{n})^n \geq y - x$ for $x < y$.)

We also have the functional equation

$$e^{x+y} = e^x e^y \quad \text{for all } x, y \in \mathbb{R}. \tag{12}$$

Since $e^0 = 1$ (which follows from (10)), we obtain in particular that

$$e^{-x} = \frac{1}{e^x} \quad \text{for all } x \in \mathbb{R}. \tag{13}$$

We have the following asymptotic results:

$$\lim_{x \to \infty} \frac{e^x}{x^m} = \infty \quad \text{for all } m \in \mathbb{N} \cup \{0\}. \tag{14}$$

(In terms of sequences, this limit means that for every sequence $(x_n)_{n \in \mathbb{N}}$ with $\lim_{n \to \infty} x_n = \infty$, we have $\lim_{n \to \infty} \frac{e^{x_n}}{x_n^m} = \infty$.)

Proof. For $n \geq m + 1$,

$$(1 + \frac{x}{n})^n \geq \binom{n}{m+1} \frac{x^{m+1}}{n^{m+1}} \quad \text{by binomial expansion}$$

$$= \frac{n(n-1) \cdot \ldots \cdot (n-m)}{n^{m+1}} \frac{1}{(m+1)!} x^{m+1}.$$

Since $\lim_{n \to \infty} \frac{n(n-1) \cdot \ldots \cdot (n-m)}{n^{m+1}} = 1$ for fixed $m \in \mathbb{N}$, e^x asymptotically grows at least like x^{m+1} which implies (14).

(14) and the quotient rule for limits yield

$$\lim_{x \to \infty} x^m e^{-x} = 0 \quad \text{for all } m \in \mathbb{N} \cup \{0\}. \tag{15}$$

Put $y = \frac{1}{x}$, (14) also yields

$$\lim_{\substack{y \to 0 \\ y > 0}} y^m e^{1/y} = \infty \quad \text{for all } m \in \mathbb{N} \cup \{0\}. \tag{16}$$

If $(a_n)_{n \in \mathbb{N}} \subset \mathbb{R}$ converges to 0, with $a_n \neq 0$ and $a_n > -1$ for all n, then

$$\lim_{n \to \infty} (1 + a_n)^{1/a_n} = e. \tag{17}$$

More generally, if $x \in \mathbb{R}$, $\lim_{n \to \infty} a_n = 0, a_n \neq 0, xa_n > -1$ for all n, then

$$\lim_{n \to \infty} (1 + a_n x)^{1/a_n} = e^x. \tag{18}$$

(This is a slight generalization of (10), and it can be derived from (10) by elementary estimates.)

From (18), one derives the following rule (for $\alpha, x \in \mathbb{R}$)

$$
\begin{aligned}
(e^x)^\alpha &= \lim_{n \to \infty} (1 + \frac{x}{n})^{\alpha n} \text{ (w.l.o.g., we take only} \\
&\qquad \text{those } n \text{ with } \frac{x}{n} > -1 \text{ into account)} \\
&= \lim_{n \to \infty} (1 + a_n \alpha x)^{1/a_n} \text{ with } a_n := \frac{1}{\alpha n} \\
&= e^{\alpha x}.
\end{aligned}
\tag{19}
$$

The logarithm is defined as the inverse of the exponential function, i.e.

$$\log(e^x) = x \quad \text{for all } x \in \mathbb{R}, \tag{20}$$

and then also

$$e^{\log x} = x \quad \text{for all } x > 0. \tag{21}$$

The functional equation (12) implies that

$$\log(xy) = \log x + \log y \quad \text{for all } x, y > 0, \tag{22}$$

and combining this with (13), we get

$$\log \frac{x}{y} = \log x - \log y \quad \text{for } x, y > 0. \tag{23}$$

Finally

$$\log x^\alpha = \alpha \log x \quad \text{for } x > 0, \alpha \in \mathbb{R} \tag{24}$$

which follows from

$$e^{\log x^\alpha} = x^\alpha = (e^{\log x})^\alpha = e^{\alpha \log x} \quad \text{(using (19))}.$$

From the monotonicity (11) of the exponential function, we derive the monotonicity of the logarithm

$$\log x < \log y \quad \text{whenever } 0 < x < y. \tag{25}$$

In §2, when we wish to compute the derivative of the exponential function, we shall also need the limit of

$$\frac{e^{x_n} - 1}{x_n} \quad \text{for } x_n \to 0, \quad \text{assuming that all } x_n \neq 0. \tag{26}$$

In order to determine this limit, we put

$$a_n := e^{x_n} - 1.$$

Then $a_n > -1$, and also $a_n \neq 0$, since $x_n \neq 0$, for all n.

Since the logarithmm is defined as the inverse of the exponential function, we have

$$x_n = \log(a_n + 1).$$

We thus need to evaluate

$$\lim_{n \to \infty} \frac{a_n}{\log(a_n + 1)}.$$

We consider

$$\exp\left(\frac{\log(a_n + 1)}{a_n}\right) = (a_n + 1)^{\frac{1}{a_n}}.$$

This expression converges to e, by (17).

In order to deduce from this result that

$$\lim_{n \to \infty} \frac{a_n}{\log(a_n + 1)} = \frac{1}{\log e} = 1, \tag{27}$$

we need the following general result.

If $(y_n)_{n \in \mathbb{N}} \subset \mathbb{R}$ converges to $y_0 > 0$, with $y_n > 0$ for all n, then

$$\lim_{n \to \infty} \log y_n = \log y_0. \tag{28}$$

(In the terminology of §1, this is the continuity of the logarithm).

Using (23), we see that (28) is equivalent to

$$\lim_{n \to \infty} \log \frac{y_n}{y_0} = 0. \tag{29}$$

This means that we need to show that for every sequence $(b_n)_{n \in \mathbb{N}}$ that converges to 1 as $n \to \infty$, with $b_n > 0$ for all n, we have

$$\lim_{n \to \infty} \log b_n = 0. \tag{30}$$

Let $\varepsilon > 0$. Then

$$e^\varepsilon > 1$$

as one sees from (10), and hence with (13)

$$e^{-\varepsilon} < 1.$$

As (b_n) converges to 1, we may thus find $N \in \mathbb{N}$ such that for all $n \geq N$

$$e^{-\varepsilon} < b_n < e^{\varepsilon}. \tag{31}$$

The monotonicity (25) of the logarithm allows us to deduce from (31) that

$$-\varepsilon < \log b_n < \varepsilon \tag{32}$$

which implies (30).

We have thus verified (28) which in turn implies (27). From the definition of $(a_n)_{n \in \mathbb{N}}$, (27) is equivalent to

$$\lim_{\substack{x_n \to 0 \\ x_n \neq 0}} \frac{e^{x_n} - 1}{x_n} = 1. \tag{33}$$

Finally, we shall assume the concept of a vector space over $\mathbb{R}$ (and also over $\mathbb{C}$, but this is not essential) from linear algebra. In particular, we shall employ the vector space $\mathbb{R}^d (d \in \mathbb{N})$.

Exercises for § 0

1) Show that $a \in \mathbb{R}$ is an accumulation point of a sequence $(a_n)_{n \in \mathbb{N}} \subset \mathbb{R}$ precisely if for all $\varepsilon > 0$ and all $N \in \mathbb{N}$ there exists $n \geq N$ with

$$|a_n - a| < \varepsilon.$$

2) Suppose that the sequence $(a_n)_{n \in \mathbb{N}} \subset \mathbb{R}$ converges to $a \in \mathbb{R}$. Show that the sequence defined by

$$b_n := \frac{1}{n}(a_1 + \ldots + a_n)$$

converges to a as well. Does the converse hold as well, i.e. does the convergence of $(b_n)_{n \in \mathbb{N}}$ to a imply the convergence of $(a_n)_{n \in \mathbb{N}}$ to a?

3) Let $(a_n)_{n \in \mathbb{N}} \subset \mathbb{R}$ be a bounded sequence with only one accumulation point a. Show that (a_n) converges to a. Is this also true for unbounded sequences?

4) Which of the following sequences $(a_n)_{n \in \mathbb{N}}$ are convergent? What are the limits of those that converge?

a) $a_n = \dfrac{n(n+1)}{n^2+1}$

b) $a_n = \dfrac{a_2 n^2 + a_1 n + a_0}{b_2 n^2 + b_1 n + b_0}$, $a_0, a_1, a_2 \in \mathbb{R}, b_0, b_1, b_2 > 0$

c) $a_n = \dfrac{13 n^6 - 27}{26 n^5 \sqrt{n} + 39 \sqrt{n}}$

d) $a_n = \dfrac{1}{n} \displaystyle\sum_{\nu=1}^{n} \dfrac{1}{\nu^2}$

e) $a_n = \dfrac{\rho^n}{n^m}$ for $m \in \mathbb{N}, \rho \in \mathbb{R}$

(here, the answer depends on ρ).

5) Which of the following series converge?

a) $\displaystyle\sum_{n=1}^{\infty} \dfrac{1}{n(n+2)}$

b) $\displaystyle\sum_{n=1}^{\infty} \dfrac{1}{1+a^n}$ $(a \neq -1)$

c) $\displaystyle\sum_{n=1}^{\infty} \dfrac{1}{\sqrt{n(n+\mu)}}$, $\mu > 0$.

6) Let $a_n := \left(1 + \frac{1}{n}\right)^n$, $b_n := \displaystyle\sum_{\nu=0}^{n} \frac{1}{\nu!}$
Show that

a) $a_n \leq a_{n+1}$ for all n
b) $a_n \leq b_n$ for all n
c) $\displaystyle\lim_{n\to\infty} b_n = \lim_{n\to\infty} a_n (= e)$.

7) Let $\alpha > 0$. Show that

$$\lim_{x\to\infty} \frac{x^\alpha}{\log x} = \infty$$

and

$$\lim_{\substack{x\to 0 \\ x>0}} x^\alpha \log x = 0.$$

1. Limits and Continuity of Functions

We introduce the concept of continuity for a function defined on a subset of $\mathbb{R}$ (or $\mathbb{C}$). After deriving certain elementary properties of continuous functions, we show the intermediate value theorem, and that a continuous function defined on a closed and bounded set assumes its maximum and minimum there.

Definition 1.1 Let $D \subset \mathbb{R}$ (or $\mathbb{C}$) and $f : D \to \mathbb{R}$ (or $\mathbb{C}$) be a function. We say that $\lim\limits_{x \to p} f(x) = y$ if and only if for every sequence $(x_n)_{n \in \mathbb{N}} \subset D$ with $\lim\limits_{n \to \infty} x_n = p$ we have $\lim\limits_{n \to \infty} f(x_n) = y$.

Theorem 1.2 *In the notations of definition 1.1*

$$\lim_{x \to p} f(x) = y$$

if and only if the following conditions are fulfilled

$$\forall \varepsilon > 0 \ \exists \delta > 0 \ \forall x \in D \quad \text{with } |x - p| < \delta :$$
$$|f(x) - y| < \varepsilon . \tag{1}$$

Proof. " $\Leftarrow$ " Let $(x_n)_{n \in \mathbb{N}} \subset D$ be a sequence with $\lim\limits_{n \to \infty} x_n = p$. We have

$$\forall \delta \ \exists N \in \mathbb{N} \ \forall n \geq N : |x_n - p| < \delta . \tag{2}$$

For $\varepsilon > 0$ we determine $\delta > 0$ as in (1) and then N as in (2): It follows that for $n \geq N$:
$$|f(x_n) - y| < \varepsilon.$$

We have therefore shown

$$\forall \varepsilon > 0 \ \exists N \in \mathbb{N} \ \forall n \geq N : |f(x_n) - y| < \varepsilon,$$

so $\lim\limits_{n \to \infty} f(x_n) = y$ and therefore, by definition, $\lim\limits_{x \to p} f(x) = y$.

" $\Rightarrow$ " If (1) is not fulfilled then

$$\exists \varepsilon > 0 \; \forall \delta > 0 \; \exists x \in D \quad \text{with } |x - p| < \delta$$
$$\text{but } |f(x) - y| > \varepsilon \,. \tag{3}$$

For $n \in \mathbb{N}$ we set $\delta = \frac{1}{n}$ and determine $x_n = x$ corresponding to δ as in (3). Then $|x_n - p| < \frac{1}{n}$ and so $\lim_{n \to \infty} x_n = p$, but for ε as in (3) we have

$$|f(x_n) - y| > \varepsilon,$$

and therefore $\lim_{n \to \infty} f(x_n) \neq y$. $\qquad\square$

Definition 1.3 Let $D \subset \mathbb{R}$ (or $\mathbb{C}$), $f : D \to \mathbb{R}$ (or $\mathbb{C}$) a function and $p \in D$. The function f is said to be continuous at p if

$$\lim_{x \to p} f(x) = f(p).$$

f is continuous in D if f is continuous at every point $p \in D$.

Theorem 1.4 *In the notations of definition 1.3 f is continuous at p precisely when the following condition is fulfilled*

$$\forall \, \varepsilon > 0 \; \exists \delta > 0 \; \forall x \in D \quad \text{with } |x - p| < \delta$$
$$|f(x) - f(p)| < \varepsilon \,. \tag{4}$$

Proof. This is a direct consequence of theorem 1.2. $\qquad\square$

Lemma 1.5 *Suppose that $g : D \to \mathbb{R}$ (or $\mathbb{C}$) is continuous at $p \in D$, and that $g(p) \neq 0$. Then there exists $\delta > 0$ with the property that for all $x \in D$ with $|x - p| < \delta$*

$$g(x) \neq 0$$

as well.

Proof. Let $\varepsilon := \frac{|g(p)|}{2} > 0$. Since g is continuous at p, we may find $\delta > 0$ such that for all $x \in D$ with $|x - p| < \delta$

$$|g(x) - g(p)| < \varepsilon = \frac{|g(p)|}{2}.$$

This implies

$$|g(x)| > \frac{|g(p)|}{2} > 0.$$

$\qquad\square$

Lemma 1.6 *Assume that the functions $f, g : D \to \mathbb{R}$ (or $\mathbb{C}$) are continuous at $p \in D$. Then so are the functions $f + g$, fg, and λf, for any $\lambda \in \mathbb{R}(\mathbb{C})$. Furthermore if $g(p) \neq 0$, then $\frac{f}{g}$ is continuous at p as well.*

Proof. This follows from the rules for the limits of sums, products, and quotients of convergent sequences described in §0.   □

Corollary 1.7 *All polynomial functions, i.e. functions of the form $f(x) = \sum_{\nu=0}^{n} a_\nu x^\nu$, are continuous.*   □

Corollary 1.8 *The functions $f : D \to \mathbb{R}(\mathbb{C})$ that are continuous at $p \in D$ form a vector space over $\mathbb{R}(\mathbb{C})$. The same holds for those functions that are continuous on all of D.*   □

Definition 1.9 Let $D \subset \mathbb{R}$ (or $\mathbb{C}$) and $f : D \to \mathbb{R}$ (or $\mathbb{C}$) a function. The function f is said to be uniformly continuous in D

$$\Longleftrightarrow \forall \varepsilon > 0 \; \exists \delta > 0 \; \forall x_1, x_2 \in D \quad \text{with } |x_1 - x_2| < \delta :$$
$$|f(x_1) - f(x_2)| < \varepsilon . \tag{5}$$

The crucial difference between the requirements of continuity and uniform continuity is that in (4), δ could depend on p, whereas in (5) it must be possible to choose δ independent of the points under consideration of D.

Example. Let $f : \mathbb{R} \to \mathbb{R}$ be the function $f(x) = x^2$. We show that f is continuous at every $p \in \mathbb{R}$. Let $\varepsilon > 0$. We set

$$\delta = \min\left(1, \frac{\varepsilon}{2|p| + 1}\right).$$

If $|x - p| < \delta$ then

$$|x^2 - p^2| = |x - p||x + p| \leq |x - p|(|x| + p) < |x - p|(2|p| + 1) < \varepsilon.$$

This shows that f is continuous. We now show that f is not uniformly continuous on $\mathbb{R}$.

For this we prove the negation of (5), namely

$$\exists \varepsilon > 0 \; \forall \delta > 0 \; \exists x_1, x_2 \in \mathbb{R} \quad \text{with } |x_1 - x_2| < \delta$$
$$\text{but } |f(x_1) - f(x_2)| > \varepsilon . \tag{6}$$

We choose $\varepsilon = 1$. For $\delta > 0$ there exist $x_1, x_2 \in \mathbb{R}$ with $|x_1 - x_2| = \frac{\delta}{2}, |x_1 + x_2| > \frac{2}{\delta}$. Therefore

$$|x_1^2 - x_2^2| = |x_1 - x_2||x_1 + x_2| > 1$$

which proves (6).

Theorem 1.10 *Let $I = [a, b]$ be a closed and bounded interval and $f : I \to \mathbb{R}$ (or $\mathbb{C}$) a continuous function. Then f is uniformly continuous on I.*

Proof. Otherwise

$$\exists \varepsilon > 0 \ \forall \delta > 0 \ \exists x_1, x_2 \in I \quad \text{with } |x_1 - x_2| < \delta,$$
$$\text{but } |f(x_1) - f(x_2)|. \tag{7}$$

For $n \in \mathbb{N}$ we choose $\delta = \frac{1}{n}$ and then points $x_{1,n} = x_1$ and $x_{2,n} = x_2$ as in (7). By the Bolzano-Weierstrass theorem, the bounded sequence $(x_{1,n})$ has a convergent subsequence which converges to, say, x_0. As I is closed the point $x_0 \in I$. As

$$|x_{1,n} - x_{2,n}| < \frac{1}{n}$$

we also have $\lim_{n \to \infty} x_{2,n} = x_0$.

Since f is continuous, it follows that

$$\lim_{n \to \infty} f(x_{1,n}) = f(x_0) = \lim_{n \to \infty} f(x_{2,n})$$

which contradicts (7).

Remark. By theorem 1.10, the function $f(x) = x^2$ is uniformly continuous on any closed and bounded interval $[a, b]$. Nevertheless, as shown above, the function $f(x) = x^2$ is not uniformly continuous on $\mathbb{R}$. Uniform continuity of a function therefore depends on its domain of definition.

Recalling corollary 1.8, we now formulate

Definition 1.11 $C^0(D, \mathbb{R})$ and $C^0(D, \mathbb{C})$ are, respectively, the vector spaces of continuous functions $f : D \to \mathbb{R}$ and $f : D \to \mathbb{C}$. We denote $C^0(D, \mathbb{R})$ also by $C^0(D)$.

Definition 1.12 Let $D \subset \mathbb{R}$ and $f : D \to \mathbb{R}$ (or $\mathbb{C}$), and $0 < \alpha < 1$. f is called α-Hölder continuous (for $\alpha = 1$ Lipschitz continuous) if for any closed and bounded interval $I \subset D$ there exists an $m_I \in \mathbb{R}$ with

$$|f(x) - f(y)| \leq m_I |x - y|^\alpha \text{ for all } x, y \in I. \tag{8}$$

One easily checks that if $f, g : D \to \mathbb{R}(\mathbb{C})$ are α-Hölder continuous, then so is their sum $f + g$, and likewise λf, for any $\lambda \in \mathbb{R}(\mathbb{C})$.

Definition 1.13 The vector space of α-Hölder continuous functions $f : D \to \mathbb{R}$ (resp. $\mathbb{C}$) will be denoted by $C^{0,\alpha}(D, \mathbb{R})$ (resp. $C^{0,\alpha}(D, \mathbb{C})$). We also write $C^{0,\alpha}(D)$ for $C^{0,\alpha}(D, \mathbb{R})$ and, for $0 < \alpha < 1, C^{0,\alpha}(D)$ as $C^\alpha(D)$.

We now come to the important *intermediate value theorem* of Bolzano

Theorem 1.14 *Let* $f : [a, b] \to \mathbb{R}$ *be continuous. Then* f *assumes any value* κ *between* $f(a)$ *and* $f(b)$. *This means that if for example* $f(a) \leq f(b)$, *and*

$$f(a) \leq \kappa \leq f(b),$$

then there exists some $x_0 \in [a, b]$ with

$$f(x_0) = \kappa.$$

Proof. By considering $f - \kappa$ in place of κ, we may assume that $\kappa = 0$. We may also assume $f(a) < f(b)$, because otherwise we may consider $-f$ in place of f. Thus, we may assume

$$f(a) < 0 < f(b)$$

(if $f(a)$ or $f(b)$ were equal to 0, we would have found the desired x_0 already). We now perform the following inductive construction:

Let $a_0 = a, b_0 = b$. If $a_n \geq a_{n-1}$ and $b_n \leq b_{n-1}$ have been determined, with $a_n < b_n$, put

$$c_n := \frac{1}{2}(b_n - a_n).$$

If $f(c_n) = 0$, put $x_0 = c_n$, and the process can be terminated, and so this case can be disregarded in the sequel.

If $f(c_n) < 0$, put $a_{n+1} = c_n, b_{n+1} = b_n$.
If $f(c_n) > 0$, put $a_{n+1} = a_n, b_{n+1} = c_n$.
We then have

$$b_{n+1} - a_{n+1} = \frac{1}{2}(b_n - a_n).$$

By the Bolzano-Weierstraß theorem, subsequences $(\alpha_\nu)_{\nu \in \mathbb{N}}$ of $(a_n)_{n \in \mathbb{N}}$ and $(\beta_\nu)_{\nu \in \mathbb{N}}$ of $(b_n)_{n \in \mathbb{N}}$ converge, and since

$$\lim_{n \to \infty} |b_n - a_n| = 0,$$

they both converge to the same point $x_0 \in [a, b]$.

(In fact since $a_n \leq a_{n+1} \leq b_{n+1} \leq b_n$ for all $n \in \mathbb{N}$, it is easy to verify that the sequences (a_n) and (b_n) themselves converge.)

Since

$$f(a_n) < 0, f(b_n) > 0 \quad \text{for all } n \in \mathbb{N},$$

the continuity of f implies that

$$f(x_0) = \lim_{\nu \to \infty} f(\alpha_\nu) = \lim_{\nu \to \infty} f(\beta_\nu) = 0.$$

Thus, the desired x_0 has been found. □

Theorem 1.15 *Let $f : [a, b] \to \mathbb{R}$ be continuous. Then f is bounded, i.e. there exists $M \geq 0$ with*

$$|f(x)| \leq M \quad \text{for all } x \in [a, b],$$

and it assumes its minimum and maximum on $[a, b]$, i.e. there exist $x_0, x_1 \in [a, b]$ with

$$f(x_0) = \inf\{f(x) : x \in [a, b]\}$$
$$f(x_1) = \sup\{f(x) : x \in [a, b]\}.$$

Proof. The second statement implies the first, because

$$|f(x)| \leq \max(|f(x_0)|, |f(x_1)|) =: M$$

by the choice of x_0, x_1. By considering $-f$ in place of f, the reasoning for the maximum is reduced to the one for the minimum, and so we only describe the latter. Let $(x_n)_{n \in \mathbb{N}} \subset [a, b]$ be a sequence with

$$\lim_{n \to \infty} f(x_n) = \inf\{f(x) : x \in [a, b]\}.$$

(At this point, we have not yet excluded that this infimum is $-\infty$, and so in this case $f(x_n)$ would approach $-\infty$.)

Since (x_n) is contained in the bounded interval $[a, b]$, the Bolzano-Weierstraß theorem implies the existence of a subsequence $(\xi_\nu)_{\nu \in \mathbb{N}}$ of $(x_n)_{n \in \mathbb{N}}$ that converges towards some $x_0 \in [a, b]$. Since f is continuous on $[a, b]$, we obtain

$$f(x_0) = \lim_{\nu \to \infty} f(\xi_\nu) = \inf\{f(x) : x \in [a, b]\}.$$

In particular, the latter expression is finite, and f assumes its minimum at x_0.

Remarks.

1)   To simplify our terminology, in the sequel we shall usually say in proofs of the type of the preceding one: "after selection of a subsequence, $(x_n)_{n \in \mathbb{N}}$ converges" in place of "there exists a subsequence $(\xi_\nu)_{\nu \in \mathbb{N}}$ of $(x_n)_{n \in \mathbb{N}}$ that converges".

2)   For the preceding theorem to hold, it is important that the interval of definition of f is closed. In fact, $f : (0, 1) \to \mathbb{R}, f(x) = x$, neither assumes its minimum nor its maximum on $(0, 1)$, and the function $g : (0, 1) \to \mathbb{R}, g(x) = \frac{1}{x}$ even is unbounded on $(0, 1)$, although both of them are continuous there.

Exercises for § 1

1)   Show the following result:
   Let $f : (a, b) \to \mathbb{R}, x_0 \in (a, b)$. Then $\lim_{x \to x_0} f(x)$ exists precisely if the limits $\lim_{\substack{x \to x_0 \\ x > x_0}} f(x)$ and $\lim_{\substack{x \to x_0 \\ x < x_0}} f(x)$ exist and coincide.

2)   Let $f, g : D \to \mathbb{R}$ be continuous. Show that $F, G : D \to \mathbb{R}$ defined by

$$F(x) := \min(f(x), g(x))$$
$$G(x) := \max(f(x), g(x))$$

are continuous as well.

3)   Are the following functions continuous?

a) $f : \mathbb{R} \setminus \{-1, 1\} \to \mathbb{R}$,

$$f(x) = \frac{x}{x^2 - 1}.$$

b) $f : (0, \infty) \to \mathbb{R}$,

$$f(x) = \frac{x^2 - 5x + 6}{x^2 - 4} \quad \text{for } x \neq 2, f(2) = 0.$$

c) $f : (0, \infty) \to \mathbb{R}$,

$$f(x) = \frac{x^2 - 5x + 6}{x^2 - 4} \quad \text{for } x \neq 2, f(2) = -\frac{1}{4}.$$

d) $f : \mathbb{R} \to \mathbb{R}$,

$$f(x) = x^3 e^{(3x^2 + 6x + 1)}.$$

e) $f : \mathbb{R} \to \mathbb{R}$,

$$f(x) = \begin{cases} \frac{1}{|x|} & \text{for } x \neq 0 \\ 0 & \text{for } x = 0 \end{cases}.$$

f) $f : \mathbb{R} \to \mathbb{R}$,

$$f(x) = \begin{cases} \frac{x}{|x|} & \text{for } x \neq 0 \\ 0 & \text{for } x = 0 \end{cases}.$$

g) $f : \mathbb{R} \to \mathbb{R}$,

$$f(x) = \begin{cases} \frac{x^2}{|x|} & \text{for } x \neq 0 \\ 0 & \text{for } x = 0 \end{cases}.$$

4)   We consider the following function $z : \mathbb{R} \to \mathbb{R}$

$$z(x) := \begin{cases} x - 4n & \text{for } x \in [4n - 1, 4n + 1] \\ -x + 4n + 2 & \text{for } x \in [4n + 1, 4n + 3] \end{cases}.$$

Show that z is uniformly continuous. (One possibility to solve this exercise is to show that z is continuous on the interval $[-1, 3]$ and to then show that every continuous function $f : \mathbb{R} \to \mathbb{R}$ that is periodic (i.e. there exists $\omega > 0$ with $f(x + \omega) = f(x)$ for all $x \in \mathbb{R}$) is uniformly continuous.)

5)   Let $f : [a, b] \to [a, b]$ (i.e. $a \leq f(x) \leq b$ for all $x \in [a, b]$) be a continuous function. Show that there exists some $x \in [a, b]$ with $f(x) = x$. (Such an x with $f(x) = x$ is called a fixed point of f.)

6)   Let $p : \mathbb{R} \to \mathbb{R}$ be a polynomial of odd order, i.e.

$$p(x) = a_k x^k + a_{k-1} x^{k-1} + \ldots + a_0,$$

where $k \in \mathbb{N}$ is odd. Show that there exists some $x_0 \in \mathbb{R}$ with $p(x_0) = 0$.

2. Differentiability

We define the notion of differentiability of functions defined on subsets of $\mathbb{R}$, and we show the basic rules for the computation of derivatives.

Definition 2.1 Let $D \subset \mathbb{R}$ and $f : D \to \mathbb{R}$ be a function. The function f is said to be differentiable at $x \in D$ if

$$f'(x) := \lim_{\substack{\xi \to x \\ \xi \in D \setminus \{x\}}} \frac{f(\xi) - f(x)}{\xi - x} = \lim_{\substack{h \to 0 \\ h \neq 0 \\ x+h \in D}} \frac{f(x + h) - f(x)}{h}$$

exists. We shall also write $\frac{df}{dx}$ in place of $f'(x)$, and we shall call this expression the derivative of f at x. We call f differentiable in D if it is differentiable at every $x \in D$.

(In this definition it is tacitly assumed that there exists a sequence $(x_n)_{n \in \mathbb{N}} \subset D \setminus \{x\}$ which converges to x).

Remark. Although D is allowed to be a subset of $\mathbb{R}$ of rather general type, in our subsequent applications, D typically will be an interval.

Theorem 2.2 *Let $D \subset \mathbb{R}$ and $f : D \to \mathbb{R}$ a function. Let $x_0 \in D$ and $(\xi_n) \subset D \setminus \{x_0\}$ a sequence with $\lim_{n \to \infty} \xi_n = x_0$. The function f is differentiable at x_0, with derivative $f'(x_0) = c$, precisely if for $x \in D$*

$$f(x) = f(x_0) + c(x - x_0) + \phi(x), \tag{1}$$

where $\phi : D \to \mathbb{R}$ is a function with

$$\lim_{\substack{x \to x_0 \\ x \neq x_0}} \frac{\phi(x)}{x - x_0} = 0 \,. \tag{2}$$

Equivalently

$$|f(x) - f(x_0) - c(x - x_0)| \leq \psi(x), \tag{3}$$

where $\psi : D \to \mathbb{R}$ is again a function with

$$\lim_{\substack{x \to x_0 \\ x \neq x_0}} \frac{\psi(x)}{x - x_0} = 0. \tag{4}$$

Proof. " $\Rightarrow$ " Let f be differentiable at x_0. We set $\phi(x) := f(x) - f(x_0) - f'(x_0)(x - x_0)$. Therefore

$$\lim_{\substack{x \to x_0 \\ x \neq x_0}} \frac{\phi(x)}{x - x_0} = \lim_{\substack{x \to x_0 \\ x \neq x_0}} \left(\frac{f(x) - f(x_0)}{x - x_0} - f'(x_0)\right) = 0.$$

" $\Leftarrow$ " Suppose (1) holds. Then

$$\lim_{\substack{x \to x_0 \\ x \neq x_0}} \frac{f(x) - f(x_0)}{x - x_0} = c + \lim_{\substack{x \to x_0 \\ x \neq x_0}} \frac{\phi(x)}{x - x_0} = c.$$

$\square$

Corollary 2.3 *If $f : D \to \mathbb{R}$ is differentiable at $x_0 \in D$ then f is continuous at x_0.*

Proof. Equation (3) implies

$$|f(x) - f(x_0)| \leq |f'(x_0)||x - x_0| + \psi(x)$$

with $\lim_{x \to x_0} \psi(x) = 0$. It follows that

$$\lim_{x \to x_0} f(x) = f(x_0)$$

and therefore f is continuous at x_0. $\square$

If $f, g : D \to \mathbb{R}$ are differentiable at $x \in D$, then it is straightforward to check that $f + g$ is likewise differentiable at x, with

$$(f + g)'(x) = f'(x) + g'(x). \tag{5}$$

Similarly, for $\lambda \in \mathbb{R}$,

$$(\lambda f)'(x) = \lambda f'(x). \tag{6}$$

In particular, the functions $f : D \to \mathbb{R}$ that are differentiable at x form a vector space.

We next have the *product rule*.

Theorem 2.4 *If $f, g : D \to \mathbb{R}$ are differentiable at $x \in D$, then so is their product fg, with*

$$(fg)'(x) = f'(x)g(x) + g'(x)f(x). \tag{7}$$

Proof. We have to compute

$$\lim_{\substack{\xi \to x \\ \xi \in D\setminus\{x\}}} \frac{f(\xi)g(\xi) - f(x)g(x)}{\xi - x}$$

$$= \lim_{\substack{\xi \to x \\ \xi \in D\setminus\{x\}}} \left\{ \frac{(f(\xi) - f(x))g(x) + (g(\xi) - g(x))f(\xi)}{\xi - x} \right\}$$

$$= \lim_{\substack{\xi \to x \\ \xi \in D\setminus\{x\}}} \left(\frac{(f(\xi) - f(x))}{\xi - x} g(x) \right) + \lim_{\substack{\xi \to x \\ \xi \in D\setminus\{x\}}} \left(\frac{g(\xi) - g(x)}{\xi - x} f(\xi) \right)$$

$$\leq f'(x)g(x) + g'(x) \lim_{\substack{\xi \to x \\ \xi \in D\setminus\{x\}}} f(\xi),$$

according to the rules for limits described in § 0

$$= f'(x)g(x) + g'(x)f(x) \text{ since } f \text{ is continuous at } x \text{ by corollary 2.3.}$$

$\square$

Similarly, we have the *quotient rule*.

Theorem 2.5 *If* $f, g : D \to \mathbb{R}$ *are differentiable at* $x \in D$, *and if* $g(x) \neq 0$, *then their quotient* $\frac{f}{g}$ *is differentiable at* x *as well, with*

$$\left(\frac{f}{g} \right)'(x) = \frac{f'(x)g(x) - g'(x)f(x)}{g^2(x)}. \tag{8}$$

Proof. It follows from the continuity of g at x (corollary 2.3) that there exists $\delta > 0$ with the property that

$$g(\xi) \neq 0 \quad \text{whenever } |\xi - x| < \delta.$$

Therefore, in the subsequent limit processes where we let ξ tend to x, we may assume that $g(\xi) \neq 0$. We now compute

$$\lim_{\substack{\xi \to x \\ \xi \in D\setminus\{x\}}} \frac{1}{\xi - x} \left(\frac{f(\xi)}{g(\xi)} - \frac{f(x)}{g(x)} \right), \quad \lim_{\substack{\xi \to x \\ \xi \in D\setminus\{x\}}} \frac{1}{\xi - x} \frac{f(\xi)g(x) - g(\xi)f(x)}{g(\xi)g(x)}$$

$$= \lim_{\substack{\xi \to x \\ \xi \in D\setminus\{x\}}} \frac{1}{g(\xi)g(x)} \lim_{\substack{\xi \to x \\ \xi \in D\setminus\{x\}}} \frac{f(\xi)g(x) - g(\xi)f(x)}{\xi - x}$$

$$= \frac{1}{g^2(x)} \lim_{\substack{\xi \to x \\ \xi \in D\setminus\{x\}}} \left(\frac{f(\xi) - f(x)}{\xi - x} g(x) - \frac{g(\xi) - g(x)}{\xi - x} f(x) \right),$$

using the continuity of g at x

$$= \frac{f'(x)g(x) - f'(x)f(x)}{g^2(x)}.$$

$\square$

The next theorem presents the important *chain rule*.

Theorem 2.6 *Let $f : D_1 \to \mathbb{R}$ be differentiable at $x \in D_1$ with $f(D_1) \subset D_2$, and let $g : D_2 \to \mathbb{R}$ be differentiable at $y := f(x)$. Then the composition $g \circ f$ is differentiable at x, with*

$$(g \circ f)'(x) = g'(f(x))f'(x). \qquad (9)$$

Proof. We set out to compute

$$\lim_{\substack{\xi \to x \\ \xi \in D_1 \setminus \{x\}}} \frac{g(f(\xi)) - g(f(x))}{\xi - x}$$

$$= \lim_{\substack{\xi \to x \\ \xi \in D_1 \setminus \{x\}}} \frac{g(f(\xi)) - g(f(x))}{f(\xi) - f(x)} \cdot \frac{f(\xi) - f(x)}{\xi - x}.$$

Here, the slight technical problem arises that we may have $f(\xi) = f(x)$ even if $\xi \neq x$, and so the first quotient may not be well defined. This, however, can easily be circumvented by considering

$$g^{(1)}(\eta) := \begin{cases} \frac{g(\eta) - g(y)}{\eta - y} & \text{for } \eta \neq y \\ \lim_{\substack{\eta \to y \\ \eta \in D_2 \setminus \{y\}}} \frac{g(\eta) - g(y)}{\eta - y} = g'(y) & \text{for } \eta = y \end{cases}$$

(note that here we are using the assumption that g is differentiable at y) and replacing the above limit by

$$\lim_{\substack{\xi \to x \\ \xi \in D_1 \setminus \{x\}}} \left(g^{(1)}(f(\xi)) \frac{f(\xi) - f(x)}{\xi - x} \right)$$

$$= g'(f(x))f'(x),$$

using the continuity of f at x (corollary 2.3) to handle the first term and the assumed differentiability of f at x to handle the second one. □

The next result allows to compute the derivative of the inverse of a function once the derivative of that function is known. To formulate that result, we call a function $f : I \to \mathbb{R}$, I an interval, monotonically increasing (decreasing) if $f(x_1) \leq (\geq)f(x_2)$ whenever $x_1 < x_2$ $(x_1, x_2 \in I)$. f is called strictly monotonically increasing (decreasing) if we have $f(x_1) < (>)f(x_2)$ whenever $x_1 < x_2$.

Theorem 2.7 *Let $I \subset \mathbb{R}$ be an open interval, and let $f : I \to \mathbb{R}$ be strictly monotonically increasing or decreasing. Then there exists a continuous function*

$$\varphi : f(I) \to \mathbb{R}$$

with

$$\varphi \circ f(x) = x \quad \text{for all } x \in I \qquad (10)$$

and

$$f \circ \varphi(y) = y \quad \text{for all } y \in f(I). \tag{11}$$

φ *is called the inverse function of f. We shall usually denote the inverse function φ of f by f^{-1}. φ is also strictly monotonically increasing or decreasing, resp. If f is differentiable at $x \in I$, with $f'(x) \neq 0$, then φ is differentiable at $y := f(x)$, with*

$$\varphi'(y) = \frac{1}{f'(x)} = \frac{1}{f'(\varphi(y))}, \tag{12}$$

or, equivalently $(f^{-1})'(y) = \frac{1}{f'(f^{-1}(y))}$.

Proof. The strict monotonicity of f implies that it is injective. Therefore, for each $y \in f(I)$, there exists a unique $x \in I$ with

$$f(x) = y,$$

and we put

$$x = \varphi(y).$$

φ then satisfies (10) and (11).

The strict monotonicity of f implies the strict monotonicity of φ. If suffices to treat the increasing case as the decreasing case is reduced to the increasing one by considering $-f$ in place of f. In the increasing case,

$$x_1 < x_2 \iff f(x_1) < f(x_2),$$

hence,

$$\varphi(y_1) < \varphi(y_2) \iff y_1 < y_2$$

for $y_i = f(x_i), i = 1, 2$. This is the strict monotonicity of φ.

We shall now verify the continuity of φ. For that purpose, let $y \in f(I), x := \varphi(y) \in I$. Since I is an open interval, there exists $\rho > 0$ with

$$(x - \rho, x + \rho) \subset I.$$

For the continuity of φ, it suffices to consider $0 < \epsilon < \rho$. For such ϵ, $x - \epsilon$ and $x + \epsilon$ belong to I, and since f is strictly increasing

$$f(x - \epsilon) < y < f(x + \epsilon).$$

Hence, there exists $\delta > 0$ with

$$f(x - \epsilon) < y - \delta < y + \delta < f(x + \epsilon).$$

If now $\eta \in f(I)$ satisfies $|\eta - y| < \delta$, the strict monotonicity of $\varphi = f^{-1}$ yields

$$x - \epsilon < f^{-1}(\eta) < x + \epsilon.$$

This means that

$$|f^{-1}(y) - f^{-1}(\eta)| < \epsilon \quad (\text{since } x = f^{-1}(y))$$

whenever $|\eta - y| < \delta$, verifying the continuity of $\varphi = f^{-1}$. For the claim about the differentiability, let $(y_n)_{n \in \mathbb{N}} \subset f(I)$ converge to $y = f(x)$. We put $x_n := \varphi(y_n)$. Since we have already seen that φ is continuous, we conclude

$$\lim_{n \to \infty} x_n = x.$$

We suppose $y_n \neq y$ for all n. By the strict monotonicity of φ, then also $x_n \neq x$ for all n. Now

$$\lim_{n \to \infty} \frac{\varphi(y_n) - \varphi(y)}{y_n - y} = \lim_{n \to \infty} \frac{x_n - x}{f(x_n) - f(x)} = \lim_{n \to \infty} \frac{1}{\frac{f(x_n) - f(x)}{x_n - x}} = \frac{1}{f'(x)},$$

since $f'(x) \neq 0$. Consequently

$$\varphi'(y) = \frac{1}{f'(x)}.$$

$\square$

The preceding results allow to compute the derivatives of many functions:

Examples.

0) $f : \mathbb{R} \to \mathbb{R}, f(x) = c \ (c \in \mathbb{R})$ satisfies $f'(x) = 0$ for all $x \in \mathbb{R}$ as is obvious from definition 2.1.

1) $f : \mathbb{R} \to \mathbb{R}, f(x) = cx, c \in \mathbb{R}$, satisfies

$$f'(x) = c \quad \text{for all } x \in \mathbb{R}.$$

This follows directly from definition 2.1.

2) $f : \mathbb{R} \to \mathbb{R}, f(x) = x^2$, satisfies

$$f'(x) = 2x \quad \text{for all } x \in \mathbb{R}.$$

This follows from 1) and the product rule, writing $x^2 = x \cdot x$.

3)   More generally, for $f : \mathbb{R} \to \mathbb{R}, f(x) = x^m, m \in \mathbb{N}$,

$$f'(x) = mx^{m-1} \quad \text{for all } x \in \mathbb{R},$$

as one verifies by induction on m, (write $x^m = x \cdot x^{m-1}$ and apply the product rule).

4) $f : \mathbb{R} \backslash \{0\} \to \mathbb{R}, f(x) = \frac{1}{x^m}, m \in \mathbb{N}$, satisfies

$$f'(x) = \frac{-m}{x^{m+1}} \quad \text{for all } x \neq 0$$

by 3) and the quotient rule.

5)   For the exponential function $\exp(x) = e^x$, we have

$$\exp'(x) = \lim_{\substack{h \to 0 \\ h \neq 0}} \frac{e^{x+h} - e^x}{h} = e^x \lim_{\substack{h \to 0 \\ h \neq 0}} \frac{e^h - 1}{h}, \quad \text{using (12) in §0}$$

$$= e^x \quad \text{by (26) in §0.}$$

Thus, the exponential function has the remarkable property that it coincides with its own derivative, i.e. satisfies

$$f'(x) = f(x) \quad \text{for all } x \in \mathbb{R}.$$

6)   More generally, $f(x) = e^{cx}, c \in \mathbb{R}$, satisfies

$$f'(x) = ce^{cx} = cf(x) \quad \text{for all } x \in \mathbb{R},$$

as follows from 1), 5) and the chain rule.

7)   For the logarithm, theorem 2.7 yields in conjunction with 5)

$$\log'(x) = \frac{1}{\exp'(\log x)} = \frac{1}{\exp(\log x)} = \frac{1}{x} \quad \text{for } x > 0.$$

8)   For a differentiable function $f : D \to \mathbb{R}$ and $x \in D$ with $f(x) > 0$, the chain rule and 7) yield

$$(\log f(x))' = \log'(f(x))f'(x) = \frac{f'(x)}{f(x)}.$$

We next define second and higher order derivatives:

Definition 2.8 Let $f : D \to \mathbb{R}$ be a differentiable function. We say that f is twice differentiable at $x \in D$ if the derivative f' of f is differentiable at x. We write

$$f''(x) := (f')'(x) \quad \text{(the derivative of } f' \text{ at } x),$$

and also $\frac{d^2 f}{dx^2}(x)$ in place of $f''(x)$. (The latter symbol is explained as $(\frac{d^2}{dx^2})f = (\frac{d}{dx})^2 f = \frac{d}{dx}(\frac{df}{dx})$.) Inductively, we say that a k-times differentiable function $f : D \to \mathbb{R}$ is $(k+1)$-times differentiable at $x \in D$ if the kth derivative $f^{(k)} = \frac{d^k}{dx^k}f$ is differentiable at x. It is $(k+1)$-times continuously differentiable in D if the $(k+1)$st derivative $f^{(k+1)}$ exists and is continuous in D.

Definition 2.9 $C^k(D, \mathbb{R}) = C^k(D)$ is the vector space of k-fold continuously differentiable real valued functions on D. $C^{k,\alpha}(D, \mathbb{R}) = C^{k,\alpha}(D)$ is the vector space of those $f \in C^k(D)$ with $f^{(k)} \in C^{0,\alpha}(D)$ $(0 < \alpha \leq 1)$. Finally, we put $C^\infty(D, \mathbb{R}) = C^\infty(D) := \bigcap_{k \in \mathbb{N}} C^k(D)$, the vector space of infinitely often differentiable functions on D.

It is clear that e^x is in $C^\infty(\mathbb{R})$, because it is differentiable and coincides with its own derivative so that the latter is differentiable as well and so inductively all derivatives are differentiable. Likewise, any polynomial is in $C^\infty(\mathbb{R})$ because the derivative of a polynomial is again a polynomial. A less trivial example of an infinitely often differentiable function is

$$f(x) := \begin{cases} e^{-\frac{1}{1-x^2}} & \text{for } |x| < 1 \\ 0 & \text{for } |x| \geq 1 \ . \end{cases}$$

Using the chain rule and induction, it is easy to verify that f is infinitely often differentiable for $|x| \neq 1$, and thus we only need to check the differentiability properties at ± 1. Since f is even, i.e. $f(x) = f(-x)$ for all x, the behaviour at $x = -1$ is the same as the one at $x = 1$, and so we only consider the latter. We put $y := \varphi(x) := \frac{1}{1-x^2}$. Thus $f(x) = e^{-y}$. Since

$$\lim_{\substack{x \to 1 \\ x < 1}} \varphi(x) = \infty,$$

(15) of §0 implies that

$$\lim_{\substack{x \to 1 \\ x < 1}} f(x) = 0.$$

Thus, f is continuous at $x = 1$. This, of course, is only the first step, and we have to investigate the differentiability properties of f at $x = 1$.

For $|x| < 1$, by the chain and quotient rules,

$$f'(x) = \frac{2x}{(1-x^2)^2} f(x) = 2xy^2 e^{-y}.$$

Using (15) of §0 again, we see that

$$\lim_{\substack{x \to 1 \\ x < 1}} f'(x) = 0.$$

Therefore, f is differentiable at $x = 1$ (see exercise 2)). Inductively, one realizes that for any $n \in \mathbb{N}$ $f^{(m)}(x)$ for $|x| < 1$ has the structure of a polynomial in x and $y = \varphi(x)$ multiplied by e^{-y}. All terms containing x are bounded for $|x| < 1$, and (15) of §0 shows that for any $m \in \mathbb{N}$

$$\lim_{y \to \infty} y^m e^{-y} = 0,$$

and therefore

$$\lim_{\substack{x \to 1 \\ x < 1}} f^{(n)}(x) = 0 \quad \text{for all } n \in \mathbb{N}.$$

Thus, f is differentiable to any order at $x = 1$, with

$$f^{(n)}(1) = 0 \quad \text{for all } n \in \mathbb{N}.$$

Exercises for § 2

1) Which of the following functions are differentiable at $x_0 = 0$?

   a) $f(x) = x|x|$

   b) $f(x) = |x|^{\frac{1}{2}}$

   c) $f(x) = |x|^{\frac{3}{2}}$

   d) $f(x) = \begin{cases} e^{-\frac{1}{|x|}} & \text{for } x \neq 0 \\ 0 & \text{for } x = 0 \end{cases}$

   e) $f(x) = \begin{cases} \frac{1}{x} e^{-\frac{1}{|x|}} & \text{for } x \neq 0 \\ 0 & \text{for } x = 0 \end{cases}$

2) Let $f : D \to \mathbb{R}$ $(D \subset \mathbb{R})$, be a function, $x \in D$.

   a)   Show that f is differentiable at x precisely if the following limits both exist and coincide:

   $$f'_-(x) := \lim_{\substack{\xi \to x \\ \xi \in D, \xi < x}} \frac{f(\xi) - f(x)}{\xi - x}, \quad f'_+(x) := \lim_{\substack{\xi \to x \\ \xi \in D, \xi > x}} \frac{f(\xi) - f(x)}{\xi - x}.$$

   b)   Show that f is differentiable at x if it is differentiable on $D \setminus \{x\}$, and if

   $$\lim_{\substack{z \to x \\ z \in D \setminus \{x\}}} f'(z)$$

   exists. That limit then yields $f'(x)$.

3) Let $I \subset \mathbb{R}$ be an interval, $f, g : I \to \mathbb{R}, x_0 \in I, f(x_0) = 0$, and assume that f is differentiable at x_0. Show that $f \cdot g$ is differentiable at x_0, and compute $(f \cdot g)'(x_0)$, provided one of the following assumptions holds:

   a: g is continuous at x_0.

   b: $f'(x_0) = 0$, and g is bounded on I.

4) Let $f : \mathbb{R} \to \mathbb{R}$ be a differentiable function, $x_0 \in \mathbb{R}$. Show that

   $$\bar{f}(x) := f(x_0) + f'(x_0)(x - x_0)$$

   is the best approximation of f at x_0 among all affine linear functions in the following sense: For any affine linear $\ell : \mathbb{R} \to \mathbb{R}$ (i.e. $\ell(x) = cx + d$), there exists $\delta > 0$ such that for all x with $|x - x_0| < \delta$

   $$|f(x) - \bar{f}(x)| \leq |f(x) - \ell(x)|.$$

3. Characteristic Properties of Differentiable Functions. Differential Equations

We treat the mean value theorems for differentiable functions, characterize interior minima and maxima of such functions in terms of properties of the derivatives, discuss some elementary aspects of differential equations, and finally show Taylor's formula.

Definition 3.1 Let $D \subset \mathbb{R}, f : D \to \mathbb{R}$ a function. We say that f has a local minimum (maximum) at $x_0 \in d$ if there exists $\varepsilon > 0$ with the property that for all $y \in D$ with $|x_0 - y| < \varepsilon$

$$f(y) \geq f(x_0) \quad (f(y) \leq f(x_0)). \tag{1}$$

Idxstrict local minimum (maximum) A local minimum (maximum) is called strict if we have " $>$ " (" $<$ ") for $y \neq x_0$ in (3.1).

$x_0 \in D$ is called a global minimum (maximum) for f in D if (3.1) holds for all $y \in D$, and not only for those with $|x_0 - y| < \varepsilon$. Often, the qualifier "global" is omitted.

Remark. x_0 is a (global) minimum for f on D precisely if

$$f(x_0) = \inf\{f(x) : x \in D\}.$$

Thus, the terminology just defined is consistent with the one implicitly employed in §1.

Theorem 3.2 *Let $f : (a, b) \to \mathbb{R}$ be a function with a local minimum or maximum at $x_0 \in (a, b)$. If f is differentiable at x_0, then $f'(x_0) = 0$.*

Proof. We only treat the case of a minimum, as the case of a maximum is reduced to the case of a minimum by considering $-f$ in place of f. Thus, we have

$$f(\xi) \geq f(x_0)$$

for all $\xi \in (a, b) \cap (x_0 - \varepsilon, x_0 + \varepsilon)$ for some $\varepsilon > 0$.

Thus, if $\xi < x_0$

$$\frac{f(x_0) - f(\xi)}{x_0 - \xi} \leq 0$$

while for $\xi > x_0$

$$\frac{f(x_0) - f(\xi)}{x_0 - \xi} \geq 0.$$

As f is differentiable at x_0, we conclude that necessarily

$$f'(x_0) = \lim_{\substack{\xi \to x_0 \\ \xi \neq x_0}} \frac{f(x_0) - f(\xi)}{x_0 - \xi} = 0.$$

$\square$

Examples.

1) $f(x) = x^2, f : \mathbb{R} \to \mathbb{R}$, has a global minimum at $x_0 = 0$.

2) $f(x) = x^3, f : \mathbb{R} \to \mathbb{R}$, satisfies $f'(0) = 0$ although it neither has a local minimum nor a local maximum at $x_0 = 0$. Thus, $f'(x_0) = 0$ is not a sufficient condition for a local minimum or maximum.

3) $f(x) = x, f : [0,1] \to \mathbb{R}$, has a local minimum at $x_0 = 0$ and a local maximum at $x_0 = 1$. However, at neither point $f'(x_0) = 0$. The reason is that the domain of definition of f, $[0,1]$, does not contain an open interval around $x_0 = 0$ or $x_0 = 1$.

4) $f(x) = 2x^3 - 3x^2, f : \mathbb{R} \to \mathbb{R}$, satisfies $f'(x_0) = 0$ for $x_0 = 0, 1$, and it has a local maximum at $x_0 = 0$, and a local minimum at $x_0 = 1$.

Theorem 3.3 (Rolle's theorem) *Let $f : [a,b] \to \mathbb{R}$ be continuous and differentiable on (a,b) $(a < b)$. Let $f(a) = f(b)$. Then there exists an $x_0 \in (a,b)$ with $f'(x_0) = 0$.*

Proof. If f is constant on $[a,b]$, then $f'(x) = 0$ for all $x \in (a,b)$. If f is not constant, we may find some $x_1 \in (a,b)$ with $f(x_1) \neq f(a)(= f(b))$. W.l.o.g., let us suppose that $f(x_1) > f(a)$. By theorem, the continuous function f assumes its maximum at some $x_0 \in [a,b]$. For a maximum point, we must have $f(x_0) \geq f(x_1) > f(a) = f(b)$, and so we must have $x_0 \in (a,b)$. Since f is differentiable on (a,b), theorem 3.2 implies $f'(x_0) = 0$. $\square$

Corollary 3.4 (2nd mean value theorem) *Let $f, g : [a,b] \to \mathbb{R}$ be continuous functions which are differentiable on (a,b). Let $g(b) \neq g(a)$ and $g'(x) \neq 0$ for all $x \in (a,b)$. Then there exists an $x_0 \in (a,b)$ with*

$$\frac{f(b) - f(a)}{g(b) - g(a)} = \frac{f'(x_0)}{g'(x_0)}.$$

Proof. Let

$$F(x) := f(x) - f(a) - \frac{f(b) - f(a)}{g(b) - g(a)}(g(x) - g(a)).$$

Now $F(a) = 0 = F(b)$ so by Rolle's theorem there exists an $x_0 \in (a, b)$ with

$$0 = F'(x_0) = f'(x_0) - \frac{f(b) - f(a)}{g(b) - g(a)} g'(x_0).$$

□

A special case is the *1st mean value theorem:*

Corollary 3.5 *Let $a < b, f : [a, b] \to \mathbb{R}$ be continuous on $[a, b]$ and differentiable on (a, b). Then there exists an $x_0 \in (a, b)$ with*

$$f'(x_0) = \frac{f(b) - f(a)}{b - a}.$$

A further consequence follows from corollary 3.5, namely

Corollary 3.6 *Let $f : [a, b] \to \mathbb{R}$ be continuous on $[a, b]$ and differentiable on (a, b) with*

$$\mu \leq f'(x) \leq m \quad \text{for all } x \in (a, b).$$

We then have for $a \leq x_1 \leq x_2 \leq b$

$$\mu(x_2 - x_1) \leq f(x_2) - f(x_1) \leq m(x_2 - x_1).$$

In particular if $M := \max(|\mu|, |m|)$ then

$$|f(x_2) - f(x_1)| \leq M|x_2 - x_1| \quad \text{for all } x_1, x_2 \in (a, b).$$

Therefore if $f'(x) \equiv 0$ then f is constant.

□

In particular, we have

Corollary 3.7 *Let $f : (a, b) \to \mathbb{R}$ be differentiable. If $f'(x) \geq 0$ for all $x \in (a, b)$, then*

$$f(x_1) \leq f(x_2) \quad \text{whenever } a < x_1 \leq x_2 < b,$$

i.e. f is monotonically increasing.
    Likewise if $f'(x) \leq 0$ on (a, b) then

$$f(x_1) \geq f(x_2) \quad \text{for } a < x_1 \leq x_2 < b,$$

i.e. f is monotonically decreasing.
    If $f'(x) > 0(< 0)$, we get the strict inequality in (2) (resp. (3)) for $x_1 < x_2$.

□

Another easy consequence of corollary 3.6 is

Corollary 3.8 *Let $f_1, f_2 : [a, b] \to \mathbb{R}$ be continuous on $[a, b]$ and differentiable on (a, b), with*

$$f_1(a) = f_2(a)$$

and

$$f_1'(x) \leq f_2'(x) \quad \text{for all } x \in (a, b).$$

Then

$$f_1(x) \leq f_2(x) \quad \text{for all } x \in [a, b].$$

Proof. The function $f = f_1 - f_2$ satisfies

$$f'(x) \leq 0 \quad \text{for all } x \in (a, b).$$

We apply corollary 3.6 with $m = 0, x_1 = a, x_2 = x$ to conclude

$$f(x) \leq 0$$

which is our claim.

□

Corollary 3.6 also implies

Corollary 3.9 *Let $f : [a, b] \to \mathbb{R}$ be continuous on $[a, b]$ and differentiable on (a, b). Suppose*

$$f'(x) \equiv \gamma$$

for some constant γ.
     Then

$$f(x) = \gamma x + c,$$

with some constant c, for all $x \in [a, b]$.

Proof. We consider

$$F(x) := f(x) - \gamma x.$$

$F : [a, b] \to \mathbb{R}$ satisfies the assumptions of corollary 3.6, and

$$F'(x) = f'(x) - \gamma \equiv 0$$

on (a, b). Therefore, $F \equiv c$ on $[a, b]$, for some constant c. The claim follows.

□

Theorem 3.10 *Let $\gamma \in \mathbb{R}$ and $f : \mathbb{R} \to \mathbb{R}$ be a function which satisfies the differential equation $f' = \gamma f$, that is, $f'(x) = \gamma f(x)$ for all $x \in \mathbb{R}$. Then*

$$f(x) = f(0)e^{\gamma x} \quad \text{for all } x \in \mathbb{R}.$$

Proof 1. We consider

$$F(x) := f(x)e^{-\gamma x}.$$

Now

$$F'(x) = f'(x)e^{-\gamma x} - \gamma f(x)e^{-\gamma x} = 0,$$

so by corollary 3.6

$$F \equiv \text{const.} = F(0) = f(0),$$

and therefore

$$f(x) = f(0)e^{\gamma x}.$$

□

Proof 2. Either $f \equiv 0$, in which case there is nothing to prove, or else there exists an $x_0 \in \mathbb{R}$ with $f(x_0) \neq 0$; say $f(x_0) > 0$ (otherwise one considers $-f$). As f is continuous, there exists a neighborhood $U(x_0)$ with $f(x) > 0$ for all $x \in U(x_0)$. For $x \in U(x_0)$ we have

$$(\log f(x))' = \frac{f'(x)}{f(x)} = \gamma,$$

so by corollary 3.9, $\log f(x) = \gamma x + c$, for some constant c. Hence

$$f(x) = c_1 e^{\gamma x} \text{ where } c_1 = e^c. \tag{1}$$

We now show that if f is not identically zero then $f(x) = 0$ has no solutions. For this, it is important to observe that (1) holds on any neighborhood $U(x_0)$ provided $f(x) \neq 0$ for all $x \in U(x_0)$. Now if f is not identically zero, say $f(x_0) > 0$ for some x_0, there exists, by continuity of f, a smallest $x_1 > x_0$ with $f(x_1) = 0$ or a greatest $x_2 < x_0$ with $f(x_2) = 0$. In the first case (1) holds for $x_0 \leq x < x_1$. But then

$$\lim_{\substack{x \to x_1 \\ x < x_1}} f(x) = c_1 e^{\gamma x_1} > 0 = f(x_1),$$

in contradiction to the continuity of f.

Therefore $f(x_1) \neq 0$ and similarly $f(x_2) \neq 0$. Consequently f has no zeroes and (1) holds for all $x \in \mathbb{R}$. Setting $x = 0$ one has $c_1 = f(0)$.   □

In the proof above we have used local solutions to rule out zeroes. Nevertheless, one can also argue abstractly.

Theorem 3.11 *Let $f : [a, b] \to \mathbb{R}$ be a differentiable function that satisfies for all $x \in [a, b]$ $|f'(x)| \leq \gamma|f(x)|$, γ a constant. If $f(x_0) = 0$ for some $x_0 \in [a, b]$ then $f \equiv 0$ on $[a, b]$.*

Proof. We may assume that $\gamma > 0$, otherwise there is nothing to prove. Set

$$\delta := \frac{1}{2\gamma}$$

and choose $x_1 \in [x_0 - \delta, x_0 + \delta] \cap [a, b] =: I$ such that

$$|f(x_1)| = \sup_{x \in I} |f(x)|$$

(such an x_1 exists by continuity of f).

By corollary 3.6

$$|f(x_1)| = |f(x_1) - f(x_0)| \le |x_1 - x_0| \sup_{\xi \in I} |f'(\xi)|$$

$$\le \gamma |x_1 - x_0| \sup_{\xi \in I} |f(\xi)| \le \gamma \delta |f(x_1)| = \frac{1}{2} |f(x_1)|,$$

and therefore $f(x_1) = 0$. It follows that

$$f(x) = 0 \quad \text{for all } x \in I.$$

We have therefore shown that there exists a $\delta > 0$ with the following property: If $f(x_0) = 0$ then $f(x) = 0$ for all $x \in [x_0 - \delta, x_0 + \delta] \cap [a, b]$. If f is not identically zero, there exists a smallest ξ_1 with $a < \xi_1 \le b$ and $f(\xi_1) = 0$, or a greatest ξ_2 with $a \le \xi_2 < b$ and $f(\xi_2) = 0$. However, this is not compatible with the statement which we just proved. $\quad\square$

Corollary 3.12 *Let $\phi : \mathbb{R} \to \mathbb{R}$ be Lipschitz continuous, $c \in \mathbb{R}$ and $[a, b] \subset \mathbb{R}$. There exists at most one solution $f : [a, b] \to \mathbb{R}$ of the differential equation*

$$f'(x) = \phi(f(x)) \text{ for all } x \in [a, b]$$

with

$$f(a) = c.$$

Proof. Let f_1 and f_2 be solutions with $f_1(a) = f_2(a) = c$. The function $F = f_1 - f_2$ satisfies

$$F(a) = 0$$

and

$$|F'(x)| = |\phi(f_1(x)) - \phi(f_2(x))|$$
$$\le L|f_1(x) - f_2(x)| = L|F(x)|$$

for a suitable constant L, as f_1 and f_2, being continuous, map the bounded interval $[a, b]$ onto a bounded interval and ϕ is Lipschitz continuous. Theorem 3.11 implies that $F \equiv 0$, that is $f_1 \equiv f_2$, whence the uniqueness of the solutions. $\quad\square$

We shall see below, in theorem 6.16, that the differential equation

$$f'(x) = \phi(x, f(x)) \tag{2}$$

indeed possesses a solution with $f(a) = c$ on some interval $[a, a+h]$, provided that ϕ is uniformly bounded and uniformly Lipschitz continuous with respect to the second variable, i.e.

$$|\phi(x, y)| \leq M \tag{3}$$

and

$$|\phi(x, y_1) - \phi(x, y_2)| \leq L|y_1 - y_2|, \tag{4}$$

for some constants M and L and all relevant x, y, y_1, y_2.

In order to see why such a condition of uniform Lipschitz continuity is needed, we shall now analyze a situation where the solution of a differential equation becomes infinite on a finite interval.

Theorem 3.13 *We consider the differential equation*

$$f'(x) = \phi(f(x)) \tag{5}$$

with

$$\phi(y) \geq \gamma y^\alpha \tag{6}$$

for some $\gamma > 0, \alpha > 1$. Let f be a solution of (5) with

$$f(a) = c > 0.$$

Then there exists some $b_1 > a$ such that $f(x)$ tends to ∞ when x approaches b_1 from below. Thus, f cannot be continued as a solution of (5) beyond b_1.

Remark. One says that f "blows up" at b_1.

Proof. Since $f(a) = c > 0$ and the right hand side of (5) is positive whenever $f(x)$ is positive, $f(x)$ is a monotonically increasing function of x. Let $n \in \mathbb{N}, n \geq c$. We now suppose that we have found some $x_n \geq a$ with

$$f(x_n) = n.$$

Since f is monotonically increasing and consequently $\phi(f(x)) \geq \gamma x^\alpha \geq \gamma x_n^\alpha$ for $x \geq x_n$, we have for $x \geq x_n$, by corollary 3.6, that

$$f(x) \geq f(x_n) + (x - x_n)\gamma x_n^\alpha$$
$$= n + (x - x_n)\gamma x_n^\alpha.$$

From this, we find x_{n+1} with $f(x_{n+1}) = n + 1$, and

$$n + 1 = f(x_{n+1}) \geq n + (x_{n+1} - x_n)\gamma x_n^\alpha$$

implies

$$x_{x+1} \leq \frac{1}{\gamma n^\alpha} + x_n.$$

Iterating this inequality yields

$$x_N \leq \sum_{n=n_0}^{N-1} \frac{1}{\gamma n^\alpha} + x_{n_0}$$

for all $N \geq n_0 \geq c$. We then choose n_0 as the smallest $n \in \mathbb{N}$ with $n \geq c$, and by the same reasoning as above, we obtain

$$x_{n_0} \leq \frac{n_0 - c}{\gamma c^\alpha} + a.$$

Altogether,

$$x_N \leq \sum_{n=n_0}^{N-1} \frac{1}{\gamma n^\alpha} + \frac{n_0 - c}{\gamma c^\alpha} + a.$$

The essential point now is that the preceding sum, when extended from n_0 to ∞, is finite for $\alpha > 1$. We put

$$b_0 := \sum_{n=n_0}^{\infty} \frac{1}{\gamma n^\alpha} + \frac{n_0 - c}{\gamma c^\alpha} + a.$$

Since then

$$x_N \leq b_0,$$

and $x_{N+1} \geq x_N$, because f is monotonically increasing, $(x_N)_{N \in \mathbb{N}}$ converges to some $b_1 \leq b_0$. Since x_N is chosen such that $f(x_N) = N$ and f is monotonically increasing, we conclude that

$$f(x) \to \infty$$

as x approaches b_1.

$\square$

In order to complete our qualitative picture of the growth behavior of solutions of ODEs, we now consider the differential inequality

$$f'(x) \leq \gamma f(x)^\beta \tag{7}$$

for a positive function f, some constant γ, and

$$0 \leq \beta < 1.$$

We assume that f solves this inequality for $x \geq a$.

For $g := f^{1-\beta}$, we then have

$$g'(x) = (1 - \beta) f^{-\beta}(x) f'(x)$$
$$\leq \gamma(1 - \beta).$$

Suppose that

$$g(a) = c.$$

Then, from corollaries 3.9 and 3.8, for $x \geq a$

$$g(x) \leq \gamma(1 - \beta)(x - a) + c.$$

In terms of our original f, we get

$$f(x) \leq (\gamma(1 - \beta)(x - a) + c)^{\frac{1}{1-\beta}}.$$

The essential feature here is that f asymptotically grows like the power

$$x^{\frac{1}{1-\beta}}.$$

We now can summarize our results about the qualitative behavior of solutions of ODEs ($\gamma > 0$):

Case A:
$$f' = \gamma f.$$

Here, $f(x)$ is proportional to $e^{\gamma x}$, i.e., it grows exponentially.

Case B:
$$f' \geq \gamma f^\alpha \quad \text{for } \alpha > 1.$$

Here, a positive solution necessarily blows up, i.e., it grows so fast that it becomes infinite on some finite interval. Sometimes, this is called hyperbolic growth.

Case C:

$$f' \leq \gamma f^\beta \quad \text{for some } \beta < 1.$$

Here, a solution is controlled by a finite power of x, namely $x^{\frac{1}{1-\beta}}$. It thus grows at most polynomially.

If we even have

$$f' \leq \gamma,$$

then f grows at most linearly.

We now return to the local discussion of functions and their derivatives.

In order to distinguish minima from maxima, it is convenient to consider second derivatives as well.

Theorem 3.14 *Let $f : (a, b) \to \mathbb{R}$ be twice differentiable, and let $x_0 \in (a, b)$, with*

$$f'(x_0) = 0, f''(x_0) > 0. \tag{8}$$

Then f has a strict local minimum at x_0. If we have $f''(x_0) < 0$ instead, it has a strict local maximum at x_0. Conversely, if f has a local minimum at $x_0 \in (a, b)$, and if it is twice differentiable there, then

$$f''(x_0) \geq 0. \tag{9}$$

Proof. We only treat the case of a local minimum. We apply the reasoning of corollary 3.7 to f' in place of f. If

$$\lim_{\substack{x \to x_0 \\ x \neq x_0}} \frac{f'(x) - f'(x_0)}{x - x_0} = f''(x_0) > 0,$$

then there exists $\delta > 0$ with

$$f'(x) < f'(x_0) = 0 \quad \text{for } x_0 - \delta < x < x_0$$

and

$$f'(x) > f'(x_0) = 0 \quad \text{for } x_0 < x < x_0 + \delta.$$

Thus, by corollary 3.7, f is strictly monotonically decreasing on $(x_0 - \delta, x_0)$, and strictly monotonically increasing on $(x_0, x_0 + \delta)$. This implies that

$$f(x) > f(x_0) \quad \text{for } 0 < |x - x_0| < \delta,$$

and consequently f has a strict local minimum at x_0.
The second half of the theorem follows from the first half. $\qquad \square$

Examples.
1) $f : \mathbb{R} \to \mathbb{R}, f(x) = x^2$ satisfies $f'(0) = 0, f''(0) > 0$. f therefore has a strict local minimum at $x_0 = 0$.

2) $f : \mathbb{R} \to \mathbb{R}, f(x) = x^4$ satisfies $f'(0) = 0, f''(0) = 0$. The condition of theorem 3.14 thus does not hold, but f nevertheless has a strict local minimum at $x_0 = 0$.

3) $f : \mathbb{R} \to \mathbb{R}, f(x) = x^3$ satisfies $f'(0) = 0, f''(0) = 0$. At $x_0 = 0$, f has neither a minimum nor a maximum.

We finally discuss the famous theorem of Taylor.

Theorem 3.15 (Taylor expansion) *Assume that the function*

$$f : [x_0, x] \to \mathbb{R} \quad (\text{or } f : [x, x_0] \to \mathbb{R},$$
$$\text{depending on whether } x_0 < x \text{ or } x < x_0)$$

possesses a continuous derivative $f^{(n)}$ on $[x_0, x]$ and is even $(n + 1)$ times differentiable on (x_0, x) (resp. on (x, x_0)). Then there exists some ξ between x_0 and x with

$$f(x) = f(x_0) + f'(x_0)(x - x_0) + \frac{1}{2!}f''(x_0)(x - x_0)^2$$

$$\tag{10}$$

$$+ \ldots + \frac{1}{n!}f^{(n)}(x_0)(x - x_0)^n + \frac{1}{(n+1)!}f^{(n+1)}(\xi)(x - x_0)^{n+1}.$$

Proof. We may find $z \in \mathbb{R}$ with

$$f(x) = \sum_{\nu=0}^{n} \frac{1}{\nu!} f^{(\nu)}(x_0)(x - x_0)^{\nu} + \frac{1}{(n+1)!}(x - x_0)^{n+1} z. \qquad (11)$$

We consider

$$\varphi(y) := f(x) - \sum_{\nu=0}^{n} \frac{1}{\nu!} f^{(\nu)}(y)(x - y)^{\nu} - \frac{1}{(n+1)!}(x - y)^{n+1} z.$$

φ is continuous on $[x_0, x]$ and differentiable on (x_0, x) (we discuss only the case $x_0 < x$, as the case $x > x_0$ is analogous). Moreover, $\varphi(x) = 0$, and also $\varphi(x_0) = 0$ by choice of z. By Rolle's theorem 3.3, there exists some $\xi \in (x_0, x)$ with

$$\varphi'(\xi) = 0.$$

As

$$\varphi'(y) = -f'(y) - \sum_{\nu=1}^{n} \left(\frac{1}{\nu!} f^{(\nu+1)}(y)(x - y)^{\nu} - \frac{1}{(\nu-1)!} f^{(\nu)}(y)(x - y)^{\nu-1} \right)$$

$$+ \frac{1}{n!}(x - y)^n z$$

$$= -\frac{1}{n!} f^{(n+1)}(y)(x - y)^n + \frac{1}{n!}(x - y)^n z,$$

$\varphi'(\xi) = 0$ implies that

$$z = f^{(n+1)}(\xi).$$

Inserting this in (11) yields (10). $\qquad \square$

Exercises for § 3

1) Let $f : \mathbb{R} \to \mathbb{R}$ be differentiable with

$$f^{n+1} \equiv 0 \quad \text{for some } n \in \mathbb{N}$$

(i.e. the $(n+1)^{\text{st}}$ derivative of f vanishes identically). Show that f is a polynomial of degree at most n.

2 Let $f : [a, b] \to \mathbb{R}$ be continuous, and let it be twice differentiable in (a, b) with

$$f'' = cf \quad \text{for some } c > 0.$$

Show that for all $x \in (a, b)$

$$|f(x)| \leq \max(|f(a)|, |f(b)|).$$

(Hint: Show that f may in (a, b) neither assume a positive maximum nor a negative minimum.)

3)   Let $f : \mathbb{R} \to \mathbb{R}$ be differentiable with

$$|f'(x)| \le cf(x) \quad \text{for all } x \in \mathbb{R} \text{ and some } c \ge 0.$$

a)   Show that $|f(x)| \le \gamma e^{c|x|}$ for some $\gamma \ge 0$.
b)   Assume in addition that $f(x_0) = 0$ for some $x_0 \in \mathbb{R}$. Show that $f \equiv 0$.

4)   Let $f : [a, b] \to [a, b]$ be differentiable, $f'(x) \ne 1$ for all $x \in [a, b]$. Show that there exists a unique $x \in [a, b]$ with $f(x) = x$. (The existence of such an x has been shown already in exercise. The point here is the uniqueness.)

5)   Let $f : (a, b) \to \mathbb{R}$ be differentiable, $x_0 \in (a, b)$, f' continuous at x_0. Let $(x_n)_{n \in \mathbb{N}}, (y_n)_{n \in \mathbb{N}} \subset (a, b)$ with $x_n \ne y_n$ for all n, $\lim\limits_{n \to \infty} x_n = \lim\limits_{n \to \infty} y_n = x_0$. Show that

$$\lim_{n \to \infty} \frac{f(x_n) - f(y_n)}{x_n - y_n} = f'(x_0).$$

6)

a)   Let $p \ge 1, f : [0, 1] \to \mathbb{R}, f(x) = \frac{(1+x)^p}{1+x^p}$. Determine the maximum and minimum of f in $[0, 1]$. Use this to conclude the following inequalities:
For all $a, b \in \mathbb{R}$

$$|a|^p + |b|^p \le (|a| + |b|)^p \le 2^{p-1}(|a|^p + |b|^p).$$

b)   Let $p > 1, q := \frac{p}{p-1}, f(x) = \frac{|a|^p}{p} + \frac{x^q}{q} - |a|x$ for $x \ge 0, a \in \mathbb{R}$. Determine the minimum of f for $x \ge 0$ and conclude the following inequality: For all $a, b \in \mathbb{R}$

$$|ab| \le \frac{|a|^p}{p} + \frac{|b|^q}{q}.$$

7)   Let $f : \mathbb{R} \to \mathbb{R}$ be differentiable, $f \not\equiv 0$, and satisfy

$$f(x + y) = f(x)f(y) \quad \text{for all } x, y \in \mathbb{R}.$$

Show that
$$f(x) = e^{\lambda x} \quad \text{for some } \lambda \in \mathbb{R}.$$

4. The Banach Fixed Point Theorem. The Concept of Banach Space

As the proper setting for the convergence theorems of subsequent §§, we introduce the concept of a Banach space as a complete normed vector space. The Banach fixed point theorem is discussed in detail.

Theorem 4.1 *Let $I \subset \mathbb{R}$ be a closed, not necessarily bounded, interval and $f : I \to \mathbb{R}$ a function with $f(I) \subset I$ and which satisfies for a fixed θ, $0 \leq \theta < 1$, the inequality*

$$|f(x) - f(y)| \leq \theta |x - y| \quad \text{for all } x, y \in I. \tag{1}$$

Then there exists exactly one fixed point of f, i.e. a $\xi \in I$ with

$$f(\xi) = \xi.$$

Proof. We choose an arbitrary $x_0 \in I$ and set iteratively

$$x_n := f(x_{n-1}) \quad \text{for } n \geq 1.$$

This is possible as $f(I) \subset I$. We now show that (x_n) forms a Cauchy sequence. For
$n > m, m \geq 1$ we have

$$
\begin{aligned}
|x_n - x_m| &\leq |x_n - x_{n-1}| + |x_{n-1} - x_{n-2}| + \ldots + |x_{m+1} - x_m| \\
&= |f(x_{n-1}) - f(x_{n-2})| + \ldots + |f(x_m) - f(x_{m-1})| \\
&\leq \theta(|x_{n-1} - x_{n-2}| + \ldots + |x_m - x_{m-1}|) \\
&\leq \sum_{\nu=m}^{n-1} \theta^\nu |x_1 - x_0| \\
&\leq \theta^m \frac{1}{1 - \theta} |x_1 - x_0| \text{ (taking into consideration that } \theta < 1).
\end{aligned}
$$

As $\theta < 1$ we deduce,

$$\forall \varepsilon > 0 \ \exists N \in \mathbb{N} \ \forall n, m \geq N : |x_n - x_m| < \varepsilon.$$

This shows that $(x_n)_{n \in \mathbb{N}}$ is a Cauchy sequence.

As $\mathbb{R}$ is complete, the sequence (x_n) converges to an $\xi \in \mathbb{R}$ and as I is closed, $\xi \in I$. Furthermore, as f is continuous on account of (1), we have:

$$f(\xi) = \lim_{n \to \infty} f(x_n) = \lim_{n \to \infty} x_{n+1} = \lim_{m \to \infty} x_m = \xi.$$

Therefore ξ is a fixed point.

For uniqueness let ξ_1, ξ_2 be fixed points, so

$$f(\xi_1) = \xi_1, f(\xi_2) = \xi_2.$$

From (1) it follows that

$$|\xi_1 - \xi_2| = |f(\xi_1) - f(\xi_2)| \le \theta|\xi_1 - \xi_2|$$

and as $\theta < 1$ we have $\xi_1 = \xi_2$, hence the uniqueness.   □

Corollary 4.2 Let $I \subset \mathbb{R}$ be a closed interval, $f : I \to \mathbb{R}$ a differentiable function with $f(I) \subset I$. Let $\theta, 0 \le \theta < 1$, be given with

$$|f'(x)| \le \theta \quad \text{for all } x \in I.$$

Then there exists exactly one fixed point ξ of f in I.

Proof. The mean value theorem implies that

$$|f(x) - f(y)| \le \theta|x - y| \quad \text{for all } x, y \in I.$$

(Notice that we have used here that I is an interval.)

Therefore, the hypotheses of theorem 4.1 are fulfilled and the assertion follows.   □

We shall now return to theorem 4.1 and analyse its proof closely. Indeed, we shall investigate which properties of the real numbers have been used in its proof. These properties are given in the following definition:

Definition 4.3 Let V be a vector space over $\mathbb{R}$ (or $\mathbb{C}$). A mapping

$$\| \cdot \| : V \to \mathbb{R}$$

is called a norm if the following conditions are fulfilled:

(i)   For $v \in V \backslash \{0\}, \|v\| > 0$ (positive definiteness)

(ii)   For $v \in V, \lambda \in \mathbb{R}$ resp. $\mathbb{C}, \|\lambda v\| = |\lambda| \|v\|$

         For $v, w \in V, \|v + w\| \le \|v\| + \|w\|$ (triangle inequality).

(iii)

A vector space V equipped with a norm $\| \cdot \|$ is called a normed vector space $(V, \| \cdot \|)$. A sequence $(v_n)_{n \in \mathbb{N}} \subset V$ is said to converge (relative to $\| \cdot \|$) to v if

$$\lim_{n\to\infty} \|v_n - v\| = 0;$$

we write $\lim_{n\to\infty} v_n = v$, when $\|\cdot\|$ is clear from the context.

Examples. The absolute value $|\cdot|$ in $\mathbb{R}$ can be generalized to a norm in $\mathbb{R}^d$ by putting for $x = (x^1, \ldots, x^d) \in \mathbb{R}^d$

$$\|x\|_p := \left(\sum_{i=1}^d (|x^i|)^p\right)^{\frac{1}{p}}, \quad \text{with } 1 \le p < \infty,$$

or by

$$\|x\|_\infty := \max_{i=1,\ldots,d} |x^i|.$$

Of particular importance is the Euclidean norm $\|x\|_2$. Sometimes, we shall write $\|x\|$, or even $|x|$, in place of $\|x\|_2$. The notation $|x|$ will be employed when we want emphasize the analogy with the usual absolute value $|\cdot|$ in $\mathbb{R}$. The examples just given are norms on finite dimensional vector spaces. In the sequel, however, we shall encounter norms on infinite dimensional vector spaces, and the properties of certain such norms will form an important object of study for us.

Remark. Of the above properties of a norm, the condition (ii) will not be used in the present §. One can therefore define more generally a metric on a space X as follows:

Definition 4.4 A metric on a space X is a mapping

$$d : X \times X \to \mathbb{R}$$

such that

(i) $\forall\, x, y \in X : d(x,y) \ge 0$
         and $d(x,y) = 0$ only then when $x = y$ (positive definiteness)
(ii) $\forall\, x, y \in X : d(x,y) = d(y,x)$ (symmetry)
(iii) $\forall\, x, y, z \in X : d(x,z) \le d(x,y) + d(y,z)$ (triangle inequality)

A sequence $(x_n)_{n\in\mathbb{N}} \subset X$ is said to converge (relative to d) to $x \in X$ if

$$\lim_{n\to\infty} d(x_n, x) = 0.$$

The fact that every Cauchy sequence in $\mathbb{R}$ is convergent was crucial to the proof of theorem 4.1. We now define next the concept of Cauchy sequence in a normed vector space:

Definition 4.5 Let V be a vector space with a norm $\| \cdot \|$. A sequence $(v_n)_{n \in \mathbb{N}} \subset V$ is called a Cauchy sequence (w.r.t. $\| \cdot \|$) if

$$\forall \, \varepsilon > 0 \, \exists \, N \in \mathbb{N} \, \forall \, n, m \geq N : \|v_n - v_m\| < \varepsilon \, .$$

Now we distinguish those vector spaces in which every Cauchy sequence is convergent.

Definition 4.6 A normed vector space $(V, \| \cdot \|)$ is called complete, or a Banach space, if every Cauchy sequence $(v_n)_{n \in \mathbb{N}} \subset V$ converges to some $v \in V$. A subset A of a Banach space is called closed if $(v_n)_{n \in \mathbb{N}} \subset A$, $\lim\limits_{n \to \infty} v_n = v$ implies $v \in A$.

Our preceding analysis gives as a result:

Theorem 4.7 (Banach fixed point theorem) *Let $(V, \| \cdot \|)$ be a Banach space, $A \subset V$ a closed subset, $f : A \to V$ a function with $f(A) \subset A$ which satisfies the inequality*

$$\|f(v) - f(w)\| \leq \theta \|v - w\| \quad \text{for all } v, w \in V,$$

θ being fixed with $0 \leq \theta < 1$.
  Then f has a uniquely determined fixed point in A.

Proof. The proof is exactly the same as that of theorem 4.1; one substitutes V for $\mathbb{R}$, A for I and $\| \cdot \|$ for $| \cdot |$.   □

Definition 4.8 In continuation of definition 4.5, a sequence $(x_n)_{n \in \mathbb{N}}$ in a metric space (X, d), that is a space X equipped with a metric d, is a Cauchy sequence if
$$\forall \, \varepsilon > 0 \, \exists \, N \in \mathbb{N} \, \forall \, n, m \geq N : d(x_n, x_m) < \varepsilon.$$

A metric space (X, d) is called complete if every Cauchy sequence has a limit in X.

Remark.   Theorem 4.7 also holds in complete metric spaces. Instead of $\|f(v) - f(w)\|$ one now has to write $d(f(v), f(w))$ etc. without changing anything in the proof.

5. Uniform Convergence. Interchangeability of Limiting Processes. Examples of Banach Spaces. The Theorem of Arzela-Ascoli

We introduce the notion of uniform convergence. This leads to Banach spaces of continuous and differentiable functions. We discuss when the limit of the derivatives of a convergent sequence of functions equals the derivative of the limit and related questions. The theorem of Arzela-Ascoli is shown, saying that an equicontinuous and uniformly bounded sequence of functions on a closed and bounded set contains a uniformly convergent subsequence.

Definition 5.1 Let K be a set and $f_n (n \in \mathbb{N})$ real (or complex) valued functions defined on K. The sequence $(f_n)_{n \in \mathbb{N}}$ converges pointwise to a function f if for every $x \in K$ $\lim_{n \to \infty} f_n(x) = f(x)$.

Remark. For every $x \in K$ we have:

$$\forall \varepsilon > 0 \ \exists N \in \mathbb{N} \ \forall n \geq N : |f_n(x) - f(x)| < \varepsilon.$$

Here N depends in general not only on ε but also on x.

Examples.

1) Let $f_n : [0, 1] \to \mathbb{R}$ be the function $f_n(x) = x^n$. The sequence $(f_n)_{n \in \mathbb{N}}$ converges pointwise to the function f defined by:

$$f(x) := \begin{cases} 0, & \text{if } 0 \leq x < 1 \\ 1, & \text{if } x = 1. \end{cases}$$

For $x = 1$ we always have $f_n(x) = 1$, whereas for $0 \leq x < 1$, given $\varepsilon > 0$ there exists an $N \in \mathbb{N}$, e.g. the smallest natural number greater than $\frac{\log \varepsilon}{\log x}$, such that

$$|f_n(x) - 0| = |f_n(x)| = x^n < \varepsilon \quad \text{for all } n \geq N.$$

We observe that the limit function f is not continuous, although all the f_n are continuous. The concept of pointwise convergence is therefore too weak to allow for continuity properties to carry over to limit functions.

2) The weakness of this convergence concept is demonstrated more drastically in the next example:

Define $f_n : [0,1] \to \mathbb{R}, n \geq 2$, by requiring f_n to be continuous and given by the following prescription:

$$f_n(x) := \begin{cases} 0 & \text{for } x = 0 \\ n & \text{for } x = \frac{1}{n} \\ 0 & \text{for } \frac{2}{n} \leq x \leq 1 \\ \text{linear} & \text{for } 0 \leq x \leq \frac{1}{n} \\ \text{linear} & \text{for } \frac{1}{n} \leq x \leq \frac{2}{n} \end{cases}$$

or, concisely

$$f_n(x) := \max(n - n^2 |x - \frac{1}{n}|, 0).$$

Now f_n converges pointwise to the function $f(x) = 0$, since $f(0) = 0$ and for $0 < x \leq 1$ there exists an $N \in \mathbb{N}$ with $\frac{2}{N} \leq x$, so for all $n \geq N$ we have $f_n(x) = 0$ and thus $\lim_{n \to \infty} f_n(x) = 0$. Consequently the sequence (f_n) converges pointwise to 0, although the f_n become unbounded as $n \to \infty$.

We now introduce a better convergence concept.

Definition 5.2 Let K be a set, and $f_n : K \to \mathbb{R}$ or $\mathbb{C}$. The sequence $(f_n)_{n \in \mathbb{N}}$ converges uniformly to the function $f : K \to \mathbb{R}$ or $\mathbb{C}$ if

$$\forall \varepsilon > 0 \; \exists N \in \mathbb{N} \; \forall n \geq N, x \in K : |f_n(x) - f(x)| < \varepsilon.$$

Symbolically: $f_n \rightrightarrows f$.

The crucial point in this formulation is that for all $x \in K$ the same N can be chosen.

Theorem 5.3 Let $K \subset \mathbb{R}$ or $\mathbb{C}$ and $f_n : K \to \mathbb{R}$ (or $\mathbb{C}$) continuous functions which converge uniformly to $f : K \to \mathbb{R}$ (resp. $\mathbb{C}$). Then the function f is continuous.

Proof. Let $x \in K, \varepsilon > 0$. By virtue of the uniform convergence of (f_n), there exists a sufficiently large $N \in \mathbb{N}$ so that for all $\xi \in K$ we have

$$|f_N(\xi) - f(\xi)| < \frac{\varepsilon}{3}.$$

Corresponding to x and ε we then determine a $\delta > 0$ so that

$$|f_N(y) - f_N(x)| < \frac{\varepsilon}{3} \quad \text{for all } y \in K \text{ with } |x - y| < \delta.$$

This is possible as the functions f_N are by assumption continuous. We then have for all $y \in K$ with $|x - y| < \delta$

$$|f(x) - f(y)| \leq |f(x) - f_N(x)| + |f_N(x) - f_N(y)| + |f_N(y) - f(y)|$$
$$< \frac{\varepsilon}{3} + \frac{\varepsilon}{3} + \frac{\varepsilon}{3} = \varepsilon,$$

whereby f is continuous at x and therefore also in K, as $x \in K$ was arbitrary. $\quad\square$

Definition 5.4 Let K be a set and $f : K \to \mathbb{R}$ (or $\mathbb{C}$) a function.

$$\|f\|_K := \sup\{|f(x)| : x \in K\}.$$

Lemma 5.5 $\|\cdot\|_K$ *is a norm on the vector space of bounded real (resp. complex) valued functions on K.*

Proof. If f is bounded then $\|f\|_K < \infty$.

(i) $\qquad\qquad\|f\|_K \geq 0$ for all f, and $\|f\|_K = 0 \Leftrightarrow f \equiv 0$ on K

(ii) $\qquad\qquad$ Let $\lambda \in \mathbb{R}$ resp. $\mathbb{C}$

$\qquad\qquad \Rightarrow \|\lambda f\|_K = \sup\{|\lambda f(x)| : x \in K\}$

$\qquad\qquad = |\lambda| \sup\{|f(x)| : x \in K\} = |\lambda| \|f\|_K$

(iii) $\quad \|f + g\|_K = \sup\{|f(x) + g(x)| : x \in K\}$

$\qquad\qquad \leq \sup\{|f(x)| + |g(x)| : x \in K\}$

$\qquad\qquad \leq \sup\{|f(y)| : y \in K\} + \sup\{|g(z)| : z \in K\}$

$\qquad\qquad = \|f\|_K + \|g\|_K$

(note that f and g could assume their maximum at different points).

Thereby all the properties of a norm are fulfilled. $\quad\square$

Theorem 5.6 $f_n : K \to \mathbb{R}$ *(or $\mathbb{C}$) converges uniformly to $f : K \to \mathbb{R}$ (resp. $\mathbb{C}$) if and only if*

$$\lim_{n \to \infty} \|f_n - f\|_K = 0.$$

Proof.

$f_n \rightrightarrows f \Leftrightarrow \forall\, \varepsilon > 0 \; \exists\, N \in \mathbb{N} \; \forall\, n \geq N, x \in K : |f_n(x) - f(x)| < \varepsilon$

$\qquad \Leftrightarrow \forall\, \varepsilon > 0 \; \exists\, N \in \mathbb{N} \; \forall\, n \geq N : \sup\{|f_n(x) - f(x)| : x \in K\} < \varepsilon$

$\qquad \Leftrightarrow \forall\, \varepsilon > 0 \; \exists\, N \in \mathbb{N} \; \forall\, n \geq N : \|f_n - f\|_K < \varepsilon$

$\qquad \Leftrightarrow \lim_{n \to \infty} \|f_n - f\|_K = 0.$ $\quad\square$

Theorem 5.7 *The space $C_b^0(K) := \{f \in C^0(K) : \|f\|_K < \infty\}$, equipped with the norm $\|\cdot\|_K$, is a Banach space. Correspondingly, so is $C_b^0(K, \mathbb{C})$.*

Proof. Let $(f_n)_{n\in\mathbb{N}} \subset C_b^0(K)$ be a Cauchy sequence relative to $\|\cdot\|_K$. So

$$\forall\, \varepsilon > 0\ \exists\, N \in \mathbb{N}\ \forall\, n, m \geq N : \|f_n - f_m\|_K < \varepsilon,$$
$$\text{and therefore}\quad \sup\{|f_n(x) - f_m(x)| : x \in K\} < \varepsilon.$$

Thereby, for every $y \in K$ $|f_n(y) - f_m(y)| < \varepsilon$, and hence $(f_n(y))_{n\in\mathbb{N}}$ is a Cauchy sequence. As $\mathbb{R}$ is complete, this sequence converges to $f(y)$. For, when we let m tend to infinity in the above inequality we have

$$\forall\, \varepsilon > 0\ \exists\, N \in \mathbb{N}\ \forall\, n \geq N : \sup\{|f_n(x) - f(x)| : x \in K\} < \varepsilon,$$

as for every $y \in K$ $|f_n(y) - f(y)| < \varepsilon$ and N is independent of y. Therefore

$$\lim_{n\to\infty} \|f_n - f\|_K = 0$$

and f is the limit of (f_n) with respect to $\|\cdot\|_K$. By theorem 5.3 the function f is continuous, which proves the completeness of $C_b^0(K)$. □

Corollary 5.8 (Weierstraß) *Let $f_n : K \to \mathbb{R}$ or $\mathbb{C}$ be functions with*

$$\sum_{n=0}^{\infty} \|f_n\|_K < \infty.$$

The series $\sum\limits_{n=0}^{\infty} f_n$ converges uniformly on K, and for every $x \in K$, the series $\sum\limits_{n=0}^{\infty} f_n(x)$ converges absolutely.

Proof. We first show that the sequence of partial sums $F_m := \sum\limits_{n=0}^{m} f_n$ is a Cauchy sequence relative to $\|\cdot\|$. Now for $m \geq m'$ we have

$$\|F_m - F_{m'}\|_K = \Big\| \sum_{n=m'}^{m} f_n \Big\|_K \leq \sum_{n=m'}^{m} \|f_n\|_K,$$

whereby the sequence (F_m) is Cauchy on account of the convergence of $\sum\limits_{n=0}^{\infty} \|f_n\|_K$. Therefore by theorem 5.7 $\sum\limits_{n=0}^{\infty} f_n$ converges uniformly on K.

The absolute convergence follows from $\Sigma|f_n(x)| \leq \Sigma\|f_n\|_K$ as the latter majorizes the former series. □

Corollary 5.9 *Let $z_0 \in \mathbb{C}, (a_n)_{n\in\mathbb{N}} \subset \mathbb{C}$. Assume that for some $z_1 \neq z_0$ the series*

$$f(z) := \sum_{n=0}^{\infty} a_n (z - z_0)^n$$

is convergent.

For $r > 0$ let $B(z_0, r) := \{z \in \mathbb{C} : |z - z_0| \le r\}$. Then for every r with $0 < r < |z_1 - z_0|$ the above power series converges absolutely and uniformly on $B(z_0, r)$ and so does the power series

$$\sum_{n=1}^{\infty} n \, a_n (z - z_0)^{n-1}.$$

Proof. For $z \in B(z_0, r)$ we have

$$|a_n(z - z_0)^n| = |a_n(z_1 - z_0)^n| \frac{|z - z_0|^n}{|z_1 - z_0|^n} \le |a_n(z_1 - z_0)^n| \vartheta^n$$

with $\vartheta := \frac{r}{|z_1 - z_0|} < 1$. As $f(z_1)$ converges, there exists an $m \in \mathbb{R}$ with

$$|a_n(z_1 - z_0)^n| \le m \quad \text{for all } n \in \mathbb{N},$$

so

$$|a_n(z - z_0)^n| \le m \vartheta^n \quad \text{for } z \in B(z_0, r).$$

Therefore the series

$$\sum_{n=0}^{\infty} \|a_n(z - z_0)^n\|_{B(z_0, r)}$$

converges and corollary 5.8 gives the absolute and uniform convergence of $\sum_{n=0}^{\infty} a_n(z - z_0)^n$ on $B(z_0, r)$.

Similarly for $z \in B(z_0, r)$

$$|n a_n(z - z_0)^{n-1}| \le n m \vartheta^{n-1}.$$

As $\sum_{n=1}^{\infty} n \vartheta^{n-1}$ converges by the ratio test, the statements about convergence of $\sum_{n=1}^{\infty} n a_n(z - z_0)^{n-1}$ for $z \in B(z_0, r)$ follow likewise from corollary 5.8. $\qquad \square$

Definition 5.10 For a power series $f(z)$ as in corollary 5.9 its radius of convergence R is

$$R := \sup\{r : \sum_{n=0}^{\infty} a_n(z - z_0)^n \text{ convergent on } B(z_0, r)\}.$$

Theorem 5.11 *Let $I = [a, b]$ be a bounded interval in $\mathbb{R}$. Let $f_n : I \to \mathbb{R}$ be continuously differentiable functions. Assume that*

(i)   *there exists a $z \in I$ for which $f_n(z)$ converges*

(ii)   *the sequence of derivatives (f_n') converges uniformly on I.*

Then the sequence (f_n) converges uniformly on I to a continuously differentiable function f and we have

$$f'(x) = \lim_{n \to \infty} f'_n(x) \quad \text{for all } x \in I.$$

Proof. Let $g(x)$ be the limit of $f'_n(x)$. Let $\eta > 0$. Because of uniform convergence of f'_n we can find an $N \in \mathbb{N}$ with the following property

$$\forall\, n, m \geq N : \sup\{|f'_n(x) - f'_m(x)| : x \in I\} < \eta \tag{1}$$

$$\text{and so } \sup\{|f'_n(x) - g(x)| : x \in I\} < \eta \quad \text{for } n \geq N \tag{2}$$

Furthermore, for all $x \in I, n, m \in \mathbb{N}$, we have, by the mean value theorem

$$|f_n(x) - f_m(x) - (f_n(z) - f_m(z))| \leq |x - z| \sup_{\xi \in I} |f'_n(\xi) - f'_m(\xi)|, \tag{3}$$

and therefore

$$|f_n(x) - f_m(x)| \leq |f_n(z) - f_m(z)| + |x - z| \sup_{\xi \in I} |f'_n(\xi) - f'_m(\xi)|,$$

wherefrom, on account of (i) and (1) it follows easily that (f_n) is a Cauchy sequence relative to $\| \cdot \|_I$. Therefore by theorem 5.7 the sequence (f_n) converges to a continuous limit function f.

In particular (i), and thereby the above considerations, hold for every $z \in I$.

In (3) we let m tend to ∞ and obtain from (2)

$$|f_N(x) - f(x) - (f_N(z) - f(z))| \leq |x - z| \eta. \tag{4}$$

For N, which depends only on η, and x we find a $\delta > 0$ with

$$|f_N(x) - f_N(z) - (x - z)f'_N(x)| \leq \eta|x - z| \text{ for } |x - z| < \delta. \tag{5}$$

This follows from our characterization of differentiability (see theorem 2.2).

It follows from (2), (4) and (5) that

$$|f(x) - f(z) - g(x)(x - z)| \leq 3\eta|x - z|, \text{ if } |x - z| < \delta.$$

Since this holds for every $x \in I$ and for all z with $|x - z| < \delta$, it follows from our characterization of differentiability, that $f'(x)$ exists and

$$f'(x) = g(x).$$

$\square$

Theorem 5.11 has the following stronger version, which we shall use later in introducing integrals.

Theorem 5.12 *Let $f_n, g_n : [a, b] \to \mathbb{R}$ be functions. Assume that the f_n are continuous and for every n there exists a countable subset $D_n \subset [a, b]$ such that for every $x \in I \backslash D_n$*

$$f_n'(x) = g_n(x).$$

Assume moreover that

(i)   *there exists a $z \in [a, b]$ for which $f_n(z)$ converges*

(ii)   *$(g_n)_{n \in \mathbb{N}}$ converges uniformly on $[a, b]$.*

Then f_n converges uniformly on $[a, b]$ to a continuous function f and for all $x \in I \backslash \underset{n \in \mathbb{N}}{\cup} D_n$, $f'(x)$ exists and

$$f'(x) = \lim_{n \to \infty} g_n(x).$$

The proof of this theorem is the same as that of theorem 5.11 if, instead of the mean value theorem one uses the following lemma:

Lemma 5.13 *Let $I := [a, b] \subset \mathbb{R}$ and $f : I \to \mathbb{R}$ continuous, D a countable subset of I such that f is differentiable on $I \backslash D$ with*

$$f'(x) \le M \text{ for all } x \in I \backslash D.$$

We then have

$$f(b) - f(a) \le M(b - a) \tag{6}$$

Proof. It suffices to show that for all $\eta > 0$

$$f(b) - f(a) \le M(b - a) + \eta(b - a + 1).$$

As the left hand side of this inequality is independent of η, inequality (6) follows immediately.

Let $n \mapsto \rho_n (n \in \mathbb{N})$ be a counting of D.

$$A := \{\xi \in I : \text{for all } \zeta \text{ with } a \le \zeta < \xi$$

$$f(\zeta) - f(a) \le M(\zeta - a) + \eta(\zeta - a) + \eta \sum_{\rho_n < \zeta} 2^{-n}\}$$

As $a \in A$, the set A is nonempty. $c := \sup A$. By continuity of f we have $c \in A$ and so

$$f(c) - f(a) \le M(c - a) + \eta(c - a) + \eta \sum_{\rho_n < c} 2^{-n}. \tag{7}$$

We claim that $c = b$, which implies the inequality (6). So we assume that $c < b$ and consider two cases:

Case 1: $c \notin D$. Then $f'(c)$ exists and from our characterization of the derivative (theorem 2.2) there exists a $\delta > 0$ with $c + \delta \le b$ such that for all ζ with $c \le \zeta \le c + \delta$

$$f(\zeta) - f(c) - f'(c)(\zeta - c) \le \eta(\zeta - c).$$

From this it follows that

$$f(\zeta) - f(c) \le f'(c)(\zeta - c) + \eta(\zeta - c) \le M(\zeta - c) + \eta(\zeta - c)$$

and with (7)

$$f(\zeta) - f(a) \le M(\zeta - a) + \eta(\zeta - a) + \eta \sum_{\rho_n < \zeta} 2^{-n}.$$

Therefore $\zeta \in A$ for $c \le \zeta \le c + \delta$ which contradicts the definition of c. So in case 1 necessarily $c = b$.

Case 2: $c \in D$, say $c = \rho_m$. By continuity of f there exists a $\delta > 0$ with $c + \delta \le b$ such that for all ζ with $c \le \zeta \le c + \delta$ we have

$$|f(\zeta) - f(c)| \le \eta 2^{-m}.$$

By (7) it follows again for $c < \zeta < c + \delta$

$$f(\zeta) - f(a) \le M(c - a) + \eta(c - a) + \eta \sum_{\rho_n \le c} 2^{-n}$$

$$\le M(\zeta - a) + \eta(\zeta - a) + \eta \sum_{\rho_n < \zeta} 2^{-n}$$

and we obtain again a contradiction to the definition of c. This completes the proof Lemma 5.13. $\square$

We now return to theorem 5.11 and draw some consequences:

Definition 5.14 For $f \in C^k(D), k = 0, 1, 2, \ldots, D \subset \mathbb{R}$, let

$$\|f\|_{C^k(D)} := \sum_{i=0}^{k} \|f^{(i)}\|_D \text{ (here } f^{(0)} := f),$$

$$C_b^k(D) := \{f \in C^k(D) : \|f\|_{C^k(D)} < \infty\}.$$

For $f \in C^{k,\alpha}(D), k = 0, 1, 2, \ldots, 0 < \alpha \le 1, D \subset \mathbb{R}$, let

$$\|f\|_{C^{k,\alpha}(D)} := \|f\|_{C^k(D)} + \sup_{\substack{x,y \in D \\ x \ne y}} \frac{|f^{(k)}(x) - f^{(k)}(y)|}{|x - y|^\alpha},$$

$$C_b^{k,\alpha}(D) := \{f \in C^{k,\alpha}(D) : \|f\|_{C^{k,\alpha}(D)} < \infty\}.$$

Lemma 5.15 $\|\cdot\|_{C^k(D)}$ and $\|\cdot\|_{C^{k,\alpha}(D)}$ define norms on $C_b^k(D)$ and $C_b^{k,\alpha}(D)$, respectively.

The proof is elementary, if one draws on Lemma 5.5.

Theorem 5.16 $(C_b^k(D), \|\cdot\|_{C^k(D)})$ is a Banach space for $k = 0, 1, 2, \ldots$

Proof. We deal only with the case $k = 1$, as the general case then follows easily by induction. The case $k = 0$ is theorem 5.7. Let $(f_n)_{n\in\mathbb{N}}$ be a Cauchy sequence relative to $\|\cdot\|_{C^1(D)}$. This means that (f_n) and (f_n') are Cauchy sequences relative to $\|\cdot\|_D$. As by theorem 5.7 $C_b^0(D)$ is a Banach space, there exist limit functions, so

$$f_n \rightrightarrows f, \quad f_n' \rightrightarrows g.$$

By theorem 5.11 (note that here even a stronger condition than (i) of theorem 5.11 is fulfilled) we have

$$f' = g$$

on every bounded interval contained in D, and so, $f \in C_b^1(D)$ and $\|f_n - f\|_{C^1(D)} \to 0$. Therefore $C_b^1(D)$ is complete and hence a Banach space. $\quad\square$

Theorem 5.17 For $k = 0, 1, 2, \ldots, 0 < \alpha \le 1$ $(C_b^{k,\alpha}(D), \|\cdot\|_{C^{k,\alpha}(D)})$ is a Banach space.

Proof. We consider only the case $k = 0$, as the general case follows easily inductively by using theorem 5.16. So let $(f_n)_{n\in\mathbb{N}}$ be a Cauchy sequence relative to $\|\cdot\|_{C^{0,\alpha}(D)}$. Then $(f_n)_{n\in\mathbb{N}}$ is also a Cauchy sequence relative to $\|\cdot\|_D$ and by theorem 5.16, the sequence (f_n) converges uniformly to a limit function f. Moreover $\forall\, \varepsilon > 0 \ \exists\, N \in \mathbb{N} \ \forall\, n, m \ge N$:

$$\sup_{\substack{x,y\in D \\ x\neq y}} \frac{|f_n(x) - f_m(x) - (f_n(y) - f_m(y))|}{|x - y|^\alpha} < \varepsilon.$$

For every pair $x, y \in D, x \neq y$, we can let m tend to ∞ and obtain

$$\frac{|f_n(x) - f(x) - (f_n(y) - f(y))|}{|x - y|^\alpha} < \varepsilon.$$

This implies firstly that

$$\frac{|f(x) - f(y)|}{|x - y|^\alpha} \le \varepsilon + \frac{|f_n(x) - f_n(y)|}{|x - y|^\alpha}$$

and so $f \in C_b^{0,\alpha}(D)$ and secondly $\lim_{n\to\infty} \|f_n - f\|_{C^{0,\alpha}(D)} = 0$. Therefore $C_b^{0,\alpha}(D)$ is complete, hence a Banach space. $\quad\square$

We now come to

Definition 5.18 Let $K \subset \mathbb{R}$ or $\mathbb{C}$ and $f_n : K \to \mathbb{R}$ or $\mathbb{C}$ functions $(n \in \mathbb{N})$.

a)   The sequence $(f_n)_{n \in \mathbb{N}}$ is called uniformly bounded in K

$$:\Leftrightarrow \exists\, M \in \mathbb{R} \,\forall\, n \in \mathbb{N}, x \in K : |f_n(x)| \leq M$$
$$(\text{so for all } n : \|f_n\|_K \leq M)$$

b)   The sequence $(f_n)_{n \in \mathbb{N}}$ is called equicontinuous on K

$$\Leftrightarrow \forall\, \varepsilon > 0 \,\exists\, \delta > 0 \,\forall\, n \in \mathbb{N}, x, y \in K \text{ with } |x - y| < \delta : \quad (8)$$
$$|f_n(x) - f_n(y)| < \varepsilon.$$

If the sequence $(f_n)_{n \in \mathbb{N}}$ is equicontinuous in K then in particular all the f_n are there uniformly continuous. The important point in (8) is that δ is independent of n.

Theorem 5.19 *Let I be an interval in $\mathbb{R}$. Let*

$$(f_n)_{n \in \mathbb{N}} \subset C^k(I), k \geq 1$$
$$or \quad \subset C^{k,\alpha}(I), 0 < \alpha \leq 1, k \geq 0,$$
$$with \quad \|f_n\|_{C^k(I)} \; resp. \; \|f_n\|_{C^{k,\alpha}(I)} \leq M \quad for \; all \; n.$$

Then (f_n) is uniformly bounded and equicontinuous.

Proof. Uniform boundedness of (f_n) is clear. For equicontinuity, it suffices to consider the cases $k = 1$ and $k = 0, 0 < \alpha \leq 1$. First, let $(f_n) \subset C^1(I), \|f_n\|_{C^1(I)} \leq M$. By the mean value theorem, for all $x, y \in I$ and $n \in \mathbb{N}$ we have:

$$|f_n(x) - f_n(y)| \leq |x - y| \sup_{\xi \in I} |f'_n(\xi)| \leq M|x - y|,$$

and equicontinuity follows.

Similarly now let $(f_n) \subset C^{0,\alpha}(I), \|f_n\|_{C^{0,\alpha}(I)} \leq M$. Then for all $x, y \in I$ and $n \in \mathbb{N}$

$$|f_n(x) - f_n(y)| \leq M|x - y|^\alpha$$

and equicontinuiuty follows again. $\qquad\qquad\qquad\qquad\qquad\qquad\square$

Theorem 5.20 (Arzela-Ascoli). *Let $I = [a, b]$ be a closed, bounded interval in $\mathbb{R}$, $f_n : I \to \mathbb{R}$ or $\mathbb{C}$ a sequence of uniformly bounded and equicontinuous functions. Then the sequence $(f_n)_{n \in \mathbb{N}}$ contains a uniformly convergent subsequence.*

Proof. We consider a sequence $(x_n)_{n\in\mathbb{N}}$, which is dense in I. So every $x \in I$ is a limit of a subsequence of (x_n). For $I = [0,1]$ we can take for example the following sequence:

$$\left(0, 1, \frac{1}{2}, \frac{1}{4}, \frac{3}{4}, \frac{1}{8}, \frac{3}{8}, \frac{5}{8}, \frac{7}{8}, \frac{1}{16}, \ldots\right).$$

Since (f_n) is uniformly bounded, $|f_n(x_1)|$ is, in particular, bounded, independently of n. Therefore, there exists a subsequence $(f_{1,n})_{n\in\mathbb{N}}$ of (f_n) for which $f_{1,n}(x_1)$ is convergent. Iteratively, we find a subsequence $(f_{k,n})_{n\in\mathbb{N}}$ of $(f_{k-1,n})_{n\in\mathbb{N}}$, for which $f_{k,n}(x_k)$ converges. As $(f_{k,n})$ is a subsequence of $(f_{k-1,n})$, inductively the sequences $(f_{k,n}(x_1)), \ldots, (f_{k,n}(x_{k-1}))$ also converge. We now construct the diagonal sequence $(f_{kk})_{k\in\mathbb{N}}$. Then $f_{kk}(x_i)$ converges for every fixed $i \in \mathbb{N}$ as $k \to \infty$. Namely, if $k \geq i$, then $f_{kk} \in (f_{i,n})_{n\in\mathbb{N}}$, and $f_{i,n}(x_i)$ converges. The above process is called the Cantor diagonal process. We now show that $(f_{kk})_{k\in\mathbb{N}}$ is a Cauchy sequence relative to $\|\cdot\|_I$. Let $\varepsilon > 0$. We choose

1) $\delta > 0$ such that

$$\forall\, x, y \in I \text{ with } |x - y| < \delta, k \in \mathbb{N} : |f_{kk}(x) - f_{kk}(y)| < \frac{\varepsilon}{3}.$$

This is possible because of the equicontinuity of the sequence $(f_n)_{n\in\mathbb{N}}$ and therefore of its subsequence $(f_{kk})_{k\in\mathbb{N}}$.

2) $M \in \mathbb{N}$ such that:

$$\forall\, x \in [0,1] \,\exists\, i \in \{1, \ldots, M\} : |x - x_i| < \delta.$$

It is in this step that the boundedness of I is used.

3) $N \in \mathbb{N}$ such that

$$\forall\, i \in \{1, \ldots, M\}, n, m \geq N : |f_{nn}(x_i) - f_{mm}(x_i)| < \frac{\varepsilon}{3}.$$

As M has already been chosen, this is a question of finitely many points $x_i, i = 1, \ldots, M$, and as $(f_{nn}(x_i))_{n\in\mathbb{N}}$ converges for all i, such an N can be found. So N depends on M, M on δ and δ ond ε. Now for every $x \in [0,1]$ and $n, m \geq N$

$$|f_{nn}(x) - f_{mm}(x)| \leq |f_{nn}(x) - f_{nn}(x_i)| + |f_{nn}(x_i) - f_{mm}(x_i)|$$
$$+ |f_{mm}(x_i) - f_{mm}(x)|$$
$$< \frac{\varepsilon}{3} + \frac{\varepsilon}{3} + \frac{\varepsilon}{3} = \varepsilon$$

where x_i $(1 \leq i \leq M)$ has been chosen as in 2).
Therefore

$$\forall\, \varepsilon > 0 \,\exists\, N \in \mathbb{N} \,\forall\, n, m \geq N : \sup\{|f_{nn}(x) - f_{mm}(x)| : x \in [0,1]\} < \varepsilon,$$

as N is independent of x.

Thereby $(f_{nn})_{n \in \mathbb{N}}$ has been shown to be a Cauchy sequence. $\qquad \square$

Corollary 5.21 *Let* $(f_n)_{n \in \mathbb{N}} \subset C^k(I), k \geq 1$ *or* $\subset C^{k,\alpha}(I), k \geq 0, 0 < \alpha \leq 1$ *with*

$$\|f_n\|_{C^k(I)} \leq M \text{ or } \|f_n\|_{C^{k,\alpha}(I)} \leq M,$$

where I is a closed bounded interval in $\mathbb{R}$. Then $(f_n)_{n \in \mathbb{N}}$ has a uniformly bounded convergent subsequence.

Proof. This follows directly from theorems 5.19 and 5.20. $\qquad \square$

Remarks.

1) Quite analogously one can prove the theorem of Arzela-Ascoli for example for a sequence (f_n) of uniformly bounded and equicontinuous functions defined on $B(z_o, r) \subset \mathbb{C}, 0 < r < \infty$.

2) We want to show by an example that in the theorem of Arzela-Ascoli, the assumption on the interval of definition I being bounded, is essential. For this we set

$$f_n(x) := \begin{cases} \sin x & \text{for } 2\pi n \leq x \leq 2\pi(n+1) \\ 0 & \text{otherwise} \end{cases}$$

The functions f_n are uniformly bounded and equicontinuous and converge pointwise, but not uniformly, to 0.

Exercises for § 5

1) Which of the following sequences $(f_n)_{n \in \mathbb{N}}$ of functions $f_n : \mathbb{R} \to \mathbb{R}$ converge for $n \to \infty$? What are the limits of the convergent ones? Find all intervals where the convergence is uniform.

   a) $f_n(x) = \frac{nx}{1+nx^2}$.

   b) $f_n(x) = \frac{nx}{1+n^2 x^2}$.

   c) $f_n(x) = \frac{n^2 x}{1+nx^2}$.

2) Which of the following sequences $(f_n)_{n \in \mathbb{N}}$ is uniformly convergent? If it is, what is the limit?

   a) $f_n : \mathbb{R} \to \mathbb{R}, f_n(x) := \begin{cases} |x - n| - 1 & \text{for } n - 1 \leq x \leq n + 1 \\ 0 & \text{otherwise} \end{cases}$.

   b) $f_n : \mathbb{R} \to \mathbb{R}, f_n(x) := \frac{1}{1+(\frac{x}{n})^2}$.

   c) $f_n : \mathbb{R} \to \mathbb{R}, f_n(x) = \sin(\frac{x}{n})$.

d) $f_n : \mathbb{R} \to \mathbb{R}, \ f_n(x) := \begin{cases} \frac{1}{n}\sin x & \text{for } 2\pi n \leq x \leq 2\pi(n+1) \\ 0 & \text{otherwise} \end{cases}$.

3)   Show that $f_n(x) = x^n$ converges uniformly on every interval $[a,b] \subset (0,1)$.

4)   Let $D \subseteq \mathbb{R}, f : D \to \mathbb{R}$ a function, $(a_n)_{n \in \mathbb{N}} \subset \mathbb{R}$ a sequence that converges to 0, and put $f_n(x) := a_n f(x)$ for $x \in D$. Show that $(f_n)_{n \in \mathbb{N}}$ converges to 0 pointwise on D, and the convergence is uniform precisely if f is bounded on D.

5)

a)   Let $g_n : K \to \mathbb{R}$ be functions that converge uniformly to a function g. Let $f : K \to \mathbb{R}$ be bounded. Then $g_n f$ converges uniformly to gf.

b)   Let $g_n \rightrightarrows g$ on K as in a), and suppose

$$|g_n(x)| \geq c > 0 \quad \text{for all } x \in K, n \in \mathbb{N}.$$

Then

$$\frac{1}{g_n} \rightrightarrows \frac{1}{g} \quad \text{on } K.$$

6)   For $f \in C^\infty(\mathbb{R})$, put

$$\|f\| := \sup\{|f(x)| : 0 \leq x \leq 1\}.$$

Does $\|\cdot\|$ define a norm on $C^\infty(\mathbb{R})$? (Hint: Consider the function

$$f(x) := \begin{cases} e^{-\frac{1}{x^2}} & \text{for } x \leq 0 \\ 0 & \text{elsewhere.} \end{cases}$$

(Of course, one needs to verify that $f \in C^\infty(\mathbb{R})$.))

7)   Let $0 < \beta < \alpha \leq 1, I = [a,b]$ a closed and bounded interval in $\mathbb{R}$, $(f_n)_{n \in \mathbb{N}} \subset C^\alpha(I)$ with

$$\|f_n\|_{C^\alpha(I)} \leq M \quad \text{for some constant } M.$$

show that a subsequence of $(f_n)_{n \in \mathbb{N}}$ converges w.r.t. the norm $\|\cdot\|_{C^\beta(I)}$.

8)   Let $\mathcal{P}$ be the space of polynomials $p : \mathbb{R} \to \mathbb{R}$, i.e. $p(x) = a_n x^n + a_{n-1} c^{n-1} + \ldots + a_0$, with $a_0, \ldots, a_n \in \mathbb{R}$. For $p \in \mathcal{P}$, we put

$$\|p\| := \sup\{|p(x)| : 0 \leq x \leq 1\}.$$

a)   Show that $\|\cdot\|$ defines a norm on $\mathcal{P}$.

b)   Is $(\mathcal{P}, \|\cdot\|)$ a Banach space?

9)   Let $f_n : [a, b] \to \mathbb{R}$ be continuously differentiable functions. Assume that $(f_n)_{n \in \mathbb{N}}$ converges pointwise to some function f. Which – if any – of the following assertions can be concluded from these assumptions?

a) f is continuously differentiable.

b)   The sequence $(f_n')_{n \in \mathbb{N}}$ converges pointwise.

c)   If f is continuously differentiable, then $\lim_{n \to \infty} f_n'(x) = f'(x)$ for all $x \in [a, b]$.

d)   If f is continuously differentiable, and if $\lim_{n \to \infty} f_n'(x)$ exists for all $x \in [a, b]$, then $\lim_{n \to \infty} f_n'(x) = f'(x)$ for all x.

6. Integrals and Ordinary Differential Equations

A continuous function g is called a primitive of another function f if the derivative of g exists and coincides with f. A primitive thus is an indefinite integral. We derive the basic rules for the computation of integrals. We use the Banach fixed point theorem to derive the Picard-Lindelöf theorem on the local existence of solutions or integrals of ordinary differential equations (ODEs).

Definition 6.1 Let I be a closed interval in $\mathbb{R}$, $f : I \to \mathbb{R}$ a function. A continuous function $g : I \to \mathbb{R}$ is called a primitive of f on I if there exists a countable subset D of I such that for all $\xi \in I \backslash D$, g is differentiable in ξ and $g'(\xi) = f(\xi)$. In that case, f is called (Riemann) integrable on I.

Remark. Starting with §12, we shall develop Lebesgue integration theory, and we shall introduce a more general notion of integrability.

Lemma 6.2 *If g_1, g_2 are primitives of f on I then $g_1 - g_2$ is constant on I.*

Proof. This is a direct consequence of lemma 5.13. □

Remark. In order to verify whether $f : I \to \mathbb{R}$ has a primitive, it suffices to consider bounded, closed subintervals of I. We have

$$I = \bigcup_{n \in \mathbb{N}} J_n$$

and the J_n are so chosen that some $x_0 \in J_n$ for all n and J_n are closed and bounded. Let g_n be a primitive of f on J_n with $g_n(x_0) = 0$. Then by lemma 6.2 the function g defined by

$$g(x) := g_n(x) \text{ for } x \in J_n$$

is a primitive of f on I.

Definition 6.3 A function $f : I \to \mathbb{R}$ is called admissible if it is the uniform limit of step functions. Here, $I \to \mathbb{R}$ is called a step function if I is the disjoint union of finitely many intervals on each of which t is constant.

Lemma 6.4 *Let I be a closed, bounded interval in $\mathbb{R}$ and $f : I \to \mathbb{R}$ continuous. Then f is admissible.*

Proof. Let $\varepsilon = \frac{1}{n}$. For ε we determine a $\delta > 0$ such that for all $x, y \in I$ with $|x - y| < \delta$, we have $|f(x) - f(y)| < \varepsilon$. Let $I = [a, b]$ and m the greatest natural number with $a + m\delta < b$. For an integer μ with $0 \le \mu \le m$ we set

$$x_\mu := a + \mu\delta \text{ and } x_{m+1} := b$$

and define a step function by

$$t_n(x) = f(x_\mu) \text{ for } x_\mu \le x < x_{\mu+1}, t_n(b) = f(b). \tag{1}$$

Then for all $x \in [a, b]$

$$|f(x) - t_n(x)| = |f(x) - f(x_\mu)|, \ x_\mu \text{ as in (1)}$$
$$< 1/n$$

as $|x - x_\mu| < \delta$.

Therefore $(t_n)_{n \in \mathbb{N}}$ converges uniformly to f. □

Theorem 6.5 *Let I be a closed interval, $f : I \to \mathbb{R}$ a admissible function. Then f has a primitive on I.*

Proof. By the remark after lemma 6.2 we may assume that I is bounded. We consider first the case where f is a step function. So let $I = [a, b], x_0 = a < x_1 < \ldots < x_n = b, f(\xi) = c_i$ for $x_i < \xi < x_{i+1}, i = 0, \ldots, n - 1$. We set

$$g(\xi) := c_i(\xi - x_i) + \sum_{\nu=0}^{i-1} c_\nu(x_{\nu+1} - x_\nu) \text{ for } x_i \le \xi < x_{i+1}.$$

Then g is continuous and $g'(\xi) = f(\xi)$ for $x \notin \{x_0, \ldots, x_n\}$ and therefore g is a primitive of f.

In the general case of an admissible function f, there exists a sequence $(t_n)_{n \in \mathbb{N}}$ of step functions which converges uniformly to f. By what has already been shown, every t_n has a primitive g_n which we can normalize by setting $g_n(a) = 0$. Since $g_n' = t_n$ on $I \backslash D_n$, D_n countable, by theorem 5.12 the g_n converge uniformly to a continuous function g with $g' = f$ on $I \backslash \bigcup_{n \in \mathbb{N}} D_n$.

Hence g is a primitive of f. □

Theorem 6.6 *Let $f : I \to \mathbb{R}$ be continuous with primitive g. Then g is differentiable on I and $g'(x) = f(x)$ for all $x \in I$.*

Proof. By lemma 5.13 applied to the function $\phi(h) = g(x + h) - f(x)h$ we have

$$|g(x + h) - g(x) - f(x)h| \le h \sup_{|\eta| \le |h|} |f(x + \eta) - f(x)|.$$

As f is continuous, we have for every $x \in I$

$$\lim_{h \to 0} \frac{1}{h}(h \sup_{|\eta| \le |h|} |f(x+\eta) - f(x)|)$$

$$= \lim_{h \to 0} \sup_{|\eta| \le |h|} |f(x+\eta) - f(x)| = 0,$$

and therefore by theorem 2.2 it follows that $g'(x) = f(x)$ for all $x \in I$.   □

Definition 6.7 Let g be a primitive of f on I and $x_1, x_2 \in I$. Then the integral of f from x_1 to x_2 is defined as

$$\int_{x_1}^{x_2} f(\xi)d\xi := g(x_2) - g(x_1).$$

Notice that by lemma 6.2, this expression is independent of the choice of the primitive g.

Notation. If g is a primitive of f, we write $g' = f$, although there could be a countable subset on which g is either not differentiable or on which g' does not agree with f.

Lemma 6.8 *Let I be a closed bounded interval in $\mathbb{R}$, $f : \mathbb{R} \to \mathbb{R}$ continuous and $\phi : I \to \mathbb{R}$ admissible. Then $f\phi : I \to \mathbb{R}$ is admissible.*

Proof. Let $(t_n)_{n \in \mathbb{N}}$ be a sequence of step functions which converge uniformly to ϕ and let $M := \sup\{\|f\|_I, \|t_n\|_I; n \in \mathbb{N}\} < \infty$. For a given $\varepsilon > 0$ there exists an $N \in \mathbb{N}$ and a $\delta > 0$ such that:

$$\forall\, n \ge N : \|t_n - \phi\|_I < \frac{\varepsilon}{2M}$$

$$\forall\, x, y \in I \text{ with } |x - y| < \delta : \quad |f(x) - f(y)| < \frac{\varepsilon}{2M}.$$

Let $I = [a, b]$ and $m \in \mathbb{N}$ maximal such that $a + m\delta < b$. For an integer μ with $0 \le \mu \le m$ we set $x_\mu := a + \mu\delta$ and $x_{m+1} = b$. We define

$$\tau_n(x) := t_n(x)f(x_\mu) \quad \text{for } x_\mu \le x < x_{\mu+1}, \quad \tau_n(b) := t_n(b)f(b). \qquad (2)$$

Now with x_μ as in (2) above we have

$$|\tau_n(x) - f(x)\phi(x)|$$
$$\le |t_n(x)(f(x_\mu) - f(x))| + |f(x)(t_n(x) - \phi(x))|,$$
$$< M\frac{\varepsilon}{2M} + M\frac{\varepsilon}{2M} = \varepsilon \quad \text{for } n \ge N.$$

Therefore the sequence of step functions $(\tau_n)_{n \in \mathbb{N}}$ converges uniformly to $f\phi$ and $f\phi$ is admissible.   □

Theorem 6.9 (Transformation law) *Let I, J be closed intervals, h a primitive of an admissible function on I with $h(I) \subset J$ and $f : J \to \mathbb{R}$ continuous. Then for all $x_1, x_2 \in I$*

$$\int_{x_1}^{x_2} f(h(\xi))h'(\xi)d\xi = \int_{h(x_1)}^{h(x_2)} f(y)dy. \tag{3}$$

Proof. By lemma 6.8 the function $f(h(\xi))h'(\xi)$ is admissible on $[x_1, x_2]$. Let g be a primitive of f. By the chain rule $g(h(\xi))$ is a primitive for $f(h(\xi))h'(\xi)$. Therefore the left as well as the right hand side of (3) are equal to

$$g(h(x_2)) - g(h(x_1)).$$

□

Theorem 6.10 (Integration by parts) *Let f and g be primitives of admissible functions defined on a closed interval I and $x_1, x_2 \in I$. Then*

$$\int_{x_1}^{x_2} f(\xi)g'(\xi)d\xi = f(x_2)g(x_2) - f(x_1)g(x_1) - \int_{x_1}^{x_2} f'(\xi)g(\xi)d\xi.$$

Proof. By lemma 6.8 the function $fg' + f'g$ is admissible on $[x_1, x_2]$, and its primitive, by the product formula, is gf.   □

Lemma 6.11 (Monotonicity of the integral) *Let $f_1, f_2 : [a, b] \to \mathbb{R}$ be admissible functions with primitives g_1, g_2. Assume that for all $x \in [a, b]$*

$$f_1(x) \leq f_2(x).$$

Then

$$g_1(b) - g_1(a) \leq g_2(b) - g_2(a).$$

Proof. This follows directly from lemma 5.13.   □

Theorem 6.12 (Mean value theorem) *Let $f : [a, b] \to \mathbb{R}$ be continuous. There exists an $x \in [a, b]$ such that*

$$\int_a^b f(\xi)d\xi = f(x)(b - a).$$

Proof. Let

$$\lambda := \inf\{f(\xi) : \xi \in [a, b]\}$$
$$\mu := \sup\{f(\xi) : \xi \in [a, b]\}$$

and g a primitive of f.

As $\lambda\xi$ and $\mu\xi$ are primitives of the constant functions λ and μ, respectively, it follows from lemma 6.11 that

$$\lambda(b - a) \leq g(b) - g(a) = \int_a^b f(\xi)d\xi \leq \mu(b - a).$$

Therefore, there exists an m with $\lambda \leq m \leq \mu$ and

$$\int_a^b f(\xi)d\xi = m(b - a).$$

By the intermediate value theorem 1.14, there exists an $x \in [a, b]$ with $f(x) = m$. □

More generally, the following mean value inequality holds:

Theorem 6.13 *Let $f : [a, b] \to \mathbb{R}$ be admissible. Then*

$$\left| \int_a^b f(\xi)d\xi \right| \leq \int_a^b |f(\xi)|d\xi \leq (b - a) \sup_{\xi \in [a,b]} |f(\xi)|.$$

Proof. With f the function $|f|$ is also admissible, as one verifies easily. Now if $(t_n)_{n \in \mathbb{N}}$ converges uniformly to f then $(|t_n|)_{n \in \mathbb{N}}$ converges uniformly to $|f|$ as $||t_n| - |f|| \leq |t_n - f|$. Let g be a primitive of f and h that of $|f|$. Then, by lemma 6.11 applied $f_1 = \pm f$ and $f_2 = |f|$, we obtain

$$|g(b) - g(a)| \leq h(b) - h(a) \leq (b - a) \sup_{\xi \in [a,b]} |f(\xi)|.$$

□

Theorem 6.14 *Let I be a closed, bounded interval, $f_n : I \to \mathbb{R}$ admissible with $f_n \rightrightarrows f$ on I. Then for all $x_1, x_2 \in I$ we have*

$$\int_{x_1}^{x_2} f_n(\xi)d\xi \to \int_{x_1}^{x_2} f(\xi)d\xi.$$

Proof. One can see easily by the Cantor diagonal process that f is admissible. Namely, if the sequence of step functions $(t_{mn})_{m \in \mathbb{N}}$ converges uniformly to

f_n, then one can find a subsequence $(t_{m(n),n})_{n\in\mathbb{N}}$ which converges uniformly to f. The formal proof goes as follows:

$$\forall \varepsilon > 0 \; \exists \, N \in \mathbb{N} \; \forall \, n \geq N : \|f_n - f\|_I < \frac{\varepsilon}{2}.$$

Furthermore for every n there exists an $m(n)$ such that

$$\|t_{m(n),n} - f_n\|_I < \frac{\varepsilon}{2}.$$

It follows that for $n \geq N : \|t_{m(n),n} - f\|_I < \varepsilon$. Therefore f is admissible. Now the proof follows from theorem 5.12.   □

Theorem 6.15 *Let I and f_n be as in theorem 6.14, $\sum\limits_{n=1}^{\infty} \|f_n\|_I < \infty, F :=$ $\sum\limits_{n=1}^{\infty} f_n$ (which exists by corollary 5.8). Then*

$$\int_{x_1}^{x_2} F(\xi)d\xi = \sum_{n=1}^{\infty} \int_{x_1}^{x_2} f_n(\xi)d\xi,$$

and the convergence is absolute.

Proof. The convergence follows as in theorem 6.14 and the absolute convergence follows from theorem 6.13.   □

We now want to describe briefly the connection of the integral with calculations of area.

For this, let $f : I \to \mathbb{R}$ be a function defined on an interval $I = [a, b]$; we shall assume that $f(x) \geq 0$ for all $x \in I$, so that we do not have to worry about the sign of area.

First, as in the proof of theorem 6.5, let f be a step function. So there exists a subdivision $x_0 = a < x_1 < \ldots < x_n = b$ of I with

$$f(\xi) = c_i \text{ for } x_i < \xi < x_{i+1}, i = 0, \ldots, n - 1.$$

By our assumption all $c_i \geq 0$. Furthermore, as in the proof of theorem 6.5

$$g(\xi) := c_i(\xi - x_i) + \sum_{\nu=0}^{i-1} c_\nu(x_{\nu+1} - x_\nu) \text{ for } x_i \leq \xi < x_{i+1}$$

is a (continuous) primitive of f.

An elementary geometric consideration shows directly that the area bounded by the ξ-axis and the graph of f (where the different levels of the step function are connected by perpendicular lines as in the figure below) is exactly the area $g(b) - g(a)$.

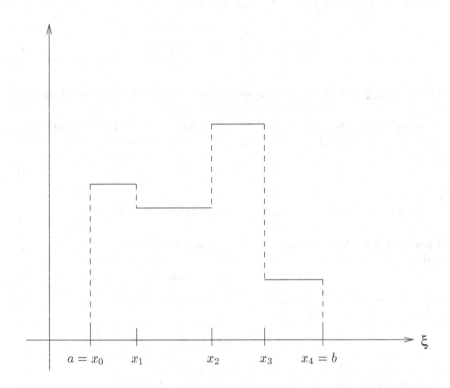

For an arbitrary nonnegative admissible function $f : [a, b] \to \mathbb{R}$ it is now natural to define the area bounded by the ξ-axis and the graph of f to be $g(b) - g(a)$, where g is a primitive of f. This is meaningful because if a sequence $(t_n)_{n \in \mathbb{N}}$ of step functions converges uniformly to f on $[a, b]$ then the corresponding primitive functions converge (uniformly) to a primitive function of f, as we have seen in the proof of theorem 6.5.

Historically, of course, one has gone the opposite way. Namely, the definite integral of f over $[a, b]$ was defined to be the area bounded by the ξ-axis and the graph of f, and then the so-called *Fundamental Theorem of Differential and Integral Calculus* was derived, which says that the derivative of a primitive of a, say, continuous function gives again the original function. We have just now taken this so-called Fundamental Theorem as definition of the integral and then we have derived the connection of the integral with area.

We now wish to study *ordinary differential equations (ODEs)* of the form

$$f'(x) = \phi(x, f(x)) \tag{4}$$

on some interval $I \subset \mathbb{R}$. Here, $\phi(x, y)$ is assumed to be continuous in both arguments. More precisely, we assume that ϕ is continuous for $x \in I$ and for $y \in J$, for some interval $J \subset \mathbb{R}$. A continuous function f on I then is a solution of (4) on I if $f(x) \in J$ for all $x \in I$, and

$$f(\xi_2) - f(\xi_1) = \int_{\xi_1}^{\xi_2} \phi(\xi, f(\xi)) d\xi \tag{5}$$

whenever $\xi_1, \xi_2 \in I$. This is a consequence of theorem 6.6 since for continuous f, $\varphi(x) := \phi(x, f(x))$ is a continuous function of x.

We wish to study the initial value problem for the ODE (ordinary differential equation) (4), i.e. given $x_0 \in I, y_0 \in J$, we wish to find a solution of

$$f'(x) = \phi(x, f(x)) \tag{6}$$

with $f(x_0) = y_0$.

Such a solution is provided by the *theorem of Picard-Lindelöf*.

Theorem 6.16 *Suppose $\phi(x, y)$ is continuous for $|x - x_0| \le \rho$, $|y - y_0| \le \eta$, with*

$$|\phi(x, y)| \le M \quad \text{for all such } x, y. \tag{7}$$

Moreover, suppose that $\phi(x, y)$ is Lipschitz continuous in y, i.e. suppose there exists some $L < \infty$ with

$$|\phi(x, y_1) - \phi(x, y_2)| \le L|y_1 - y_2| \tag{8}$$

whenever $|x - x_0| \le \rho, |y_1 - y_0| \le \eta, |y_2 - y_0| \le \eta$.

Then there exists some $h > 0$ with the property that (6) possesses a unique solution on $[x_0 - h, x_0 + h] \cap I$.

Proof. Putting $\xi_2 = x, \xi_1 = x_0$ in (5), we see that we need to solve the integral equation

$$f(x) = y_0 + \int_{x_0}^{x} \phi(\xi, f(\xi)) d\xi \tag{9}$$

on $I' := [x_0 - h, x_0 + h] \cap I$.

We shall achieve this by applying the Banach fixed point theorem 4.7 to

$$A := \{f \in C^0(I') : \|f - y_0\|_{C^0(I')} \le \eta\}.$$

Here, we have to choose the parameter $h > 0$ so small that

$$hM \le \eta \tag{10}$$

and

$$\theta := hL < 1. \tag{11}$$

We put $H(f)(x) := y_0 + \int_{x_0}^{x} \phi(\xi, f(\xi))d\xi$, and we shall now verify that with this choice of h, A and H satisfy the assumptions of theorem 4.7. Indeed, H maps the closed subset A of the Banach space $C^0(I')$ to itself: namely, if $f \in A$, then

$$\|Hf - y_0\|_{C^0(I')} = \max_{x \in I'} \left| \int_{x_0}^{x} \phi(\xi, f(\xi))d\xi \right|$$

$$\leq h \max_{\substack{\xi \in I' \\ |y-y_0| \leq \eta}} |\phi(\xi, y)|$$

$$\text{since } |f(\xi) - y_0| \leq \eta \text{ for } f \in A, \text{ and } \xi \in I'$$

$$\leq hM$$

$$\leq \eta \quad \text{by (10)}.$$

Also, H satisfies the contraction property: namely, for $f, g \in A$

$$\|Hf - Hg\|_{C^0(I')} = \max_{x \in I'} \left| \int_{x_0}^{x} \{\phi(\xi, f(\xi)) - \phi(\xi, g(\xi))\}d\xi \right|$$

$$\leq hL \max_{\xi \in I'} |f(\xi) - g(\xi)| \quad \text{by (8)}$$

$$= \theta \|f - g\|_{C^0(I')},$$

and we recall $\theta < 1$ by (11).

Therefore, by the Banach fixed point theorem,

$$H : A \to A$$

has a unique fixed point f, and this fixed point f then provides the desired solution of (9).   □

We have already seen examples of ODEs in §3, and we shall see further examples in subsequent §§. Here, we just recall from theorem 3.10 that the unique solution of

$$f'(x) = \gamma f(x) \quad (\gamma \in \mathbb{R})$$
$$f(0) = y_0$$

is given by

$$f(x) = y_0 e^{\gamma x}.$$

Exercises for § 6

1)   Let $f : [0, b] \to \mathbb{R}$ be continuous, $0 \le x \le b$. Then

$$\int\limits_0^x \left(\int\limits_0^y f(\xi)d\xi \right) dy = \int\limits_0^x (x-y)f(y)dy.$$

(Hint: Write the left integral as

$$\int\limits_0^x \varphi(y)\psi(y)dy$$

with $\varphi(y) \equiv 1, \psi(y) = \int\limits_0^y f(\xi)d\xi$ and integrate by parts.)

2)   Compute $\int\limits_a^b xe^x dx, \int\limits_a^b x^2 e^x dx$, and derive a recursion formula for

$\int\limits_a^b x^n e^x dx, n \in \mathbb{N}$.

3)   For which $\lambda \in \mathbb{R}$ is $\sin^\lambda(x)$ integrable on $(0, \pi)$? Compute

$$\int\limits_0^\pi \sin^n(x)dx \quad \text{for } n \in \mathbb{N}$$

and

$$\int\limits_{\frac{\pi}{4}}^{\frac{3\pi}{4}} \frac{dx}{\sin x}.$$

4)   Compute

$$\int\limits_{-1}^1 \frac{x+1}{x^2-2} dx.$$

5)   Compute

$$\lim_{n\to\infty} \int\limits_{-2}^2 \frac{dx}{1+x^{2n}}.$$

6)   Let

$$\varphi, \psi : [a, b] \to \mathbb{R} \quad \text{be differentiable,}$$
$$f : [\alpha, \beta] \to \mathbb{R} \quad \text{continuous,}$$

with

$$\alpha \leq \varphi(x) < \psi(x) \leq \beta \quad \text{for all } x \in [a, b].$$

Show that

$$F(x) := \int_{\varphi(x)}^{\psi(x)} f(\xi) d\xi$$

is differentiable on $[a, b]$.

7)

a)   Find a sequence $(f_n)_{n \in \mathbb{N}}$ of integrable functions $f_n : [0, 1] \to \mathbb{R}$ such that $f(x) = \lim\limits_{n \to \infty} f_n(x)$ exists on $[0, 1]$ and is integrable, but

$$\lim_{n \to \infty} \int_0^1 f_n(x) dx \neq \int_0^1 f(x) dx.$$

b)   Find a sequence $(f_n)_{n \in \mathbb{N}}$ of integrable functions $f_n : [0, 1] \to \mathbb{R}$ such that $f(x) = \lim\limits_{n \to \infty} f_n(x)$ exists on $[0, 1]$, but is not integrable.

8)   Determine all solutions of the differential equation

$$f'(x) = f(x) + c$$

$(c \in \mathbb{R}, f : \mathbb{R} \to \mathbb{R})$.

9)   Let $I, J \subset \mathbb{R}$ be open intervals, $0 \in J$. Let $\psi : I \to \mathbb{R}$ be continuous, $y_0 \in I$. We call $f : J \to I$ a solution of the differential equation

$$f' = \psi(f)$$

with initial condition

$$f(0) = y_0$$

if it satisfies

$$\frac{df}{dx}(x) = \psi(f(x)) \quad \text{for all } x \in J \quad \text{and } f(0) = y_0.$$

Find suitable values of $\alpha > 0$ such that for $\psi : \mathbb{R} \to \mathbb{R}$, $\psi(y) = y^\alpha$ and some (or all) initial conditions for the differential equation $f' = \psi(f)$,

a)   a solution f does not exist on $J = \mathbb{R}$

b)   a solution f is not unique.

10)

a)   Compute, for $n, m \in \mathbb{Z}$,

$$\int_0^{2\pi} \sin nx \sin mx dx, \quad \int_0^{2\pi} \cos nx \cos mx dx, \quad \int_0^{2\pi} \sin nx \cos mx dx.$$

b) $f : \mathbb{R} \to \mathbb{R}$ is called periodic, with period w, if $f(x+w) = f(x)$ for all $x \in \mathbb{R}$. Show that for such a periodic f that is piecewise continuous, and any $a, b \in \mathbb{R}$

$$\int_a^b f(x)dx = \int_{a+w}^{b+w} f(x)dx.$$

c)   Let $f : \mathbb{R} \to \mathbb{R}$ be periodic with period 2π, and piecewise continuous. Put

$$a_n := \frac{1}{\pi} \int_0^{2\pi} f(x) \cos nx dx, \quad b_n := \frac{1}{\pi} \int_0^{2\pi} f(x) \sin nx dx$$

for $n = 0, 1, 2, \dots$ These numbers a_n, b_n are called the Fourier coefficients of f. Show that if f is an even function, i.e. $f(x) = f(-x)$ for all $x \in \mathbb{R}$, then

$$b_n = 0, \ a_n = \frac{2}{\pi} \int_0^{\pi} f(x) \cos nx dx \quad \text{for all } n.$$

Similarly, if f is odd, i.e. $f(x) = -f(-x)$ for all $x \in \mathbb{R}$, then

$$a_n = 0, \ b_n = \frac{2}{\pi} \int_0^{\pi} f(x) \sin nx dx \quad \text{for all } n.$$

d)   With the notations established in c), prove all steps of the following sequence of assertions: For $x \in \mathbb{R}, n \in \mathbb{N}$, we put

$$f_n(x) := \frac{a_0}{2} + \sum_{\nu=1}^{n} (a_\nu \cos \nu x + b_\nu \sin \nu x),$$

and we have

$$\int_0^{2\pi} f(x) f_n(x)dx = \int_0^{2\pi} |f_n(x)|^2 dx$$

$$= \pi \left(\frac{a_0^2}{2} + \sum_{\nu=1}^{n} (a_\nu^2 + b_\nu^2) \right).$$

With $\varphi_n := f - f_n$, we have

$$\int_0^{2\pi} f_n(x)\varphi_n(x)dx = 0,$$

and hence

$$\int_0^{2\pi} |f(x)|^2 dx = \int_0^{2\pi} |f_n(x)|^2 dx + \int_0^{2\pi} |\varphi_n(x)|^2 dx.$$

This yields Bessel's inequality

$$\frac{a_0^2}{2} + \sum_{\nu=1}^{\infty}(a_\nu^2 + b_\nu^2) \le \frac{1}{\pi}\int_0^{2\pi} |f(x)|^2 dx,$$

and equality holds if and only if

$$\lim_{n\to\infty}\int_0^{2\pi} |\varphi_n(x)|^2 dx = 0.$$

Chapter II.

Topological Concepts

7. Metric Spaces: Continuity, Topological Notions, Compact Sets

We introduce the elementary topological concepts for metric spaces, like open, closed, and compact subsets. Continuity of functions is also expressed in topological terms. At the end, we also briefly discuss abstract topological spaces.

Definition 7.1 A metric space is a pair (X, d) where X is a set and d a metric on X, that is, the following conditions are satisfied

(i) $\forall\, x, y \in X : d(x, y) \geq 0$, and $d(x, y) = 0 \iff x = y$
   (positive definiteness)

(ii) $\forall\, x, y \in X : d(x, y) = d(y, x)$ (symmetry)

(iii) $\forall\, x, y, z \in X : d(x, z) \leq d(x, y) + d(y, z)$ (triangle inequality)

When the metric d is clear from the context we often simply write X instead of (X, d). $d(x, y)$ is called the distance of x from y.

The metric spaces with which we shall work later will always carry some extra structure. However, the metric structure suffices for the basic topological concepts to which we now want to turn, and extra structures would only obscure the contents of the concepts and statements.

Theorem 7.2 A normed vector space $(X, \|\cdot\|)$, in particular a Banach space, becomes a metric space by defining

$$d(x, y) := \|x - y\|.$$

Proof.
(i) $d(x, y) = \|x - y\| \geq 0$, and $d(x, y) = 0 \iff \|x - y\| = 0 \iff x - y = 0 \iff x = y$.

(ii) $d(x, y) = \|x - y\| = \|(-1)(y - x)\| = |-1|\|y - x\| = \|y - x\| = d(y, x)$.

(iii) $d(x, z) = \|x - z\| = \|(x - y) + (y - z)\| \leq \|x - y\| + \|y - z\| = d(x, y) + d(y, z)$.

which follow from the corresponding properties of the norm.   □

Example. A particular example is $\mathbb{R}$ with

$$d(x, y) := |x - y|$$

and similarly $\mathbb{C}$.

Definition 7.3 In general we introduce on $\mathbb{R}^d$ a norm by

$$\|x\|_p := (\sum_{i=1}^{d} |x^i|^p)^{1/p} \quad (x = (x^1, \ldots, x^d)) \quad (p - \text{norm})$$

and

$$\|x\|_\infty := \max(|x^1|, \ldots, |x^d|) \quad (\text{maximum norm}).$$

Remark. Of particular importance is the Euclidean norm $\| \cdot \|_2$ – which we shall sometimes simply denote by $\|\cdot\|$ or even $|\cdot|$. By theorem 7.2 these norms define corresponding metrics on $\mathbb{R}^d$. One can naturally define the same norms and metrics on $\mathbb{C}^n$.

Of course, there exist also metric spaces that are not induced by a norm, like the following trivial one: For any set X, put $d(x, x) = 0$ and $d(x, y) = 1$ if $x \neq y \in X$.

Finally, we further note

Lemma 7.4 *Let (X, d) be a metric space and $B \subset X$. Then B is also a metric space where we define a metric d_B on B by $d_B(x, y) = d(x, y)$ $(x, y \in B)$.*

The proof is obvious.

Lemma 7.5 *Let $(X, \| \cdot \|)$ be a normed space and let $(x_n)_{n \in \mathbb{N}} \subset X$ converge to $x \in X$. Then*

$$\lim_{n \to \infty} \|x_n\| = \|x\|$$

i.e. the norm is continuous.

Proof. By the triangle inequality we have

$$|\|x_n\| - \|x\|| \leq \|x_n - x\|,$$

and by assumption, the right hand side tends to 0. $\qquad \square$

Definition 7.6 Let (X, d) be a metric space and $(x_n)_{n \in \mathbb{N}} \subset X$ a sequence. The sequence (x_n) is said to converge to $x \in X$, symbolically $x_n \to x$, if

$$\forall \, \varepsilon > 0 \, \exists \, N \in \mathbb{N} \, \forall \, n \geq N : d(x_n, x) < \varepsilon.$$

A sequence $(x_n)_{n \in \mathbb{N}} \subset X$ is called a Cauchy sequence if $\forall \, \varepsilon > 0 \, \exists \, N \in \mathbb{N} \, \forall \, n, m \geq N : d(x_n, x_m) < \varepsilon$. (X, d) is called complete if every Cauchy sequence $(x_n)_{n \in \mathbb{N}} \subset X$ converges to some $x \in X$.

The above concepts are direct generalizations of the corresponding concepts for $\mathbb{R}$ and $\mathbb{C}$. We shall now consider these concepts for $\mathbb{R}^d$. We shall equip $\mathbb{R}^d$ here, and in what follows, always with the Euclidean metric. The following considerations however continue to hold for all the metrics introduced above, and also for $\mathbb{C}^d$, which the reader may verify as an exercise.

Theorem 7.7 *Let* $(x_n)_{n \in \mathbb{N}} \subset \mathbb{R}^d, x_n = (x_n^1, \ldots, x_n^d)$ *for all* n. *Then*

$$x_n \to x = (x^1, \ldots, x^d) \iff i = 1, \ldots, d : x_n^i \to x^i.$$

In particular, $\mathbb{R}^d$ *is complete as* $\mathbb{R}$ *is complete.*

Proof. $\Rightarrow$: $\forall \, \varepsilon > 0 \, \exists \, N \in \mathbb{N} \, \forall \, n \geq N : \|x_n - x\| < \varepsilon$. As $|x_n^i - x^i| \leq \|x_n - x\|$ for $i = 1, \ldots, d$, we have

$$|x_n^i - x^i| < \varepsilon, \quad \text{for } n \geq N$$

This implies $x_n^i \to x^i$.

$\Leftarrow$: Let $\varepsilon > 0$. We choose $N \in \mathbb{N}$ such that for $n \geq N$ and $i = 1, \ldots, n$

$$|x_n^i - x^i| < \frac{\varepsilon}{\sqrt{d}} \, .$$

We then have

$$\|x_n - x\| = \left(\sum_{i=1}^{d} |x_n^i - x^i|^2 \right)^{\frac{1}{2}} < \varepsilon$$

for $n \geq N$, so $x_n \to x$.

The last assertion follows, since (x_n) is a Cauchy sequence precisely when (x_n^i) is, for $i = 1, \ldots, d$. $\qquad \square$

Definition 7.8 *Let* X *and* Y *be metric spaces. A function* $f : X \to Y$ *is said to be continuous at* $x_0 \in X$ *if*

$$\lim_{x \to x_0} f(x) = f(x_0).$$

f *is said to be continuous on* $\Omega \subset X$ *if it is continuous at every point* $x_0 \in \Omega$.

Theorem 7.9 *Let* (X, d_X) *and* (Y, d_Y) *be metric spaces. A function* $f : X \to Y$ *is continuous at* x_0 *precisely if*

$$\forall \, \varepsilon > 0 \, \exists \, \delta > 0 \, \forall \, x \in X \text{ with } d_X(x, x_0) < \delta : d_Y(f(x), f(x_0)) < \varepsilon.$$

Proof. The proof is exactly the same as that for functions between subsets of $\mathbb{R}$; in the proof one has only to substitute everywhere $|x - x_0|$ etc. by $d(x, x_0)$.   □

Definition 7.10 Let X and Y be metric spaces, $\Omega \subset X$. A function $f : \Omega \to Y$ is said to be uniformly continuous if

$$\forall \varepsilon > 0 \, \exists \, \delta > 0 \, \forall \, x, y \in \Omega \text{ with } d(x, y) < \delta : d(f(x), f(y)) < \varepsilon.$$

Example. Let X be a metric space, $x_0 \in X$,

$$f(x) := d(x, x_0) \quad (f : X \to \mathbb{R}).$$

Then f is uniformly continuous on X.

Namely, for a prescribed $\varepsilon > 0$ we can choose $\delta = \varepsilon$ and obtain for $x, y \in X$ with $d(x, y) < \varepsilon$

$$|f(x) - f(y)| = |d(x, x_0) - d(y, x_0)| \leq d(x, y) < \varepsilon,$$

on account of the triangle inequality.

Theorem 7.11 *Let X be a metric space. A function $f = (f_1, \ldots, f_d) : X \to \mathbb{R}^d$ is (uniformly) continuous precisely if all the f_i ($i = 1, \ldots, d$) are (uniformly) continuous.*

Proof. The continuity is a simple consequence of theorem 7.7. The assertion about uniform continuity is also quite easy.   □

Theorem 7.12 *Let X be a metric space, $f, g : X \to \mathbb{R}$ continuous. The functions $f + g, f \cdot g, \max(f, g), \min(f, g), |f| : X \to \mathbb{R}$ are also continuous. (Here we have set $\max(f, g)(x) := \max(f(x), g(x))$, etc.) If $g(x) \neq 0$ for all $x \in X$, then f/g is also continuous.*

Proof. The proof is the same as for functions $f : K \to \mathbb{R}$ with $K \subset \mathbb{R}$.   □

Theorem 7.13 *Let X, Y, Z be metric spaces, $f : X \to Y$ continuous at x_0 and $g : Y \to Z$ continuous at $f(x_0)$. Then $g \circ f$ is continuous at x_0.*

Proof. Similar to that for functions of a single variable.   □

Definition 7.14 Let $f, f_n : X \to Y$ be functions ($n \in \mathbb{N}$), X, Y being metric spaces. The sequence $(f_n)_{n \in \mathbb{N}}$ converges uniformly to f (notation: $f_n \rightrightarrows f$), if

$$\forall \varepsilon > 0 \, \exists \, N \in \mathbb{N} \, \forall \, n \geq N, x \in X : d(f_n(x), f(x)) < \varepsilon.$$

Theorem 7.15 *Assumptions being as in definition 7.14, let all f_n be continuous and $f_n \rightrightarrows f$. Then f is also continuous.*

Proof. The proof goes again as for functions of one variable. We give it again for the sake of completeness.

Let $x_0 \in X, \varepsilon > 0$. By uniform convergence, there exists $m \in \mathbb{N}$ with

$$d(f_m(x), f(x)) < \frac{\varepsilon}{3} \text{ for all } x \in X.$$

Since f_m is continuous, there exists $\delta > 0$ with

$$d(f_m(x), f_m(x_0)) < \frac{\varepsilon}{3}, \text{ when } d(x, x_0) < \delta .$$

Therefore for $d(x, x_0) < \delta$ we have

$$d(f(x), f(x_0)) \leq d(f(x), f_m(x)) + d(f_m(x), f_m(x_0)) + d(f_m(x_0), f(x_0))$$
$$< \frac{\varepsilon}{3} + \frac{\varepsilon}{3} + \frac{\varepsilon}{3} = \varepsilon. \qquad \square$$

We now want to consider linear maps, also called linear operators, between normed vector spaces.

Definition 7.17 V, W being normed vector spaces with norms $\| \cdot \|_V, \| \cdot \|_W$, a linear map $L : V \to W$ is called bounded if there exists a $c \in \mathbb{R}$ with

$$\|L(x)\|_W \leq c\|x\|_V \text{ for all } x \in V.$$

Theorem 7.18 *A linear operator between normed vector spaces is continuous precisely when it is bounded. In particular, every linear map $L : \mathbb{R}^n \to \mathbb{R}^m$ is continuous.*

Proof. Let $L : V \to W$ be bounded, say

$$\|L(x)\|_W \leq c\|x\|_V \text{ for all } x \in V.$$

Let $\varepsilon > 0$. We may assume $c > 0$, otherwise the assertion is trivial, and set $\delta = \frac{\varepsilon}{c}$. Now for $x, y \in V$ with $\|x - y\|_V < \delta$ we have

$$\|L(x) - L(y)\|_W \leq c\|x - y\|_V < \varepsilon.$$

Hence L is even uniformly continuous. Conversely, let now $L : V \to W$ be continuous. As L is, in particular, continuous at 0, there exists a $\delta > 0$ with

$$\|L(\xi)\|_W < 1 \quad \text{for all } \xi \text{ with } \|\xi\|_V < \delta.$$

We choose $c := \frac{2}{\delta}$. For $x \in V, x \neq 0$, we set $\xi := \frac{x}{c\|x\|_V}$. Now $\|\xi\|_V < \delta$, so $\|L(\xi)\|_W < 1$ by choice of δ; consequently

$$\|L(x)\|_W = \| \ c\|x\|_V L(\xi) \ \|_W < c\|x\|_V.$$

Finally, it is obvious that a linear map between finite dimensional vector spaces is bounded, hence also continuous by what has just been proved. ☐

Remark. We have even shown more than what is stated, namely, L is continuous at $0 \Rightarrow L$ is bounded $\Rightarrow L$ is uniformly continuous. Naturally, the assumption that L is linear is essential here.

Definition 7.19 Let $L : V \to W$ be a continuous linear map of normed vector spaces. The norm $\|L\|$ of L is defined by

$$\|L\| := \sup\{\|L(x)\|_W : \|x\|_V = 1\}$$

$$(= \sup\{\|L(x)\|_W : \|x\|_V \le 1\} = \sup\{\frac{\|L(x)\|_W}{\|x\|_V} : x \ne 0\}).$$

(By theorem 7.18, $\|L\|$ is finite since L is assumed to be continuous).

That $\|L\|$ indeed defines a norm follows very easily.

Examples.

1) $\delta_x : C^0([0,1]) \to \mathbb{R}, \ \delta_x(f) := f(x) \ (x \in [0,1])$ is continuous with $\|\delta_x\| = 1$, for

$$|\delta_x(f)| = |f(x)| \le \sup\{|f(\xi)| : \xi \in [0,1]\} = \|f\|_{C^0},$$

and equality holds for $f \equiv 1$.

2) $D : C^1([0,1]) \to C^0([0,1]), Df := f'$ is continuous with $\|D\| \le 1$, because

$$\|Df\|_{C^0} = \|f'\|_{C^0} \le \|f\|_{C^0} + \|f'\|_{C^0} = \|f\|_{C^1}.$$

But if we equip $C^1([0,1])$ not with the C^1-norm but with the supremum norm (so $C^1([0,1])$ would then be considered as a subspace of $C^0([0,1])$), then D is no longer continuous, because for $f_n(x) := \sin(nx)$ we have (for $n \ge 2$) $\|f_n\|_{C^0} = 1$, $\|Df_n\|_{C^0} = \|f'_n\|_{C^0} = n$, and therefore there is no $c \in \mathbb{R}$ with

$$\|Df\|_{C^0} \le c\|f\|_{C^0} \quad \text{for all } f \in C^1.$$

The continuity of a linear operator can, therefore, also depend on which norms are being used.

Definition 7.20 Let V and W be normed vector spaces. We denote the space of continuous linear maps $L : V \to W$ by $B(V, W)$ and equip it with the norm $\| \cdot \|$.

Theorem 7.21 *Let W be a Banach space, V a normed vector space (e.g. likewise a Banach space). Then $B(V,W)$ is also a Banach space.*

Proof. Let $(L_n)_{n\in\mathbb{N}} \subset B(V,W)$ be a Cauchy sequence relative to $\|\cdot\|$, so

$$\forall\, \varepsilon > 0 \ \exists\, N \in \mathbb{N} \ \forall\, n,m \geq N : \|L_n - L_m\| < \varepsilon.$$

For every $x \in V$ we then have for $n,m \geq N$:

$$\|L_n(x) - L_m(x)\|_W = \|(L_n - L_m)(x)\|_W \leq \varepsilon \|x\|_V,$$

which follows directly from the definition of $\|L_n - L_m\|$. Therefore, for every $x \in V$, $(L_n(x))_{n\in\mathbb{N}}$ forms a Cauchy sequence in W. As W is complete, this sequence has a limit which we shall denote by $L(x)$. Thereby we have defined a map $L : V \to W$ which is linear, since e.g. for $x,y \in V$

$$\begin{aligned}
&\|L(x+y) - L(x) - L(y)\| \\
&\leq \|L(x+y) - L_N(x+y)\| + \|L_N(x+y) - L_N(x) - L_N(y)\| \\
&\qquad + \|L_N(x) - L(x)\| + \|L_N(y) - L(y)\|,
\end{aligned}$$

and the second term on the right side vanishes by linearity of L_N, whereas the other terms can be made arbitrarily small for sufficiently large N and therefore

$$L(x+y) = L(x) + L(y)$$

and similarly

$$L(\alpha x) = \alpha L(x) \quad \text{for } \alpha \in \mathbb{R}.$$

L_n also converges to L in $B(V,W)$ since for $n,m \geq N$

$$\sup\left\{\frac{\|L_n(x) - L_m(x)\|_W}{\|x\|_V} : x \neq 0\right\} < \varepsilon$$

and therefore also

$$\frac{\|L(x) - L_m(x)\|_W}{\|x\|_V} = \lim_{n\to\infty} \frac{\|L_n(x) - L_m(x)\|_W}{\|x\|_V} < \varepsilon$$

for all $x \neq 0$. Moreover L is bounded on account of

$$\frac{\|L(x)\|_W}{\|x\|_V} \leq \frac{\|(L - L_N)(x)\|_W}{\|x\|_V} + \frac{\|L_N(x)\|_W}{\|x\|_V},$$

and we can choose N so that $\frac{\|(L-L_N)(x)\|_W}{\|x\|_V} < 1$ by the previous inequality. Hence $L \in B(V,W)$. $\qquad\square$

For later purposes, we insert the following result about invertible linear maps between Banach spaces.

Lemma 7.22 *Let $L_0 : V \to W$ be a bijective continuous linear map between Banach spaces, with a continuous inverse L_0^{-1}. If $L \in B(V,W)$ satisfies*

$$\|L - L_0\| < \left\|L_0^{-1}\right\|^{-1},$$

then L is also bijective with a continuous inverse.

Proof. $L = L_0 \left(Id - L_0^{-1}(L_0 - L)\right)$. As for the geometric series, we see that the inverse of L is given by

$$\left(\sum_{i=0}^{\infty} \left(L_0^{-1}(L_0 - L)\right)^i\right) L_0^{-1},$$

if we can show that that series converges. We have

$$\|\sum_{i=m}^{n} \left(L_0^{-1}(L_0 - L)\right)^i\| \le \sum_{i=m}^{n} \| \left(L_0^{-1}(L_0 - L)\right)^i \|$$

$$\le \sum_{i=m}^{n} \left(\|L_0^{-1}\|\|L_0 - L\|\right)^i.$$

By our assumption, $\|L_0^{-1}\| \|L_0 - L\| < 1$, and so the series satisfies the Cauchy property, and it therefore converges to a linear operator with finite norm. As observed, this operator is the inverse L^{-1}, and Theorem 7.18 implies its continuity.

$\square$

Definition 7.23 Let (X, d) be a metric space, $x_0 \in X, r > 0$.

$$U(x_0, r) := \{x \in X : d(x, x_0) < r\}$$

and

$$B(x_0, r) := \{x \in X : d(x, x_0) \le r\}$$

are called the open and closed balls centered at x_0 of radius r. $U(x_0, r)$ is also called an r-neighborhood of x_0.

Remark. Let $x_0, y_0 \in X$, $x_0 \ne y_0$. For $0 < \varepsilon \le \frac{1}{2}d(x_0, y_0)$, the open balls $U(x_0, \varepsilon)$ and $U(y_0, \varepsilon)$ are disjoint.

A metric space therefore satisfies the so called Hausdorff separation property, that for every pair of distinct points, there exist disjoint ε-neighborhoods of the points. For later purposes, we formulate this as

Lemma 7.24 *Let $x, y \in X$ (a metric space), $x \ne y$. Then there exists an $\varepsilon > 0$ with $U(x, \varepsilon) \cap U(y, \varepsilon) = \emptyset$.* $\square$

Definition 7.25 Let (X, d) be a metric space. A subset $\Omega \subset X$ is called open if for every $x \in \Omega$ there exists an $\varepsilon > 0$ with

$$U(x, \varepsilon) \subset \Omega,$$

that is, when together with every point it also contains an ε-neighborhood of the point. An open subset of X containing $x \in X$ is called an open neighborhood of x. A subset $A \subset X$ is called closed if $X \backslash A$ is open.

Examples.

1) $X = \mathbb{R}$, $a, b \in \mathbb{R}$, $(a < b)$. Then the intervals (a, b), (a, ∞), $(-\infty, b)$ are open. For example, let $x_0 \in (a, b)$, $\varepsilon = \min(|a - x_0|, |b - x_0|)$. Then $U(x_0, \varepsilon) =$
$\{\xi \in \mathbb{R} : |x_0 - \xi| < \varepsilon\} \subset (a, b)$. Hence $[a, b]$, $[a, \infty)$, $(-\infty, b]$ are closed, for $X \backslash [a, b] = (-\infty, a) \cup (b, \infty)$ is open. The latter intervals are not open, since e.g. there exists no $\varepsilon > 0$ with $U(a, \varepsilon) \subset [a, b]$. The point $a \in [a, b]$ has therefore no ε-neighborhood contained in $[a, b]$. Similarly (a, b) is not closed. By similar considerations, one sees that $\{x \in \mathbb{R} : a \leq x < b\}$ is neither open nor closed.

2)   Let (X, d) be a metric space, $x_0 \in X, r > 0$. Then $U(x_0, r)$ is open. For, if $x \in U(x_0, r)$ then with $\varepsilon := r - d(x, x_0) > 0$, we have

$$U(x, \varepsilon) \subset U(x_0, r),$$

on account of the triangle inequality.

We formulate this example as

Lemma 7.26 *Let (X, d) be a metric space, $x_0 \in X, r > 0$. Then $U(x_0, r)$ is open.*

Theorem 7.27 *Let (X, d) be a metric space.*

(i)   *$\emptyset$ and X are open.*

(ii)   *For open Ω_1, Ω_2 their intersection $\Omega_1 \cap \Omega_2$ is also open.*

(iii)   *If $(\Omega_i)_{i \in I}$ (I any index set) is a family of open subsets of X, then $\bigcup_{i \in I} \Omega_i$ is also open.*

Proof.

(i) $\emptyset$ is open, because there is no $x \in \emptyset$ and therefore no condition that must be verified. X is open, since for every $x \in X$ and all $r > 0$, $U(x, r) \subset X$.

(ii)   Let $x_0 \in \Omega_1 \cap \Omega_2$. Since Ω_1 and Ω_2 are open, there exist $r_1, r_2 > 0$ with $U(x_0, r_i) \subset \Omega_i, i = 1, 2$. For $r := \min(r_1, r_2)$ we then have

$$U(x_0, r) \subset \Omega_1 \cap \Omega_2.$$

Hence $\Omega_1 \cap \Omega_2$ is open.

(iii)   If $x_0 \in \bigcup_{i \in I} \Omega_i$, then there exists an $i_0 \in I$ with $x_0 \in \Omega_{i_0}$. Since Ω_{i_0} is open, there exists an $r > 0$ with $U(x_0, r) \subset \Omega_{i_0} \subset \bigcup_{i \in I} \Omega_i$, whereby the openness of the latter has been shown. $\qquad\qquad\square$

Remark. From (ii) one obtains inductively: if $\Omega_1, \ldots, \Omega_n$ are open, then so is $\bigcap_{\nu=1}^{n} \Omega_\nu$. This does not hold for infinite intersections as the example $\Omega_\nu = (-\frac{1}{\nu}, \frac{1}{\nu}) \subset \mathbb{R}$ shows, for $\bigcap_{\nu \in \mathbb{N}} \Omega_\nu = \{0\}$, which is not open.

Example. For the example before lemma 7.4, i.e. where $d(x, y) = 1$ for all $x \neq y$, every subset of X is open, and hence also every subset is closed as its complement is open.

Theorem 7.28 *Let X be a metric space. A subset $A \subset X$ is closed precisely when the following condition is fulfilled: if $(x_n)_{n \in \mathbb{N}} \subset A$ and (x_n) converges to $x \in X$ then x is already in A.*

Proof. $\Rightarrow$: Let A be closed and assume that the sequence $(x_n)_{n \in \mathbb{N}} \subset A$ converges to $x \in X$. If $x \in X \backslash A$, there exists an $\varepsilon > 0$ with $U(x, \varepsilon) \subset X \backslash A$. This, however, contradicts $x_n \to x$. Hence, necessarily $x \in A$.

$\Leftarrow$: If A were not closed, i.e. $X \backslash A$ not open, then there would be an $x \in X \backslash A$ with the property that for no $n \in \mathbb{N}$ $U(x, \frac{1}{n}) \subset X \backslash A$ holds. So for every $n \in \mathbb{N}$ there exists an $x_n \in A \cap U(x, \frac{1}{n})$. Now as $d(x_n, x) < \frac{1}{n}$, the sequence $(x_n) \subset A$ converges to x, and by assumption we would have $x \in A$. This contradiction proves that $X \backslash A$ is open, so A is closed. $\qquad\square$

A direct consequence is

Corollary 7.29 *Every finite subset, in particular every one element subset, of a metric space is closed.*

$\qquad\qquad\qquad\qquad\qquad\qquad\qquad\qquad\qquad\qquad\qquad\qquad\qquad\qquad\square$

Definition 7.30 Let X be a metric space, $M \subset X$. An element $x \in X$ is called a boundary point of M if every open neighborhood of x contains (at least) one point of M as well as of its complement $X \backslash M$. ∂M denotes the set of boundary points of M. $\overset{\circ}{M} := M \backslash \partial M$ denotes the interior and $\bar{M} := M \cup \partial M$ the closure of M.

The previous definition is justified by the following theorem

Theorem 7.31 *Let X be a metric space and $M \subset X$. We have*

(i)   *∂M is closed.*

(ii) $M \backslash \partial M$ is open.

(iii) $M \cup \partial M$ is closed.

Proof.

(i)   Let $x \in X \backslash \partial M$. Then there exists an ε neighborhood $U(x, \varepsilon)$ which either contains no points of M or no points of $X \backslash M$. Now if there were a $y \in \partial M$ with $y \in U(x, \varepsilon)$ then by lemma 7.26, there would exist an $\eta > 0$ with $U(y, \eta) \subset U(x, \varepsilon)$. Since $y \in \partial M$, the neighborhood $U(y, \eta)$ contains a point of M as well as of $X \backslash M$, the same holds for $U(x, \varepsilon)$, which would be a contradiction. It follows that $U(x, \varepsilon) \subset X \backslash \partial M$ and so $X \backslash \partial M$ is open and ∂M is closed.

(ii)   If $x \in M \backslash \partial M$ then again there exists an ε-neighborhood $U(x, \varepsilon)$ which either contains no points of M or none of $X \backslash M$. Now since $x \in M \cap U(x, \varepsilon)$, only the latter is possible, and it follows, using (i), that $U(x, \varepsilon) \subset M \backslash \partial M$. Hence $M \backslash \partial M$ is open.

(iii)   If $x \in X \backslash (M \cup \partial M)$, then again there exists an ε-neighborhood $U(x, \varepsilon)$ which either contains no points of M or no points of $X \backslash M$, because in particular $x \in X \backslash \partial M$. Since $x \in X \backslash M \cap U(x, \varepsilon)$, this time the first possibility must occur and it follows as before that $U(x, \varepsilon) \subset X \backslash (M \cup \partial M)$. Therefore $X \backslash (M \cup \partial M)$ is open and $M \cup \partial M$ is closed. $\qquad \square$

The next result appears at first sight perhaps somewhat surprising.

Theorem 7.32 *Let $f : X \to Y$ be a function, X, Y being metric spaces. The function f is continuous precisely if for every open set $V \subset Y$, the set $f^{-1}(V)$ is open in X.*

Proof. $\Rightarrow$: Let $V \subset Y$ be open and $x_0 \in f^{-1}(V)$, so $f(x_0) \in V$. Then there exists an $\varepsilon > 0$ with $U(f(x_0), \varepsilon) \subset V$. As f is continuous, there exists a $\delta > 0$ with $d(f(x), f(x_0)) < \varepsilon$ whenever $d(x, x_0) < \delta$. This, however, means that for $x \in U(x_0, \delta)$, $f(x) \in U(f(x_0), \varepsilon) \subset V$, and therefore $U(x_0, \delta) \subset f^{-1}(V)$. Hence $f^{-1}(V)$ is open.

$\Leftarrow$: Let $x_0 \in X, \varepsilon > 0$. The neighborhood $U(f(x_0), \varepsilon)$ is open and by assumption $f^{-1}(U(f(x_0), \varepsilon))$ is also open. Therefore there exists a $\delta > 0$ with $U(x_0, \delta) \subset f^{-1}(U(f(x_0), \varepsilon))$, so $f(U(x_0, \delta)) \subset U(f(x_0), \varepsilon)$. But this is equivalent to the implication

$$d(x, x_0) < \delta \Rightarrow d(f(x), f(x_0)) < \varepsilon.$$

Hence f is continuous. $\qquad \square$

Example. For the trivial metric space (X, d) introduced before lemma 7.4, every function $f : X \to Y$ into a metric space Y is continuous, because all subsets of X are open.

Caution: The continuity of f is not equivalent to the requirement that the image $f(U)$ of an open set $U \subset X$ be open in Y. An example is $f : \mathbb{R} \rightarrow \mathbb{R}, f(x) = |x|$. Here $f(\mathbb{R}) = \{x \in \mathbb{R} : x \geq 0\}$, which is not open in $\mathbb{R}$.

Definition 7.33 A subset C of a vector space is called convex if for all $x, y \in C$,
$x + t(y - x) \in C \ \forall \ t \in [0, 1]$.
A convex set therefore contains for any two points in it the segment joining the two points.

Example. Let B be a Banach space, $x_0 \in B$. Then for every $r > 0$ the sets $U(x_0, r)$ and $B(x_0, r)$ are convex. For example, let $x, y \in U(x_0, r)$, so $\|x - x_0\| < r$ and $\|y - x_0\| < r$. Now for $0 \leq t \leq 1$ $\|x - x_0\| < r$ and

$$\|x + t(y - x) - x_0\| \leq t\|y - x_0\| + (1 - t)\|x - x_0\| < r,$$

so $x + t(y - x) \in U(x_0, r)$.

The proof of the convexity of $B(x_0, r)$ is exactly the same.

Definition 7.34 A metric space B is said to be connected, if for every pair of open sets $\Omega_1, \Omega_2 \subset B$ with

$$\Omega_1 \cup \Omega_2 = B$$
$$\Omega_1 \cap \Omega_2 = \emptyset$$

either

$$\Omega_1 = \emptyset \text{ or } \Omega_2 = \emptyset.$$

Example. We continue to discuss the example of a metric space (X, d) with $d(x, y) = 1$ for all $x \neq y$. If X has more than one element, then X is not connected because every subset of X is open and so we may find two nonempty disjoint open sets whose union is X.

The following result is often of use in proofs.

Lemma 7.35 *Let B be connected. If Ω is a nonempty subset of B which is both open and closed then $\Omega = B$.*

Proof. $\Omega_1 = \Omega$ and $\Omega_2 = B \backslash \Omega$ fulfil the requirements of definition 7.34. Since B is connected and $\Omega \neq \emptyset$, it follows $B \backslash \Omega = \emptyset$, so $\Omega = B$. $\qquad \square$

Lemma 7.36 *Every normed vector space is connected. More generally, every convex subset of a normed vector space is connected.*

Proof. Let B be a normed vector space and Ω_1, Ω_2 nonempty open subsets of B with $B = \Omega_1 \cup \Omega_2$, $\Omega_1 \cap \Omega_2 = \emptyset$. Let $x \in \Omega_1$ and $y \in \Omega_2$. We set

$$t := \sup\{\tau \in [0,1] : \text{ for all } s \text{ with } 0 \leq s \leq \tau \text{ we have } sy + (1-s)x \in \Omega_1\}$$

(as $x \in \Omega_1$, the set under consideration is nonempty because it contains 0).
    We shall show that

$$\xi := ty + (1-t)x$$

can lie neither in Ω_1 nor in Ω_2, which is a contradiction to $\Omega_1 \cup \Omega_2 = B$, so that one of Ω_1 and Ω_2 must be empty, as asserted. To prove this assertion we shall show that every open neighborhood of ξ contains points of $\Omega_2 = B\backslash\Omega_1$ as well as points of $\Omega_1 = B\backslash\Omega_2$. Since Ω_1 and Ω_2 have both been assumed open, the point ξ can then indeed lie neither in Ω_1 nor in Ω_2. So let U be an open neighborhood of ξ. If $t = 1$ then $\xi = y \notin \Omega_1$. If $t < 1$, there exists a t' with $t < t' \leq 1$ and $\xi' := t'y + (1-t')x \in U$, but $\xi' \notin \Omega_1$. Now as U is open, there exists an $\varepsilon > 0$ with $sy + (1-s)x \in U$ for $t - \varepsilon < s < t + \varepsilon$ and by definition of t there exists for every $\varepsilon > 0$ a t' with $t < t' < t + \varepsilon$ and $t'y + (1-t')x \notin \Omega_1$, as claimed. If $t = 0$, then $\xi = x \notin \Omega_2$. If $t > 0$, there exists a t'' with $t - \varepsilon < t'' < t$, so $\xi'' := t''y + (1-t'')x \in U$, and as $t'' < t$, by definition of t the point $\xi'' \in \Omega_1 = B\backslash\Omega_2$. Therefore U contains a point $\xi' \in \Omega_2$ as well as a point $\xi'' \in \Omega_1$.
    Exactly the same proof shows that convex subsets of B are connected.   □

Lemma 7.37 *The connected subsets of $\mathbb{R}$ are precisely the intervals (open, closed or half open ($\{a < x \leq b\}$ or $\{a \leq x < b\}$), bounded or unbounded (including $\mathbb{R}$). Here a single element subset of $\mathbb{R}$ is also regarded as a closed interval).*

Proof. Let $B \subset \mathbb{R}$. As a subset of a metric space, B itself is a metric space. A neighborhood $U(x, \varepsilon)$ ($\varepsilon > 0$) of a point $x \in B$ relative to the metric of B is then

$$U(x, \varepsilon) := \{y \in B : d(x,y) < \varepsilon\} = B \cap \{y \in \mathbb{R}^n : d(x,y) < \varepsilon\}.$$

From this it follows that the open sets of B relative to the metric of B are of the form $B \cap \Omega$, Ω open in $\mathbb{R}$.
    After these preliminaries, we come to the actual proof. So let B be a non-empty connected subset of $\mathbb{R}$. If B contains just one point, there is nothing to prove. So let $x_1, x_2 \in B$, say $x_1 < x_2$. We assume that there exists an $\xi \in \mathbb{R}$ with $x_1 < \xi < x_2, \xi \notin B$. We then set

$$\Omega_1 := B \cap \{x \in \mathbb{R} : x < \xi\}$$
$$\Omega_2 := B \cap \{x \in \mathbb{R} : x > \xi\}.$$

Now Ω_1, Ω_2 are open in B, $\Omega_1 \cap \Omega_2 = \emptyset$ and $\Omega_1 \cup \Omega_2 = B$, as $\xi \notin B$. On account of $x_1 \in \Omega_1, x_2 \in \Omega_2$ none of these sets is empty and B can therefore not be connected. Therefore, if B is connected, it contains together with any two points all the points lying between them. If we set

$$M := \sup\{x \in B\} \in \mathbb{R} \cup \{\infty\}$$
$$m := \inf\{x \in B\} \in \mathbb{R} \cup \{-\infty\},$$

then B contains all points x with $m < x < M$, as well as possibly m and M, in case they are finite. Hence B is an interval. ☐

Definition 7.38 A subset B of a metric space is called bounded if there exist an $x_0 \in X$ and an $r > 0$ with

$$B \subset U(x_0, r).$$

B is said to be totally bounded if for every $\varepsilon > 0$ there exist an $n \in \mathbb{N}$ and points $x_1, \ldots, x_n \in B$ with

$$B \subset \bigcup_{i=1}^{n} U(x_i, \varepsilon).$$

Definition 7.39 A subset K of a metric space is said to be compact if for any open cover $(U_i)_{i \in I}$ of K (i.e. all U_i are open and $K \subset \bigcup_{i \in I} U_i$) there exists a finite subfamily $(U_i)_{i \in E}$ ($E \subset I$ and E finite) which covers K, that is, $K \subset \bigcup_{i \in E} U_i$. K is said to be sequentially compact if every sequence $(x_n)_{n \in \mathbb{N}} \subset K$ contains a subsequence which converges to some $x \in K$.

It may well be argued that the concept of compactness is the single most important concept of mathematical analysis. In its sequential version it embodies the existence of limits of subsequences. The concept of compactness will be fundamental for many of the subsequent §§.

Theorem 7.40 *Let K be a subset of a metric space X. Then the following three conditions are equivalent:*

(i)   *K is compact.*

(ii)   *K is sequentially compact.*

(iii)   *K is complete (i.e. if $(x_n)_{n \in \mathbb{N}} \subset K$ is a Cauchy sequence then it converges to $x \in K$) and totally bounded.*

Proof.

(i) $\Rightarrow$ (ii). Let $(x_n)_{n \in \mathbb{N}} \subset K$. If there is no subsequence of (x_n) which converges to a point in K, then every $x \in K$ has a neighborhood $U(x, r(x))$ which contains only finitely many terms of the sequence. For, if there were an $x \in X$ for which for every $\varepsilon > 0$ there were infinitely many $n \in \mathbb{N}$ with $x_n \in U(x, \varepsilon)$, the same would hold in particular for $\varepsilon = \frac{1}{m}, m \in \mathbb{N}$. We could then find inductively for $m \in \mathbb{N}$ $n_m \in \mathbb{N}$ with $n_m > n_{m-1}$ and $x_{n_m} \in U(x, \frac{1}{m})$. But then the sequence $(x_{n_m})_{m \in \mathbb{N}} \subset (x_n)_{n \in \mathbb{N}}$ would converge to x, which would be contrary to our assumption. Hence any neighborhood of an element $x \in K$ contains only finitely many terms of the sequence. Now clearly

$$K \subset \bigcup_{x \in K} U(x, r(x)),$$

and as K is compact, this cover has a finite subcover, so there exist $\xi_1, \ldots, \xi_N \in K$ with

$$K \subset \bigcup_{i=1}^{N} U(\xi_i, r(\xi_i)).$$

Since every $U(\xi_i, r(\xi_i))$ contains only finitely many sequence terms, therefore K also contains only finitely many sequence terms. Consequently, the sequence $(x_n)_{n \in \mathbb{N}}$ is contained in a finite subset of K and so a subsequence is convergent, which contradicts our assumption. Hence K must be sequentially compact.

(ii) $\Rightarrow$ (iii). First of all any sequence in K, in particular a Cauchy sequence, contains a subsequence which is convergent in K. Therefore every Cauchy sequence itself converges in K, as a Cauchy sequence is convergent if and only if any subsequence of it is convergent. So K is complete.

We assume that K is not totally bounded. There is then an $\eta > 0$ such that K cannot be covered by finitely many $U(x_i, \eta)$. We now define inductively a sequence $(x_n)_{n \in \mathbb{N}} \subset K$ as follows: choose an arbitrary $x_1 \in K$; if $x_1, \ldots, x_{n-1} \in K$ have been chosen with $d(x_i, x_j) \geq \eta$ for $i \neq j$, $1 \leq i \leq n-1, 1 \leq j \leq n-1$, then K is not covered by $\bigcup_{i=1}^{n-1} U(x_i, \eta)$. Therefore there exists an $x_n \in K \backslash \bigcup_{i=1}^{n-1} U(x_i, \eta)$, and this x_n satisfies $d(x_i, x_n) \geq \eta$ for $1 \leq i \leq n-1$. The sequence $(x_n)_{n \in \mathbb{N}}$ can contain no convergent subsequence, for otherwise this subsequence would have the Cauchy property, and in particular there would be x_{n_1} and x_{n_2} with $d(x_{n_1}, x_{n_2}) < \eta$. That (x_n) contains no convergent subsequence contradicts the assumed sequential compactness of K. Hence K must be totally bounded.

(iii) $\Rightarrow$ (i). Assume that there exists an open cover $(U_i)_{i \in I}$ of K which contains no finite subcover. We then define inductively a sequence (B_n) of closed balls $B(\xi_n, r_n)$ as follows:

Set $B_0 = X$ and let $B_{n-1} = B(\xi_{n-1}, \frac{1}{2^{n-1}})$ with $\xi_{n-1} \in K$ such that $B_{n-1} \cap K$ is not covered by any finite subfamily of $(U_i)_{i \in I}$. Since K is assumed to be totally bounded, K can be covered by finitely many closed balls $K_j = B(y_j, \frac{1}{2^n}), y_j \in K$. Amongst the balls $K_j \cap B_{n-1} \neq \emptyset$ there exists at least one for which $K_j \cap K$ cannot be covered by finitely many of the $(U_i)_{i \in I}$, for otherwise $B_{n-1} \cap K$ would itself be covered by finitely many of the U_i's. We choose $B_n = B(\xi_n, \frac{1}{2^n})$ as one K_j with this property. Therefore as $B_n \cap B_{n-1} \neq \emptyset$ we have

$$d(\xi_n, \xi_{n-1}) \leq \frac{1}{2^{n-1}} + \frac{1}{2^n} \leq \frac{1}{2^{n-2}}.$$

For $N \in \mathbb{N}$ and $n, m \geq N$ it follows (assuming $n < m$)

$$d(\xi_n, \xi_m) \leq d(\xi_n, \xi_{n+1}) + \ldots + d(\xi_{m-1}, \xi_m) \leq \frac{1}{2^{n-1}} + \ldots + \frac{1}{2^{m-2}} \leq \frac{1}{2^{N-2}}.$$

Therefore $(\xi_n)_{n \in \mathbb{N}}$ is a Cauchy sequence which converges by assumption to some $x \in K$. Now $x \in U_{i_0}$ (such an $i_0 \in I$ exists as the $(U_i)_{i \in I}$ cover K). Since U_{i_0} is open, there exists an $\varepsilon > 0$ with $U(x, \varepsilon) \subset U_{i_0}$. Because (ξ_m) converges to x, there exists some $N \in \mathbb{N}$ with $d(x, \xi_N) < \frac{\varepsilon}{2}$ and $\frac{1}{2^N} < \frac{\varepsilon}{2}$. It follows from the triangle inequality that

$$B(\xi_N, \frac{1}{2^N}) \subset U(x, \varepsilon) \subset U_{i_0}.$$

This, however, contradicts the assumption that $B_N = B(\xi_N, \frac{1}{2^N})$ cannot be covered by any finite subfamily of $(U_i)_{i \in I}$. This contradiction proves the compactness of K. □

Lemma 7.41 *Every closed subset A of a compact set K is compact.*

Proof. We can use any of the three criteria of theorem 7.40, for example that of the sequential compactness (the reader may carry out a proof using the other critieria as an exercise). So let $(x_n)_{n \in \mathbb{N}} \subset A$ be a sequence. Since $A \subset K$ and K is compact, a subsequence converges to some $x \in K$ and this $x \in A$ because A is closed (by theorem 7.28). Hence A is compact. □

Corollary 7.42 (Heine-Borel) *A subset K of $\mathbb{R}^d$ is compact precisely if it is closed and bounded.*

Proof. The implication follows one way from theorem 7.40 (iii), as a complete set is closed and a totally bounded set is bounded. Now assume that K is closed and bounded. As K is bounded, there is an $R > 0$ for which K is contained in the cube

$$W_R := \{x = (x^1, \ldots, x^d) \in \mathbb{R}^d, \ \max_{i=1,\ldots,d} |x^i| \leq R\}.$$

Since K is closed, it suffices, by lemma 7.41, to show that W_R is compact. Because $\mathbb{R}^d$ is complete (theorem 7.7) and W_R is closed (this follows likewise from theorem 7.7 using the fact that the interval $[-R, R] \subset \mathbb{R}$ is closed), it is also complete, for if $(x_n)_{n \in \mathbb{N}} \subset W_R$ is a Cauchy sequence, it converges to an $x \in \mathbb{R}^d$, which must already be in W_R, since W_R is closed. Moreover, W_R is totally bounded. For, let $\varepsilon > 0$. We choose $m \in \mathbb{N}$ with $m > \frac{R}{\varepsilon}$. We then have

$$W_R \subset \bigcup \{U((\frac{k_1 R}{m}, \frac{k_2 R}{m}, \ldots, \frac{k_n R}{m}), \varepsilon) :$$
$$k_1, \ldots, k_n \in \mathbb{Z}, -m \leq k_i \leq m \quad \text{for } i = 1, \ldots, d\}$$

So W_R is totally bounded. Therefore condition (iii) of theorem 7.40 holds and the assertion follows. □

Remark. The assertion in corollary 7.42 is no longer true for infinite dimensional spaces. As an example, consider

$$S := \{f \in C^0([0,1]) : \|f\|_{C^0} = 1\}$$

S is obviously bounded and it is closed, because if $(f_n)_{n \in \mathbb{N}}$ converges uniformly to f, then $\|f\|_{C^0} = 1$, so $f \in S$. However, S is not compact: we consider the sequence

$$f_n = x^n.$$

As we have seen earlier, $(f_n)_{n \in \mathbb{N}}$ contains no convergent subsequence, because the pointwise limit of any subsequence is

$$f(x) = \begin{cases} 0 & \text{for } 0 \leq x < 1 \\ 1 & x = 1 \end{cases} \quad \text{(see Example 1) in §5)},$$

and since f is discontinuous, it cannot be the uniform limit of continuous functions. Therefore S is not sequentially compact.

Corollary 7.43 (Bolzano-Weierstrass) *Every bounded sequence in $\mathbb{R}^d$ has a convergent subsequence.*

Proof. This follows directly from corollary 7.42 and theorem 7.40, applied to the closure of the set of all points in the sequence. □

Theorem 7.44 *Let K be compact and $f : K \to \mathbb{R}$ be continuous. Then f assumes its maximum and minimum on K, i.e. there exist $y_1, y_2 \in K$ with*

$$f(y_1) = \sup\{f(x) : x \in K\}$$
$$f(y_2) = \inf\{f(x) : x \in K\}.$$

Proof. We shall prove that f takes its maximum value on K. Let $(x_n)_{n \in \mathbb{N}} \subset K$ be a sequence with

$$\lim_{n \to \infty} f(x_n) = \sup\{f(x) : x \in K\}.$$

Since K is sequentially compact, the sequence (x_n) converges, after choosing a subsequence, to some $y_1 \in K$. Since f is continuous it follows that

$$f(y_1) = \lim_{n \to \infty} f(x_n) = \sup\{f(x) : x \in K\}.$$

$\square$

Theorem 7.44 can be proved as well using the covering criterion for compactness. For this, consider for example

$$M := \sup\{f(x) : x \in K\}.$$

If f does not assume its maximum on K, then there exists, by continuity of f, for any $x \in K$ a $\delta > 0$ with

$$f(\xi) < r(x) := \frac{f(x) + M}{2} < M$$

for $d(x, \xi) < \delta$ (this is the $\varepsilon - \delta$–criterion used for $\varepsilon = \frac{M - f(x)}{2}$). Since δ can depend on x, write $\delta(x)$ instead of δ.

Since K is compact and $(U(x, \delta(x))_{x \in K}$ is an open cover of K, it has a finite subcover, so there exist points $x_1, \ldots, x_n$ with

$$K \subset \bigcup_{i=1}^{n} U(x_i, \delta(x_i)).$$

We set

$$r := \max_{i=1,\ldots,n} r(x_i) < M.$$

Then for every $\xi \in K$ we have $f(\xi) < r$, for there is an i with $\xi \in U(x_i, \delta(x_i))$ and so $f(\xi) < r(x_i) \le r$. Since $r < M$, the supremum of f on K cannot be M. This contradiction proves that the maximum is, after all, achieved on K.

Theorem 7.45 *Let K be compact, Y a metric space and $f : K \to Y$ continuous. Then f is uniformly continuous on K.*

Proof. Let $\varepsilon > 0$. For every $x \in K$ there exists, by continuity of f, a $\delta(x) > 0$ with

$$f(U(x, 2\delta(x))) \subset U(f(x), \frac{\varepsilon}{2}).$$

The $(U(x, \delta(x)))_{x \in K}$ form an open cover of K. Since K is compact, this cover has a finite subcover, so there exist points $x_1, \ldots, x_m \in K$ such that

$$K \subset \bigcup_{i=1}^{m} U(x_i, \delta(x_i)).$$

Let $\delta := \min_{i=1,\ldots,m} \delta(x_i) > 0$. Now if $d(x, y) < \delta$ then there exists a $j \in \{1, \ldots, m\}$ with $x, y \in U(x_j, 2\delta(x_j))$. It follows that $f(x), f(y) \in U(f(x_j), \frac{\varepsilon}{2})$

and therefore $d(f(x), f(y)) < \varepsilon$. As this holds for arbitrary $x, y \in K$ with $d(x, y) < \delta$, we see that f is uniformly continuous on K. $\qquad \square$

Theorem 7.46 *Let X, Y be metric spaces, $K \subset X$ compact and $f : K \to Y$ continuous. Then $f(K)$ is compact.*

(So a continuous image of a compact set is again compact.)

Proof. Let $(U_i)_{i \in I}$ be an open cover of $f(K)$. We set $V_i := f^{-1}(U_i), i \in I$. By theorem 7.32 V_i is open. Moreover the $(V_i)_{i \in I}$ clearly form an open cover of K. Hence there exists a finite subcover $f^{-1}(U_i), i \in E$ (E finite) of K, since K is compact. But then $f(K)$ is covered by the sets $f(V_i) \subset U_i (i \in E)$. So $U_i (i \in E)$ is a finite subcover and $f(K)$ is therefore compact. $\qquad \square$

Naturally, theorem 7.46 can also be proved by means of sequential compactness.

Theorem 7.47 *All norms in $\mathbb{R}^d$ are equivalent in the following sense: If $\| \cdot \|_0$ and $\| \cdot \|_1$ are two norms on $\mathbb{R}^d$ then there exist $\lambda, \mu > 0$ with*

$$\forall\, x \in \mathbb{R}^d : \lambda \|x\|_0 \leq \|x\|_1 \leq \mu \|x\|_0.$$

Proof. We prove the left inequality: the right one follows analogously by interchanging the roles of $\| \cdot \|_0$ and $\| \cdot \|_1$.

We set

$$\lambda := \inf\{\|y\|_1 : y \in \mathbb{R}^d, \|y\|_0 = 1\}.$$

We equip $\mathbb{R}^d$ with the metric induced by $\| \cdot \|_0$. Now $\{y \in \mathbb{R}^d : \|y\|_0 = 1\}$ is bounded and closed, for if $y_n \to y$ then $\|y_n - y\|_1 \to 0$ and because $|\|y_n\|_0 - \|y\|_0| \leq \|y_n - y\|_0$ (triangle inequality) we have $\|y_n\|_0 \to \|y\|_0$ and therefore $\|y\|_0 = 1$ in case $\|y_n\|_0 = 1$ for all n. By corollary 7.42, $\{y \in \mathbb{R}^d : \|y\|_0 = 1\}$ is compact.

Now we consider the continuity of $\| \cdot \|_1 : \mathbb{R}^d \to \mathbb{R}$. First, let $v \in \mathbb{R}^d, v \neq 0$. For a sequence $(\lambda_n)_{n \in \mathbb{N}} \subset \mathbb{R}$ and $\lambda \in \mathbb{R}$ we have

$$\|\lambda_n v - \lambda v\|_0 \to 0$$
$$\Leftrightarrow |\lambda_n - \lambda| \|v\|_0 \to 0$$
$$\Leftrightarrow |\lambda_n - \lambda| \|v\|_1 \to 0$$
$$\Leftrightarrow \|\lambda_n v - \lambda v\|_1 \to 0.$$

Now let $e_1, \ldots, e_d$ be a basis of $\mathbb{R}^d$, $x = x^1 e_1 + \ldots x^d e_d \in \mathbb{R}^d$. By substituting for v the basis vectors $e_1, \ldots, e_d$, it follows that for any sequence $(x_n)_{n \in \mathbb{N}} \subset \mathbb{R}^d$, $x_n = x_n^1 e_1 + \ldots x_n^n e_d$

$$\|x_n - x\|_0 \to 0 \quad \Leftrightarrow \quad \|x_n - x\|_1 \to 0.$$

From this the continuity of $\|\cdot\|_1$ follows because we have chosen the metric on $\mathbb{R}^d$ so that $x_n \to x$ is equivalent to $\|x_n - x\|_0 \to 0$.

As $\{y \in \mathbb{R}^d : \|y\|_0 = 1\}$ is compact and $\|\cdot\|_1 : \mathbb{R}^d \to \mathbb{R}$ is continuous, there exists an x_0 with $\|x_0\|_0 = 1$ and $\|x_0\|_1 = \lambda$. Since $x_0 \neq 0$, it follows that $\lambda > 0$. Therefore for $x \in \mathbb{R}^d$

$$\|x\|_1 = \|x\|_0 \cdot \|\frac{x}{\|x\|_0}\|_1 \geq \|x\|_0 \cdot \lambda \|\frac{x}{\|x\|_0}\|_0, \text{ as } \|\frac{x}{\|x\|_0}\|_0 = 1$$

$$= \lambda \|x\|_0.$$

The assertion follows. □

Definition 7.48 A continuous function $f : X \to Y$ from a metric space X to a metric space Y is said to be compact if for every bounded sequence $(x_n)_{n \in \mathbb{N}} \subset X$, the sequence $f(x_n)_{n \in \mathbb{N}}$ has a convergent subsequence.

Examples.

1)   Every continuous function $f : \mathbb{R}^d \to \mathbb{R}^m$ is compact. This follows from corollary 7.42, as for a bounded sequence $(x_n)_{n \in \mathbb{N}} \subset \mathbb{R}^d$, the sequence $(f(x_n))_{n \in \mathbb{N}} \subset \mathbb{R}^m$ is again bounded, on account of continuity of f.

2)   The function $i : C^{0,\alpha}([0,1]) \to C^0([0,1])$, $i(g) = g$, $(0 < \alpha \leq 1)$ is compact. This follows from the theorem of Arzela-Ascoli (5.21), for if $(g_n)_{n \in \mathbb{N}}$ is bounded in $C^{0,\alpha}([0,1])$, it is also equicontinuous and contains therefore a uniformly convergent subsequence. For the same reason $i : C^k([0,1]) \to C^{k-1}([0,1])$ $(k \geq 1)$ is compact.

At the end of this §, we should like to briefly discuss the abstract concept of a topological space. This concept will not be needed in the reminder of this textbook, but it plays a basic rôle in many mathematical theories, and its inclusion at the present location seems natural. The idea simply is to use the contents of theorems 7.27 and 7.32 as axioms.

Thus

Definition 7.49 Let X be a set. A topology on X is given by a collection $\mathcal{U}$ of subsets of X, called open subsets of X, that satisfy

(i) $\emptyset, X \in \mathcal{U}$.

(ii)   If $U, V \in \mathcal{U}$, then also $U \cap V \in \mathcal{U}$.

(iii)   If $U_i \in \mathcal{U}$ for all i in some index set I, then also

$$\bigcup_{i \in I} U_i \in \mathcal{U}.$$

A topological space $(X, \mathcal{U})$ is a set X equipped with a topology $\mathcal{U}$. $A \subset X$ is called closed if $X \backslash A$ is open.

An open subset U of X containing $x \in X$ is called an open neighborhood of x.

By theorem 7.27, any metric space (X, d) becomes a topological space if we take $\mathcal{U}$ as the collection of all sets that are open in the sense of definition 7.49. The resulting topology is called the topology induced by the metric.

We have also seen (lemma 7.24) that every metric space satisfies

Definition 7.50 A topological space $(X, \mathcal{U})$ is called a Hausdorff space if for any $x, y \in X$ with $x \neq y$ there exist open neighborhoods $U(x)$ of x, $U(y)$ of y with

$$U(x) \cap U(y) = \emptyset.$$

This property, however, is not satisfied by every topological space.

Examples.

1)   Let X be a set with at least two elements, $\mathcal{U} = \{\emptyset, X\}$ then defines a topology on X that is not Hausdorff.

2)   We equip $\mathbb{R}^2$ with the following topology:
     The open subsets are $\emptyset, \mathbb{R}^2$, and all the complements of straight lines as well as the finite intersections and arbitrary unions of such sets. In other words, the closed subsets of $\mathbb{R}^2$ with this topology are precisely the affine linear subspaces and their finite unions. This topology is not Hausdorff. Namely, let $x, y \in \mathbb{R}^2, x \neq y$, and let $U(x)$ and $U(y)$ be arbitrary neighborhoods of x and y, resp. The complement of $U(x) \cap U(y)$ then is a finite union of affine linear subspaces (different from $\mathbb{R}^2$ since neither $U(x)$ nor $U(y)$ is empty). Thus, this complement is a proper subset of $\mathbb{R}^2$, and hence $U(x) \cap U(y) \neq \emptyset$.
     (A generalization of this topology yields the so called Zariski topology that is quite important in algebraic geometry.)

A remark on notation: Except for this example, $\mathbb{R}^d$ is always understood to be equipped with its Euclidean metric and the induced topology.

The next definition axiomatizes the content of theorem 7.32:

Definition 7.51 A map $f : (X, \mathcal{U}) \to (Y, \mathcal{V})$ between topological spaces is called continuous if for every open $V \subset Y, f^{-1}(V)$ is open in X.

Examples.

1)   Consider again the topology $\mathcal{U} = \{\emptyset, X\}$ on a set X. Let $(Y, \mathcal{V})$ be a Hausdorff topological space. Then the only maps $f : (X, \mathcal{U}) \to (Y, \mathcal{V})$ that are continuous are the constant ones. Namely, let $y_0 \in f(X)$. Then for every neighborhood $U(y_0)$ of $y_0, f^{-1}U(y_0)$ is nonempty, and open if f is continuous. Thus, $f^{-1}(U(y_0)) = X$. For any other point $y \neq y_0$ in Y, by the Hausdorff property, we may find disjoint

neighborhoods $U(y_0)$ and $U(y)$ of y_0 and y, resp. Then $f^{-1}(U(y_0))$ and $f^{-1}(U(y))$ are also disjoint, and open if f is continuous. Therefore, the set $f^{-1}(U(y))$ has to be empty. Since this holds for every $y \neq y_0$, $f \equiv y_0$.

2)   Consider again a metric space (X, d) with metric $d(x, y) = 1$ for $x \neq y$, with the induced topology. Since now every subset of X is open, any map $f : X \to (Y, \mathcal{V})$ into any topological space is continuous.

The concept of compactness extends to topological spaces as well, and in fact the formulation given in definition 7.39 can be taken over verbatim. However, a sequential characterization of compactness is not possible in general.

Exercises for § 7

1)   Provide two proofs, one using sequences and one using open coverings, of the following result.
Let $K_n \neq \emptyset$ be compact subsets of some metric space, and

$$K_{n+1} \subset K_n \quad \text{for all } n \in \mathbb{N}.$$

Then

$$\bigcap_{n \in \mathbb{N}} K_n \neq \emptyset.$$

Give an example to show that this result ceases to be true if the K_n are merely supposed to be closed.

2)   Let (X, d) be a metric space, $f, g : X \to \mathbb{R}$ continuous functions. Then

$$\varphi(x) := \max(f(x), g(x))$$
$$\psi(x) := \min(f(x), g(x))$$

define continuous functions as well. Use this to conclude that $|f|$ is continuous if f is.

3)   Determine

$$S_p := \{y \in \mathbb{R}^2 : \|y\|_p = 1\}$$

for $p \geq 1 (p \in \mathbb{R})$ and $p = \infty$. Draw pictures for $p = 1, 2, \infty$.

3)   Let K be a compact subset of some metric space, $(U_i)_{i \in I}$ an open covering of K. Show that there exists $\delta > 0$ with the property that for every $x \in K$ there is some $i \in I$ with

$$U(x, \delta) \subset U_i.$$

5)   Let $I, J \subset \mathbb{R}$ be compact intervals, $f : I \times J \to \mathbb{R}$ continuous. Define $F : I \to \mathbb{R}$ by $F(x) := \sup\{f(x, y) : y \in J\}$. Show that F is continuous.

6)   Determine the boundaries δA of the following sets $A \subset X$:

   a) $A = \mathbb{Q}$ (rational numbers), $X = \mathbb{R}$

   b) $A = \mathbb{R}\backslash\mathbb{Q}, X = \mathbb{R}$

   c) $A = \{x \in \mathbb{R} : 0 < x \leq 1\}, X = \mathbb{R}$

   d) $A = \{x \in \mathbb{R}^2 : \|x\| < 1\}, X = \mathbb{R}^2$

   e) $A = C^1([0, 1]), X = C^0([0, 1])$

   f) X a metric space, $A = (x_n)_{n \in \mathbb{N}}$, with $(x_n) \subset X$ converging to $x \in X$.

7)   Let M be a subset of some metric space with closure $\bar{M}$ and interior $\overset{\circ}{M}$. Show that $\bar{M}$ is the smallest closed set containing M (i.e. if A is closed and $M \subset A$, then also $\bar{M} \subset A$), and that $\overset{\circ}{M}$ is the largest open set contained in M (i.e. if Ω is open and $\Omega \subset M$, then also $\Omega \subset \overset{\circ}{M}$).

8)   Let $\mathrm{Gl}(d, \mathbb{R})$ be the space of invertible $d \times d$ matrices (with real entries). Show that $\mathrm{Gl}(d, \mathbb{R})$ is not connected.
   (Hint: Consider $\det : \mathrm{Gl}(d, \mathbb{R}) \to \mathbb{R}$ where $\det(A)$ denotes the determinant of the matrix A.)

9)   Construct a subset of some metric space that is bounded, but not totally bounded.

10)   Which of the following subsets of $\mathbb{R}^2$ are connected?

   a) $\Omega_1 := \{(x, y) : x \neq 0 : y = \sin\frac{1}{x}\}$

   b) $\Omega_2 := \Omega_1 \cup \{(0, 0)\}$

   c) $\Omega_3 := \Omega_1 \cup \{(0, y) : y \in \mathbb{R}\}$

11)   Construct a Banach space B and a linear map $L : B \to B$ that is not continuous.

Chapter III.

Calculus in Euclidean and Banach Spaces

8. Differentiation in Banach Spaces

We introduce the concept of differentiability for mappings between Banach spaces, and we derive the elementary rules for differentiation.

Definition 8.1 Let V, W be Banach spaces, $\Omega \subset V$ open, $f : \Omega \to W$ a map and $x_0 \in \Omega$. The map f is said to be differentiable at x_0 if there is a continuous linear map $L =: Df(x_0) : V \to W$ such that

$$\lim_{\substack{x \to x_0 \\ x \neq x_0}} \frac{\|f(x) - f(x_0) - L(x - x_0)\|}{\|x - x_0\|} = 0 \tag{1}$$

$Df(x_0)$ is called the derivative of f at x_0. f is said to be differentiable in Ω if it is differentiable at every $x_0 \in \Omega$.

Notice that the continuity of $Df(x_0)$ as a linear map does not mean that $Df(x_0)$ depends continuously on x_0.

Lemma 8.1 *Let f be differentiable at x_0 (assumptions being as in Def. 8.1). Then $Df(x_0)$ is uniquely determined.*

Proof. Let $L_1, L_2 : V \to W$ be linear maps that fulfil (1) above. It follows easily that

$$\lim_{\substack{x \to x_0 \\ x \neq x_0}} \frac{\|L_1(x - x_0) - L_2(x - x_0)\|}{\|x - x_0\|} = 0.$$

For $\varepsilon > 0$ there exists therefore $\delta > 0$ with

$$\|(L_1 - L_2)(\frac{x - x_0}{\|x - x_0\|})\| = \frac{\|(L_1 - L_2)(x - x_0)\|}{\|x - x_0\|} < \varepsilon$$

for $\|x - x_0\| < \delta$.

Now for any $y \in V$ with $\|y\| = 1$ there exists an $x \in V$ with $\|x - x_0\| < \delta$ and $\frac{x - x_0}{\|x - x_0\|} = y$. So for an arbitrary $\varepsilon > 0$ and all $y \in V$ with $\|y\| = 1$ we have

$$\|(L_1 - L_2)(y)\| < \varepsilon$$

so that $\|L_1 - L_2\| \leq \varepsilon$ and as ε was arbitrary, $L_1 = L_2$. $\qquad\square$

Lemma 8.2 *Let f be differentiable at x_0 (the assumptions being as in Def. 8.1). Then f is continuous at x_0.*

Proof. On account of equality (1) above, there is an $\eta > 0$ such that for $\|x - x_0\| < \eta$,
$$\|f(x) - f(x_0) - L(x - x_0)\| \leq \|x - x_0\|$$
holds, so
$$\|f(x) - f(x_0)\| \leq (1 + \|L\|)\|x - x_0\|.$$
For $\varepsilon > 0$ we choose $\delta = \min(\eta, \frac{\varepsilon}{1+\|L\|}) > 0$ and for $\|x - x_0\| < \delta$ we then have
$$\|f(x) - f(x_0)\| < \varepsilon,$$
and the continuity of f at x_0 follows. $\square$

Examples.

1)   Let $f : \Omega \to W$ be constant. Then f is differentiable in Ω and its derivative $Df(x_0)$ is the zero element of $B(V, W)$ for every $x_0 \in \Omega$.

2)   Any $L \in B(V, W)$ is differentiable on V with
$$DL(x_0) = L$$
for all $x_0 \in V$, because
$$L(x) - L(x_0) - L(x - x_0) = 0 \text{ for all } x \in V.$$

3)   Let $\Omega \subset \mathbb{R}^n$ be open. If $f : \Omega \to \mathbb{R}^m$ is differentiable at $x_0 \in \Omega$, its derivative $Df(x_0)$ is a linear map from $\mathbb{R}^n$ into $\mathbb{R}^m$ and so can be identified with an
$(m \times n)$-matrix. Recall that a linear map between finite dimensional vector spaces is automatically continuous.

4)   In particular, let $\Omega \subset \mathbb{R}, f : \Omega \to \mathbb{R}$. Then f is differentiable at $x_0 \in \Omega$ precisely if $\lim_{x \to x_0} \frac{f(x) - f(x_0)}{x - x_0} = f'(x_0)$. (Here, we observe $f'(x_0) = Df(x_0)(1)$.) Notice that a linear map $L : \mathbb{R} \to \mathbb{R}$ is simply given by the real number $L(1)$.

5)   Let $W = W_1 \times \ldots \times W_n$ be a product of Banach spaces. Setting
$$\|y\| := \max_{i=1,\ldots,n} \|y^i\|_{W_i} \text{ for } y = (y^1, \ldots, y^n) \in W,$$
the space W itself becomes a Banach space.
For $L = (L^1, \ldots, L^n) \in L(V, W)$ we have, correspondingly
$$\|L\| = \max_{i=1,\ldots,n} \|L^i\|.$$

Let Ω be an open subset of a Banach space V.

The map $f = (f^1, \ldots, f^n) : \Omega \to W$ is differentiable at $x_0 \in \Omega$ precisely if every $f^i(i = 1, \ldots, n)$ is differentiable there and we have then

$$Df(x_0) = (Df^1(x_0), \ldots, Df^n(x_0)).$$

Lemma 8.3 *Let f and g be differentiable at x_0 (assumptions as in definition 8.1), $\lambda \in \mathbb{R}$. Then $f + g$, λf are also differentiable at x_0.*

The proof is obvious.

Theorem 8.4 (Chain rule) *Let V, W, U be Banach spaces, $\Omega \subset V$ open, $x_0 \in \Omega$, $f : \Omega \to W$ differentiable at x_0 and further $\Sigma \subset W$ open with $y_0 := f(x_0) \in \Sigma$ and $g : \Sigma \to U$ differentiable at y_0.*

Then $g \circ f$ is defined in an open neighborhood of x_0 and it is differentiable at x_0 with

$$D(g \circ f)(x_0) = Dg(y_0) \circ Df(x_0).$$

Proof. As f, by lemma (8.2), is continuous at x_0, there exists an open neighborhood $\Omega' \subset \Omega$ with $f(\Omega') \subset \Sigma$. The function $g \circ f$ is then defined on Ω'.

Now for $x \in \Omega'$

$$
\begin{aligned}
&\|g(f(x)) - g(f(x_0)) - Dg(y_0) \circ Df(x_0)(x - x_0)\| \\
&\leq \|g(f(x)) - g(f(x_0)) - Dg(y_0)(f(x) - f(x_0))\| \\
&\quad + \|Dg(y_0)(f(x) - f(x_0) - Df(x_0)(x - x_0))\|.
\end{aligned}
\tag{2}
$$

For $\varepsilon > 0$, we choose $\sigma > 0$ in such a way that for $\|y - y_0\| < \sigma$

$$\|g(y) - g(y_0) - Dg(y_0)(y - y_0)\| \leq \frac{\varepsilon}{2(1 + \|Df(x_0)\|)} \|y - y_0\|. \tag{3}$$

This works, as g is differentiable at y_0. Next we choose $\eta_1 > 0$ so that for $\|x - x_0\| < \eta_1$

$$\|f(x) - f(x_0) - Df(x_0)(x - x_0)\| \leq \frac{\varepsilon}{2\|Dg(y_0)\|} \|x - x_0\|. \tag{4}$$

Finally, we choose $0 < \eta_2 < \frac{\sigma}{1 + \|Df(x_0)\|}$ in such a way that for $\|x - x_0\| < \eta_2$

$$\|f(x) - f(x_0)\| \leq (1 + \|Df(x_0)\|)\|x - x_0\| \tag{5}$$

holds. This is possible as f is differentiable at x_0.

By (5), we have for $\|x - x_0\| < \eta_2$

$$\|f(x) - f(x_0)\| \leq \sigma. \tag{6}$$

We set $\delta := \min(\eta_1, \eta_2)$. Then for $\|x - x_0\| < \delta$ we have

$$\|g(f(x)) - g(f(x_0)) - Dg(y_0) \circ Df(x_0)(x - x_0)\| \leq \varepsilon \|x - x_0\|$$

using (2), (3), (5), (6), and (4).

As $\varepsilon > 0$ was arbitrary, it follows that

$$\lim_{\substack{x \to x_0 \\ x \neq x_0}} \frac{\|g(f(x)) - g(f(x_0)) - Dg(y_0) \circ Df(x_0)(x - x_0)\|}{\|x - x_0\|} = 0,$$

hence the theorem.   □

Theorem 8.5 (Mean value theorem) *Let $I = [a, b]$ be a compact interval in $\mathbb{R}$, W a Banach space and $f : I \to W$ a function which is continuous on I and differentiable in the interior $\overset{\circ}{I}$ of I with*

$$\|Df(x)\| \leq M \quad \text{for all } x \in \overset{\circ}{I}. \tag{7}$$

Then

$$\|f(b) - f(a)\| \leq M(b - a). \tag{8}$$

Proof. We show that for every $\eta > 0$

$$\|f(b) - f(a)\| \leq M(b - a) + \eta(b - a) \tag{9}$$

holds. As $\eta > 0$ is arbitrary, the inequality (8) follows.

We set

$$A := \{\xi \in I : \text{for all } z \text{ with } a \leq z < \xi,$$
$$\|f(z) - f(a)\| \leq M(z - a) + \eta(z - a)\}.$$

The set A is non-empty as $a \in A$. We set

$$c := \sup A.$$

As f is continuous, it follows that $c \in A$, so $A = [a, c]$. In order to show (9), it suffices to prove that $c = b$.

We assume that $c < b$. Then there exists a $\delta > 0$ with $c + \delta \leq b$ and

$$\forall z \in [c, c + \delta) : \|f(z) - f(c) - Df(c)(z - c)\| \leq \eta(z - c),$$

by definition of differentiation. From this, it follows that for $z \in [c, c + \delta)$

$$\|f(z) - f(a)\| \leq \|f(z) - f(c)\| + \|f(c) - f(a)\|$$
$$\leq M(z - c) + \eta(z - c) + M(c - a) + \eta(c - a)$$
$$= M(z - a) + \eta(z - a).$$

But here, as $z > c$ is possible, this contradicts the definition of c, unless $c = b$. Therefore $c = b$ and the assertion follows.   □

Corollary 8.6 *Let V, W be Banach spaces, $x, y \in V$, $S := \{x + t(y - x) : 0 \leq t \leq 1\}$ the line joining x and y, U an open neighborhood of S, and let $f : U \to W$ be continuous and differentiable at every point of S with*

$$\|Df(z)\| \leq M \text{ for all } z \in S.$$

Then

$$\|f(y) - f(x)\| \leq M\|y - x\|.$$

Proof. We set $I = [0, 1] \subset \mathbb{R}$ and define $g : I \to W$ by

$$g(t) = f(x + t(y - x)).$$

By the chain rule (theorem 8.4) g is differentiable at every $t \in (0, 1)$ with

$$Dg(t) = Df(x + t(y - x)) \cdot (y - x).$$

Theorem 8.5 now gives the assertion.   □

For later purposes we notice further

Lemma 8.7 *Let V, W be Banach spaces, $x, y \in V$, $S := \{x + t(y - x) : 0 \leq t \leq 1\}$, U an open neighborhood of S, $f : U \to W$ differentiable. Then for all $z \in U$ we have*

$$\|f(y) - f(x) - Df(z)(y - x)\| \leq \|y - x\| \sup_{\xi \in S} \|Df(\xi) - Df(z)\|$$

Proof. We apply corollary 8.6 to the function $g : U \to W$ defined by

$$g(\zeta) = f(\zeta) - Df(z) \cdot \zeta.$$

The function g is differentiable with

$$Dg(\zeta) = Df(\zeta) - Df(z).$$

□

Definition 8.8 *Let V, W be Banach spaces, $\Omega \subset V$ open, $f : \Omega \to W$ differentiable with derivative $Df(x)$ for $x \in \Omega$. The function f is said to be continuously differentiable in Ω if $Df(x)$ is continuous in x.*

It is important to distinguish between continuity of the linear map $Df(x_0)$ for fixed x_0, which is part of the requirement for differentiability at x_0, and the continuity of Df as a function of x in the above definition.

Definition 8.9 Let $V = V_1 \times \ldots \times V_d$ ($V_1, \ldots, V_d$ being Banach spaces), Ω open in V, $a = (a^1, \ldots, a^d) \in \Omega$. A function f from Ω to some Banach space is said to be partially differentiable at a in the j-th variable ($j \in \{1, \ldots, d\}$) if the function $x^j \mapsto f(a^1, \ldots, a^{j-1}, a^j + x^j, a^{j+1}, \ldots, a^d)$ is differentiable at $x^j = 0$. The corresponding derivative is then denoted by $D_j f(a)$ and called the j-th partial derivative of f at a.

Lemma 8.10 *Notations being as in definition 8.9, let f be differentiable at a. Then f is partially differentiable at a in all the variables, and for $v = (v^1, \ldots, v^d) \in V_1 \times \ldots \times V_d$, we have*

$$Df(a)(v^1, \ldots, v^d) = D_1 f(a) v^1 + \ldots + D_d f(a) v^d. \tag{10}$$

Proof. For $j = 1, \ldots, d$, let $i_j : V_j \to V$,

$$v^j \mapsto (0, \ldots, \underset{\substack{\uparrow \\ j\text{-th slot}}}{v^j}, \ldots, 0)$$

be the canonical injection. It is continuous and linear and therefore differentiable with derivative i_j. Therefore, by the chain rule, $D_j f(a)$ exists as the derivative of the function $f(a + i_j(x^j))$ at the point $x^j = 0$ with

$$D_j f(a) = Df(a) \circ i_j. \tag{11}$$

The equation (10) now follows from $v = \sum\limits_{j=1}^{d} i_j(v^j)$. $\square$

We shall later see by an example that the converse of lemma 8.10 does not hold, that is, partial differentiability in all the variables does not necessarily imply differentiability. However, the following holds.

Theorem 8.11 *Let $V_1, \ldots, V_d, W$ be Banach spaces, $V = V_1 \times \ldots \times V_d$, $\Omega \subset V$ open. A function $f : \Omega \to W$ is continuously differentiable on Ω precisely if it is partially differentiable on Ω with respect to the first to the n-th variable and all the partial derivatives are continuous on Ω. Of course, formula (10) then holds.*

Proof. The implication in one way follows directly from lemma 8.10 and equation (11). For the converse, we shall treat only the case $d = 2$. The general case follows easily by induction. We first show that f is differentiable.

It suffices to prove the following statement: Let $a \in \Omega$, $a = (a^1, a^2)$. For every $\varepsilon > 0$ there exists a $\delta > 0$ such that for all

$$(t^1, t^2) \in V_1 \times V_2 \text{ with } \|t^1\|, \|t^2\| \leq \delta$$

$$\|f(a^1 + t^1, a^2 + t^2) - f(a^1, a^2) - D_1 f(a^1, a^2) t^1 - D_2 f(a^1, a^2) t^2\| \quad (12)$$
$$\leq \varepsilon \sup(\|t^1\|, \|t^2\|).$$

We use the following break-up in order to show (12).

$$
\begin{aligned}
&f(a^1 + t^1, a^2 + t^2) - f(a^1, a^2) - D_1 f(a^1, a^2) t^1 - D_2 f(a^1, a^2) t^2 \\
&= f(a^1 + t^1, a^2 + t^2) - f(a^1 + t^1, a^2) - D_2 f(a^1 + t^1, a^2) t^2 \\
&+ D_2 f(a^1 + t^1, a^2) t^2 - D_2 f(a^1, a^2) t^2 \\
&+ f(a^1 + t^1, a^2) - f(a^1, a^2) - D_1 f(a^1, a^2) t^1.
\end{aligned}
\quad (13)
$$

As f is partially differentiable in the first variable, there exists a $\delta_1 > 0$ such that

$$\forall\, t^1 \text{ with } \|t^1\| \leq \delta_1 : \|f(a^1 + t^1, a^2) - f(a^1, a^2) - D_1 f(a^1, a^2) t^1\| \leq \frac{\varepsilon}{4} \|t^1\|. \quad (14)$$

By continuity of $D_2 f$ there exists a $\delta_2 > 0$ such that

$$\forall\, t = (t^1, t^2) \text{ with } \|t^1\|, \|t^2\| \leq \delta_2 : \|D_2 f(a^1 + t^1, a^2 + t^2) - D_2 f(a^1, a^2)\| \leq \frac{\varepsilon}{4}. \quad (15)$$

From this it follows first that for $\|t^1\|, \|t^2\| \leq \delta_2$

$$\|D_2 f(a^1 + t^1, a^2) t^2 - D_2 f(a^1, a^2) t^2\| \leq \frac{\varepsilon}{4} \|t^2\|. \quad (16)$$

Furthermore, lemma 8.7 implies

$$
\begin{aligned}
&\|f(a^1 + t^1, a^2 + t^2) - f(a^1 + t^1, a^2) - D_2 f(a^1 + t^1, a^2) t^2\| \\
&\leq \|t^2\| \sup_{\|\tau\| \leq \|t^2\|} \|D_2 f(a^1 + t^1, a^2 + \tau) - D_2 f(a^1 + t^1, a^2)\| \\
&\leq \frac{\varepsilon}{2} \|t^2\| \text{ again by (15) .}
\end{aligned}
\quad (17)
$$

Now (13), (14), (16), (17) give (12), provided $\|t^1\|, \|t^2\| \leq \delta := \min(\delta_1, \delta_2)$.

It also follows from (10) (lemma 8.10) that

$$Df(a^1, a^2)(t^1, t^2) = D_1 f(a^1, a^2) t^1 + D_2 f(a^1, a^2) t^2,$$

and with this, the continuity of Df follows from the continuity of $D_1 f$ and $D_2 f$. $\qquad \square$

We now come to the higher derivatives.

Definition 8.12 Let V, W be Banach spaces, $\Omega \subset V$ open, $x_0 \in \Omega$, $f : \Omega \to W$ differentiable. If the derivative Df is differentiable at x_0 then f is said to be twice differentiable at x_0 and the derivative of Df in x_0 is denoted by $D^2 f(x_0)$.

Notice that f was a function from Ω into W so Df is then a map from Ω to $B(V, W)$, which is again a Banach space. Therefore $D^2 f(x_0) \in B(V, B(V, W))$.

We shall now identify the elements of $B(V, B(V, W))$ with continuous bilinear maps of $V \times V$ into W. So let $L \in B(V, B(V, W))$, and $x, y \in V$. We set

$$L(x, y) := (L(x))(y). \tag{18}$$

(On the right side, L assigns to an element x an element $L(x) \in B(V, W)$, which is then applied to $y \in V$). Clearly, $L(x, y)$ is continuous and linear in x and y, as required. With respect to the norms, we have for $L \in B(V, B(V, W))$

$$\begin{aligned}
\|L\| &= \sup_{\|x\|=1} \|L(x)\| \\
&= \sup_{\|x\|=1} \sup_{\|y\|=1} \|(L(x))(y)\| \tag{19} \\
&= \sup_{\|x\|, \|y\|=1} \|L(x, y)\|.
\end{aligned}$$

Therefore

$$\|L(x, y)\| \le \|L\| \|x\| \|y\| \quad \text{for all } x, y \in V. \tag{20}$$

Analogous constructions and inequalities as (20) above, hold also for differential operators of order $k \ge 2$.

Theorem 8.13 *With the notations of definition 8.12, let f be twice differentiable at x_0. Then the continuous bilinear map $D^2 f(x_0)$ is symmetric, that is for $s, t \in V$*

$$D^2 f(x_0)(s, t) = D^2 f(x_0)(t, s).$$

Proof. Let $U(x_0, 2\sigma) \subset \Omega$. We consider

$$\varphi : [0, 1] \to W$$
$$\varphi(\tau) := f(x_0 + \tau s + t) - f(x_0 + \tau s) \text{ for } \|s\|, \|t\| \le \sigma.$$

By lemma 8.7 it follows that

$$\|\varphi(1) - \varphi(0) - D\varphi(0)\| \le \sup_{\tau \in [0,1]} \|D\varphi(\tau) - D\varphi(0)\| \tag{21}$$

(note that, by a small abuse of notation, we put $D\varphi(0)(1) = D\varphi(0)$ for the real number 1). The chain rule gives

$$\begin{aligned}
D\varphi(\rho) &= (Df(x_0 + \rho s + t) - Df(x_0 + \rho s))s \\
&= (Df(x_0 + \rho s + t) - Df(x_0) - D^2 f(x_0)\rho s)s \\
&\quad - (Df(x_0 + \rho s) - Df(x_0) - D^2 f(x_0)\rho s)s.
\end{aligned}$$

By assumptions, for $\varepsilon > 0$ there exists a $\delta > 0$ such that for $\|s\|, \|t\| \le \delta$ and $0 \le \rho \le 1$

$$\|Df(x_0 + \rho s + t) - Df(x_0) - D^2 f(x_0)(\rho s + t)\| \leq \varepsilon(\|s\| + \|t\|),$$
$$\|Df(x_0 + \rho s) - Df(x_0) - D^2 f(x_0)(\rho s)\| \leq \varepsilon\|s\|.$$

This gives

$$\|D\varphi(\rho) - (D^2 f(x_0)t)s\| \leq 2\varepsilon(\|s\| + \|t\|)\|s\| \tag{22}$$

and further

$$\|\varphi(1) - \varphi(0) - (D^2 f(x_0) \cdot t)s\| \leq \|\varphi(1) - \varphi(0) - D\varphi(0)\| + \|D\varphi(0) - (D^2 f(x_0)t)s\|$$
$$\leq 6\varepsilon(\|s\| + \|t\|)\|s\| \tag{23}$$

by (21) and (22) for $\rho = 0$ and $\rho = \tau$.

But $\varphi(1) - \varphi(0) = f(x_0 + s + t) - f(x_0 + s) - f(x_0 + t) + f(x_0)$ is symmetric in s and t, and we may therefore interchange s and t in (23) and obtain

$$\|(D^2 f(x_0)t)s - (D^2 f(x_0)s)t\| \leq 6\varepsilon(\|s\| + \|t\|)^2, \tag{24}$$

at first for $\|s\|, \|t\| \leq \delta$.

However, if we substitute $\lambda s, \lambda t$ ($\lambda > 0$) for s and t, then both sides of (24) are defined and are multiplied with $|\lambda|^2$. Therefore (24) holds also for $\|s\| = 1 = \|t\|$, for one has only to choose $\lambda = \delta$. Therefore

$$\|D^2 f(x_0)(t, s) - D^2 f(x_0)(s, t)\| \leq 24\varepsilon$$

for all s, t with $\|s\| = 1 = \|t\|$.

But as ε was arbitrary, it follows that

$$D^2 f(x_0)(t, s) = D^2 f(x_0)(s, t).$$

$\square$

Corollary 8.14 Let $V_1, \ldots, V_d, W$ be Banach spaces, $V = V_1 \times \ldots \times V_d, \Omega \in V$ open and $x_0 \in \Omega$. If $f : \Omega \to W$ is twice continuously differentiable at x_0, then all the partial derivatives $D_k D_j f(x_0)$, $k, j = 1, \ldots, d$, exist and for $t_\alpha = (t_\alpha^j)_{j=1,\ldots,d}, \alpha = 1, 2$, we have

$$D^2 f(x_0)(t_1, t_2) = \sum_{j,k=1,\ldots,d} D_k D_j f(x_0)(t_1^k, t_2^j);$$

in particular, we have

$$D_k D_j f(x_0) = D_j D_k f(x_0) \text{ for all } j, k.$$

f is therefore twice continuously differentiable in Ω precisely if all partial derivatives of order 2 exist and are continuous there.

Proof. This follows directly from theorems 8.11 and 8.13.

Quite similar results hold for higher order derivatives $D^k f$ and as such results can be derived easily by induction, we avoid giving further details at this point. We note only

Definition 8.15 Let V, W be Banach spaces, $\Omega \subset V$ open, $k \in \mathbb{N}$:

$$C^k(\Omega, W) := \{f : \Omega \to W, f \text{ is } k \text{ times continuously differentiable in } \Omega\}$$

$$C^\infty(\Omega, W) := \bigcap_{k \in \mathbb{N}} C^k(\Omega, W).$$

We now come to the Taylor formulae:

Theorem 8.16 *Let Ω be an open subset of a Banach space V, $x_0 \in \Omega, t \in V$ and assume that $\{x_0 + \tau t, 0 \leq \tau \leq 1\} \subset \Omega, f \in C^{k+1}(\Omega, \mathbb{R})$ $(k \in \mathbb{N})$.*
Then for a suitable $\theta \in (0, 1)$ we have:

$$f(x_0 + t) = f(x_0) + Df(x_0)t + \frac{1}{2!}D^2 f(x_0)(t, t) + \ldots + \frac{1}{k!}D^k f(x_0) \underbrace{(t, \ldots, t)}_{k\text{- times}}$$

$$+ \frac{1}{(k+1)!} D^{k+1} f(x_0 + \theta t) \underbrace{(t, \ldots, t)}_{(k+1)\text{-times}} . \tag{25}$$

Proof. We consider

$$g : [0, 1] \to \mathbb{R}, g(\tau) := f(x_0 + \tau t).$$

It follows easily from the chain rule by induction that g is $(k + 1)$-times continuously differentiable, with derivative

$$D^j g(\tau) = D^j f(x_0 + \tau t) \underbrace{(t, \ldots, t)}_{j\text{-times}} \ (1 \leq j \leq k + 1). \tag{26}$$

The Taylor series for functions of one variable (see theorem 3.13) gives

$$g(1) = \sum_{j=0}^{k} \frac{1}{j!} D^j g(0) + \frac{1}{(k+1)!} D^{k+1} g(\theta) \text{ for some } \theta \in (0, 1). \tag{27}$$

Now (25) follows from (26) and (27). □

Corollary 8.17 (Taylor expansion) *Let V be a Banach space, $\Omega \subset V$ open,*
$x_0 \in \Omega, \delta > 0, U(x_0, \delta) \subset \Omega$ and $f \in C^k(\Omega, \mathbb{R})$. Then for all $t \in V$ with $\|t\| < \delta$ we have

$$f(x_0 + t) = \sum_{j=0}^{k} \frac{1}{j!} D^j f(x_0) \underbrace{(t, \ldots, t)}_{j\text{-times}} + r_{k+1}(t),$$

where

$$\lim_{\substack{t \to 0 \\ t \neq 0}} \frac{r_{k+1}(t)}{\|t\|^k} = 0.$$

Proof. By theorem 8.16 we have, for some $\theta \in [0, 1]$,

$$f(x_0 + t) = \sum_{j=0}^{k} \frac{1}{j!} D^j f(x_0)(t, \ldots, t) + r_{k+1}(t) \text{ with}$$

$$r_{k+1}(t) = \frac{1}{k!} (D^k f(x_0 + \theta t) - D^k f(x_0)) \underbrace{(t, \ldots, t)}_{k\text{-times}}.$$

As $D^k f(x)$ is continuous by assumption, we see, as $0 \leq \theta \leq 1$, that $\lim_{t \to 0} (D^k f(x_0 + \theta t) - D^k f(x_0)) = 0$, and therefore $\lim_{t \to 0} \frac{r_{k+1}(t)}{\|t\|^k} = 0$ also.   □

Exercises for § 8

1)   On $\mathbb{R}^d$, consider

$$f_1(x) := \|x\|_1$$
$$f_2(x) := \|x\|_2$$
$$f_\infty(x) := \|x\|_\infty.$$

Where are this functions differentiable? Compute the derivatives whenever they exist.

2)   For $x = (x^1, x^2) \in \mathbb{R}^2$, consider

$$f(x) := \frac{x^1 (x^2)^2}{(x^1) + (x^2)^2} \quad \text{for } x \neq 0, f(0) := 0.$$

Show that

$$\lim_{\substack{t \to 0 \\ t \neq 0}} \frac{f(x + t\xi) - f(x)}{t} =: g(x, \xi)$$

exists for every $x, \xi \in \mathbb{R}^2$. However, the map $\xi \mapsto g(0, \xi)$ is not linear. Conclude that f is not differentiable at $0 \in \mathbb{R}^2$.

3)   Let V, W be Banach spaces, $\Omega \subset V$ open and connected, $f : \Omega \to W$ differentiable with

$$Df(x) = 0 \quad \text{for all } x \in \Omega.$$

Show that f is constant.

4)   Compute the Taylor expansion up to second order of

$$f(\xi, \eta) := \frac{\xi - \eta}{\xi + \eta} \quad (\text{for } \xi + \eta \neq 0) \quad \text{at } (1, 1) \in \mathbb{R}^2.$$

5)   Let Ω be an open subset of a Banach space $V = V_1 \times \ldots \times V_d$ as in definition 8.9, $f : \Omega \to \mathbb{R}$ a function whose partial derivatives exist and are bounded in Ω. Then f is continuous in Ω.
(Hint: Let $h = (h^1, \ldots h^d) \in V_1 \times \ldots \times V_d, \kappa_i = (h^1, \ldots, h^i, 0, \ldots, 0)$, for $i = 1, \ldots, d$, and $\kappa_0 = 0$. For $x \in \Omega$, write

$$f(x + h) - f(x) = \sum_{i=1}^{d} (f(x + \kappa_i) - f(x + \kappa_{i-1}))$$

and apply the mean value theorem.)

9. Differential Calculus in $\mathbb{R}^d$

The results of the previous paragraph are specialized to Euclidean spaces. Again, as in §3, interior extrema of differentiable functions are studied. We also introduce some standard differential operators like the Laplace operator.

A. Scalar Valued Functions

In this paragraph, we shall consider the following situation: Ω is an open subset of $\mathbb{R}^d$ and $f : \Omega \to \mathbb{R}$ is a function.

Definition 9.1 The graph Γ_f of f is defined to be the set
$\Gamma_f := \{(x, y) \in \Omega \times \mathbb{R} : f(x) = y\} \subset \mathbb{R}^{d+1}$. Furthermore, for $c \in \mathbb{R}$, the set
$N_f(c) := \{x \in \Omega : f(x) = c\} \subset \mathbb{R}^d$ is called the level set of f for the value c.
For $d = 2$ the level sets are also called contour lines.

Examples. We shall consider some examples in $\mathbb{R}^2$; for $x \in \mathbb{R}^2$ we shall write
$x = (\xi, \eta)$ instead of $x = (x^1, x^2)$. In the examples, Ω will always be $\mathbb{R}^2$.

1) $\qquad f(\xi, \eta) = \xi^2 + \eta^2$

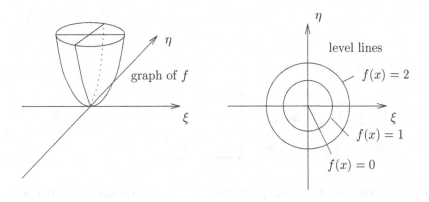

The graph of f is a paraboloid which opens upwards, the contour lines being circles centered at the origin.

2) $f(\xi, \eta) = \xi^2 - \eta^2$

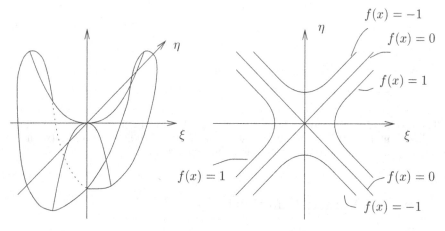

Already in this example, we see that the contours can degenerate to a point or have self intersections.

3) $f(x) = \xi^2 + \eta^3$

Here the graph rises in three directions and falls only in the direction of the negative η-axis. The contour $\{f(x) = 0\}$ has a vertex at the origin of the coordinates.

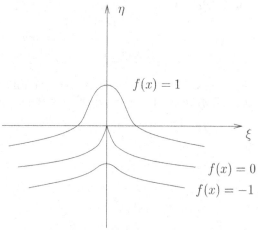

We now always split $\mathbb{R}^d$ as:

$$\mathbb{R}^d = \underbrace{\mathbb{R} \times \ldots \times \mathbb{R}}_{n\text{-times}}.$$

In what follows we always equip $\mathbb{R}^n$ with the Euclidean norm $\|\cdot\|_2$. We shall usually omit the index 2. Furthermore, we shall let e^i be the unit vector in the direction of the positive x^i-axis. So, in our current notations, we have $x = \sum_{i=1}^{d} x^i e_i$. If $y = \sum_{i=1}^{d} y^i e_i$, we have

$$\langle x, y \rangle := \sum_{i=1}^{d} x^i y^i,$$

the usual scalar product in $\mathbb{R}^d$. In particular,

$$\langle e_i, e_j \rangle = \delta_{ij} := \begin{cases} 1 & \text{for } i = j \\ 0 & \text{for } i \neq j \end{cases},$$

and

$$\|e_i\|_2 = 1 \text{ for } i = 1, \ldots, d.$$

The ith partial derivative of f, in case it exists, is then given by

$$D_i f(x) = \lim_{\substack{h \to 0 \\ h \neq 0}} \frac{f(x + h e_i) - f(x)}{h} =: \frac{\partial f(x)}{\partial x^i}.$$

Generally, we formulate the following definition.

Definition 9.2 Let Ω be open in $\mathbb{R}^d$, $f : \Omega \to \mathbb{R}$ a function, $v \in \mathbb{R}^d$, $\|v\|_2 = 1$. The directional derivative of f in the direction v is given by

$$D_v f(x) := \frac{d}{dh} f(x + hv)_{|h=0} = \lim_{\substack{h \to 0 \\ h \neq 0}} \frac{f(x + hv) - f(x)}{h},$$

provided this limit exists.

Furthermore,

Definition 9.3 Let $f : \Omega \to \mathbb{R}$ (Ω open in $\mathbb{R}^d$) be partially differentiable at $x \in \mathbb{R}^n$. The gradient of f at x is the vector field

$$\nabla f(x) := \text{grad } f(x) := (\frac{\partial f}{\partial x^1}(x), \ldots, \frac{\partial f}{\partial x^d}(x)).$$

Lemma 9.4 Let $f : \Omega \to \mathbb{R}$ be differentiable at x, $v \in \mathbb{R}^d$ with $\|v\|_2 = 1$. Then

$$D_v f(x) = Df(x)(v) = \langle \nabla f(x), v \rangle. \tag{1}$$

Proof. As f is differentiable, we have

$$\lim_{\substack{h \to 0 \\ h \neq 0}} \frac{\|f(x + hv) - f(x) - Df(x)hv\|}{\|hv\|} = 0.$$

It follows that

$$D_v f(x) = \lim_{\substack{h \to 0 \\ h \neq 0}} \frac{f(x + hv) - f(x)}{h} = Df(x)(v),$$

and formula (10) of the previous chapter (with $(v = (v^1, \dots, v^d))$ gives

$$Df(x)(v) = \sum_{i=1}^{d} D_i f(x) v^i = \langle \operatorname{grad} f(x), v \rangle.$$

$\square$

The derivative $Df(x)$ is therefore given by taking the scalar product with $\nabla f(x)$; in other words, the vector $\nabla f(x)$ is dual to the linear function $Df(x)$. Furthermore, it follows from (1) that if $\nabla f(x) \neq 0$, $\|D_v f(x)\|$ takes its maximum value over all v with $\|v\| = 1$ precisely when $v = \frac{\nabla f(x)}{\|\nabla f(x)\|}$. This can be so interpreted that $\nabla f(x)$ gives the direction of steepest ascent of f. On the other hand, if $D_v f(x) = 0$, then

$$\langle \operatorname{grad} f(x), v \rangle = 0,$$

so such v's are perpendicular (relative to the given scalar product) to $\nabla f(x)$. We shall later interpret this to mean that $\nabla f(x)$ is perpendicular to the level sets of f.

Examples.

1)   Consider the following example:

$$f(x) = \frac{\xi\eta}{\xi^2 + \eta^2} \text{ for } (\xi, \eta) \neq (0,0)$$
$$f(0) = 0.$$

For $(\xi, \eta) \neq (0,0)$, f is partially differentiable with respect to ξ and η, with

$$\frac{\partial f(x)}{\partial \xi} = \frac{\eta(\eta^2 - \xi^2)}{(\eta^2 + \xi^2)^2}, \quad \frac{\partial f(x)}{\partial \eta} = \frac{\xi(\xi^2 - \eta^2)}{(\xi^2 + \eta^2)^2}.$$

But f is also differentiable at the origin with respect to ξ and η, for

$$f(\xi, 0) = 0 \text{ for all } \xi \text{ and therefore}$$
$$\frac{\partial f(0,0)}{\partial \xi} = 0 \text{ and similarly } \frac{\partial f(0,0)}{\partial \eta} = 0.$$

Nevertheless, f is not continuous at the origin as

$$\lim_{\substack{h \to 0 \\ h \neq 0}} f(h, h) = \lim \frac{h^2}{h^2 + h^2} = \frac{1}{2} \neq f(0).$$

2)   Consider the function

$$g(x) = \frac{\xi\eta}{\xi^4 + \eta^4}$$

$$g(0) = 0;$$

g is likewise everywhere partially differentiable, but g is not only discontinuous at the origin, it is even unbounded near the origin:

$$\lim_{\substack{h \to 0 \\ h \neq 0}} g(h, h) = \lim \frac{h^2}{h^4 + h^4} = \infty.$$

By theorem 8.11, such phenomena are no longer possible when the partial derivatives not only exist, but are also continuous.

3)   We consider now for $x \in \mathbb{R}^d$ the distance $r(x)$ of x from the origin, namely
$r(x) = \|x\| = ((x^1)^2 + \ldots + (x^d)^2)^{\frac{1}{2}}$.
For $r(x) \neq 0$ (so for $x \neq 0$) we can calculate as follows:

$$\frac{\partial r(x)}{\partial x^i} = \frac{1}{2((x^1)^2 + \ldots (x^d)^2)^{\frac{1}{2}}} 2x^i = \frac{x^i}{r}, \tag{2}$$

so in particular

$$\nabla r(x) = \frac{x}{r} \text{ for } x \neq 0. \tag{3}$$

We have

$$\|\nabla r(x)\| = \frac{\|x\|}{r} = \frac{r}{r} = 1.$$

The gradient of $r(x)$ is therefore the unit vector in the direction determined by x.
It is always perpendicular to the level sets of r, namely concentric spheres centered at the origin. At the origin, $r(x)$ is not partially differentiable, for e.g. $r(x^1, 0, \ldots, 0) = |x^1|$ is not differentiable at $x^1 = 0$.

4)   Functions $f(x)$ often depend only on $r(x)$. So let $\varphi : (0, \infty) \to \mathbb{R}$ be differentiable and let $f(x) = \varphi(r(x))$. The chain rule gives

$$\frac{\partial f(x)}{\partial x^i} = \frac{d\varphi(r(x))}{dr} \frac{\partial r(x)}{\partial x^i} = \frac{d\varphi}{dr} \cdot \frac{x^i}{r} \text{ for } x \neq 0. \tag{4}$$

We shall now investigate the partial derivatives of order 2 of functions $f : \Omega \to \mathbb{R}$.

Corollary 8.15 gives directly

Corollary 9.5 *Let $f : \Omega \to \mathbb{R}$ have continuous partial derivatives of order 2. Then for $x \in \Omega$ and $i, j = 1, \ldots, n$,*

$$\frac{\partial^2}{\partial x^i \partial x^j} f(x) = \frac{\partial^2}{\partial x^j \partial x^i} f(x)$$

where we have set

$$\frac{\partial^2 f(x)}{\partial x^i \partial x^j} := \frac{\partial}{\partial x^j} \left(\frac{\partial f(x)}{\partial x^j} \right).$$

In order to show that the continuity of second order derivatives is really necessary for this, we consider the following example

$$f(x) := \begin{cases} \xi \eta \frac{\xi^2 - \eta^2}{\xi^2 + \eta^2} & \text{for } (\xi, \eta) \neq (0, 0) \\ 0 & \text{for } (\xi, \eta) = (0, 0). \end{cases}$$

We have

$$\frac{\partial f(0, \eta)}{\partial \xi} = -\eta \text{ for all } \eta \text{ and therefore } \frac{\partial^2 f(0, 0)}{\partial \eta \partial \xi} = -1$$

and

$$\frac{\partial f(\xi, 0))}{\partial \eta} = \xi \text{ for all } \xi \text{ and so } \frac{\partial^2 f(0, 0)}{\partial \xi \partial \eta} = 1.$$

Definition 9.6 Let $f : \Omega \to \mathbb{R}$ be twice continuously differentiable. The Hessian (matrix) of f is given by

$$D^2 f(x) = \left(\frac{\partial^2 f(x)}{\partial x^i \partial x^j} \right)_{i,j=1,\dots,n}.$$

One takes corollary 8.14 into account here again. The above corollary then signifies

Corollary 9.7 *Let $f : \Omega \to \mathbb{R}$ be twice continuously differentiable. Then the Hessian matrix of f is symmetric.*   □

We now come to the investigation of local extrema of scalar valued functions.

Definition 9.8 A point $x_0 \in \Omega$ is called a local maximum (minimum) of $f : \Omega \to \mathbb{R}$ if there is a neighborhood $U \subset \Omega$ of x_0 such that

$$f(x) \leq f(x_0) \ (f(x) \geq f(x_0)) \text{ for all } x \in U. \tag{5}$$

If in (5) the symbol $<$ ($>$) occurs for $x \neq x_0$, then the local extremum is called strict or isolated.

Theorem 9.9 *Let $x_0 \in \Omega$ be a local extremum of $f : \Omega \to \mathbb{R}$, and let f be partially differentiable at x_0. Then $\nabla f(x_0) = 0$.*

Proof. We consider for $i = 1, \ldots, d$

$$\varphi_i(h) = f(x_0 + he_i).$$

The function φ_i is defined in a neighborhood of 0 in $\mathbb{R}$ and φ_i has a local extremum at $h = 0$. It follows from the corresponding theorem for functions of a real variable that

$$0 = \frac{d}{dh}\varphi_i(h)_{|h=0} = Df(x_0 + he_i) \cdot \frac{d}{dh}(x_0 + he_i)_{|h=0}$$

$$= Df(x_0) \cdot e_i = \frac{\partial}{\partial x^i}f(x_0), \text{ for } i = 1, \ldots, d.$$

The assertion follows. $\qquad\qquad\square$

We shall now find conditions with which one can conversely decide whether an x_0 for which $\nabla f(x_0) = 0$ is an extremum and, more precisely, whether it is a local maximum or a minimum. As for functions of one real variable, these conditions will involve the second derivatives of f. We shall assume that f is twice continuously differentiable at x_0 and recall that by corollary 9.7, the Hessian matrix $D^2 f(x_0)$ is then symmetric.

Definition 9.10 A symmetric $(d \times d)$-matrix A is called

(i)   positive (negative) definite, if

$$\langle v, Av \rangle > 0 \ (< 0) \text{ for all } v \in \mathbb{R}^d \backslash \{0\}$$

(ii)   positive (negative) semidefinite if

$$\langle v, Av \rangle \geq 0 \ (\leq 0) \text{ for all } v \in \mathbb{R}^d$$

(iii)   indefinite if it is neither positive nor negative semidefinite.
       So for an indefinite matrix A there exist $v_1, v_2 \in \mathbb{R}^d$ such that

$$\langle v_1, Av_1 \rangle > 0 \text{ and } \langle v_2, Av_2 \rangle < 0.$$

A is negative (semi)definite precisely if $-A$ is positive (semi)definite.

Lemma 9.11 *A symmetric $(n \times n)$-matrix A is positive (negative) definite precisely if there exists a $\lambda > 0$ such that*

$$\langle v, Av \rangle \geq \lambda \|v\|^2 \ (\leq -\lambda \|v\|^2) \text{ for all } v \in \mathbb{R}^d. \tag{6}$$

Proof.
   " $\Leftarrow$ " is trivial.

" $\Rightarrow$ " : We treat only the case where A is positive definite. We consider $S^{d-1} := \{v \in \mathbb{R}^d : \|v\| = 1\}$. By corollary 7.42 S^{d-1}, being closed and bounded, is a compact subset of $\mathbb{R}^d$. Moreover, the function

$$v \mapsto \langle v, Av \rangle = \sum_{i,j=1}^{d} v^i a_{ij} v^j \ (A = (a_{ij})_{i,j=1,\ldots,d})$$

is continuous. Therefore $\langle v, Av \rangle$ assumes on S^{d-1} its minimum value λ, i.e., there exists a $v_0 \in S^{d-1}$ with $\langle v_0, Av_0 \rangle = \lambda$. Therefore $\lambda > 0$ by assumption. Furthermore, for all $w \in S^{d-1}$, $\langle w, Aw \rangle \geq \lambda$.

It follows that for $v \in \mathbb{R}^d \backslash \{0\}$ $\langle \frac{v}{\|v\|}, A\frac{v}{\|v\|} \rangle \geq \lambda$, so $\langle v, Av \rangle \geq \lambda \|v\|^2$. Finally, (6) holds trivially for $v = 0$. $\qquad \square$

Theorem 9.12 Let $f : \Omega \to \mathbb{R}$ be twice continuously differentiable. Assume that for $x_0 \in \Omega$

$$\nabla f(x_0) = 0.$$

If the Hessian matrix $D^2 f(x_0)$ is

(i)   positive definite then f has a strict (strong) minimum at x_0, i.e., there exists a neighborhood U of x_0 such that

$$f(x) > f(x_0) \text{ for all } x \in U \backslash \{x_0\}.$$

(ii)   negative definite then f has a strict maximum at x_0.

(iii)   indefinite then f has no local extremum at x_0.

Proof. By Taylor expansion (corollary 8.17) we have, using $Df(x_0) = 0$, that for $v \in \mathbb{R}^d$ with $x_0 + \tau v \in \Omega$ ($0 \leq \tau \leq 1$),

$$f(x_0 + v) = f(x_0) + \frac{1}{2}D^2 f(x_0)(v, v) + r_3(v)$$

where

$$\lim_{\substack{v \to 0 \\ v \neq 0}} \frac{r_3(v)}{\|v\|^2} = 0. \tag{7}$$

Now if $v = (v^1, \ldots, v^d)$ we have

$$D^2 f(x_0)(v, v) = \frac{\partial^2 f(x_0)}{\partial x^i \partial x^j} v^i v^j = \langle v, D^2 f(x_0) v \rangle.$$

So if $D^2 f(x_0)$ is positive definite, by lemma 9.11 there exists a $\lambda > 0$ such that

$$\langle v, D^2 f(x_0) v \rangle \geq \lambda \|v\|^2 \text{ for all } v \in \mathbb{R}^d.$$

By (7) there exists a $\delta > 0$ such that

$$|r_3(v)| \leq \frac{\lambda}{4}\|v\|^2 \text{ for } \|v\| \leq \delta. \tag{8}$$

It follows that for $\|v\| \leq \delta$

$$f(x_0 + v) \geq f(x_0) + \frac{\lambda}{2}\|v\|^2 - \frac{\lambda}{4}\|v\|^2 = f(x_0) + \frac{\lambda}{4}\|v\|^2,$$

and f therefore has a strict minimum at x_0.

If $D^2 f(x_0)$ is negative definite then $D^2(-f(x_0))$ is positive definite and $-f$ therefore has a strict minimum and consequently f has a strict maximum at x_0.

Finally, if $D^2 f(x_0)$ is indefinite, there exist $\lambda_1, \lambda_2 > 0$ and $v_1, v_2 \in \mathbb{R}^d \backslash \{0\}$ with

$$\langle v_1, D^2 f(x_0)v_1 \rangle \geq \lambda_1 \|v_1\|^2$$

$$\tag{9}$$

$$\langle v_2, D^2 f(x_0)v_2 \rangle \leq -\lambda_2 \|v_2\|^2.$$

For $\lambda := \min(\lambda_1, \lambda_2)$, we choose $\delta > 0$ as in (8), and λ remains invariant under the scaling $v \mapsto tv, t \geq 0$. We can assume that $\|v_1\| = \delta = \|v_2\|$. It follows that

$$f(x_0 + v_1) \geq f(x_0) + \frac{\lambda}{4}\|v_1\|^2$$

$$f(x_0 + v_2) \leq f(x_0) - \frac{\lambda}{4}\|v_2\|^2.$$

Therefore, f can have no extremum at x_0, as we can decrease $\delta > 0$ arbitrarily.

$\square$

Examples. We want to investigate the local extrema of functions in the examples at the beginning of this chapter.

1) $f(\xi, \eta) = \xi^2 + \eta^2$.
Here $\nabla f(x) = (2\xi, 2\eta)$. So the gradient vanishes only at the point $(0, 0)$. Furthermore

$$D^2 f(0) = \begin{pmatrix} 2 & 0 \\ 0 & 2 \end{pmatrix}.$$

So f has a strict minimum at 0.

2) $f(\xi, \eta) = \xi^2 - \eta^2, \nabla f(x) = (2\xi, -2\eta)$, so again $\nabla f(0) = 0$. Now

$$D^2 f(0) = \begin{pmatrix} 2 & 0 \\ 0 & -2 \end{pmatrix}$$

is, however, indefinite and therefore there is no extremum at hand.

3) $f(x) = \xi^2 + \eta^3, \nabla f(x) = (2\xi, 3\eta^2)$. Now

$$D^2 f(0) = \begin{pmatrix} 2 & 0 \\ 0 & 0 \end{pmatrix},$$

is positive semidefinite. But again there is no local extremum at the origin.

4) $f(x) = \xi^2 + \eta^4, \nabla f(x) = (2\xi, 4\eta^3)$, so again $\nabla f(0) = 0$. Now

$$D^2 f(0) = \begin{pmatrix} 2 & 0 \\ 0 & 0 \end{pmatrix},$$

so it is again positive semidefinite. But now there is a strict minimum at the origin.

5) $f(x) = \xi^2, \nabla f(x) = (2\xi, 0)$, so again $\nabla f(0) = 0$. We have once more

$$D^2 f(0) = \begin{pmatrix} 2 & 0 \\ 0 & 0 \end{pmatrix},$$

so positive semidefinite. A local minimum occurs at the origin, which, however, is not a strict minimum.

We give now, without a proof, the following criterion for positive definiteness of a symmetric matrix.

Lemma 9.13 *Let $A = (a_{ij})_{i,j=1,\ldots,d}$ be a symmetric matrix, so $a_{ij} = a_{ji}$ for all i, j. A is positive definite if and only if for all $\nu = 1, \ldots, d$*

$$\det \begin{pmatrix} a_{11} & \cdots & a_{1\nu} \\ \vdots & & \vdots \\ a_{\nu 1} & \cdots & a_{\nu\nu} \end{pmatrix} > 0.$$

A is negative definite when these determinants are negative for all odd ν and positive for even ν.

Definition 9.14 The Laplace operator Δ is defined as follows: For a twice differentiable function $f : \Omega \to \mathbb{R}$ we set

$$\Delta f := \frac{\partial^2 f}{(\partial x^1)^2} + \cdots \frac{\partial^2 f}{(\partial x^d)^2}. \tag{10}$$

A function $f \in C^2(\Omega)$ which satisfies the equation

$$\Delta f = 0 \text{ in } \Omega \tag{11}$$

is said to be harmonic (in Ω). The equation (11) is called the Laplace equation.

Examples. Constants and linear functions are trivially harmonic. In $\mathbb{R}^2$, the function $f(x) = \xi^2 - \eta^2$ is harmonic. We shall now consider some less trivial examples. We consider again functions of the form $f(x) = \varphi(r(x))$. By the chain rule (cf. examples after 9.4) we have

$$\frac{\partial f}{\partial x^i} = \frac{d\varphi(r(x))}{dr} \frac{x^i}{r},$$

and therefore

$$\Delta\varphi(r(x)) = \sum_{i=1}^{n} \left(\frac{d^2\varphi}{dr^2} \cdot \frac{x^i}{r} \cdot \frac{x^i}{r} + \frac{d\varphi}{dr} \cdot \frac{1}{r} - \frac{d\varphi}{dr} \cdot \frac{x^i}{r^3} x^i \right)$$

$$= \frac{d^2\varphi}{dr^2} + \frac{(n-1)}{r} \frac{d\varphi}{dr}, \tag{12}$$

for $\sum_{i=1}^{d} x^i x^i = r^2$. It follows that, in particular,

$$\text{for } d = 2: \ \Delta\log(r(x)) = 0 \text{ for } x \neq 0$$

$$\text{and for } d \geq 3: \ \Delta\frac{1}{r^{d-2}(x)} = 0 \text{ for } x \neq 0.$$

These functions, that is, $\log(r(x))$ and $r^{2-d}(x)$, respectively, are called fundamental solutions of the Laplace equation.

Definition 9.15

(i)   Let $k > 0$ be a constant (heat conductivity). The differential equation defined for functions $f : \Omega \times \mathbb{R}_+ \to \mathbb{R}, (x, t) \in \Omega \times \mathbb{R}_+$, f being twice differentiable in x and once in t, by

$$\frac{1}{k} \frac{\partial f(x, t)}{\partial t} - \Delta f(x, t) = 0$$

(Δ as in (10)) is called the heat equation.

(ii)   Let $c > 0$ be likewise a constant (wave propagation speed); $f \in C^2(\Omega \times \mathbb{R}_+, \mathbb{R}), (x, t) \in \Omega \times \mathbb{R}_+$. The differential equation

$$\frac{1}{c^2} \frac{\partial^2 f(x, t)}{\partial t^2} - \Delta f(x, t) = 0$$

(Δ as in (10)) is called the wave equation.

We consider the function

$$k(x, t) := \frac{1}{t^{\frac{d}{2}}} \exp\left(-\frac{\|x\|^2}{4t} \right) \text{ for } x \in \mathbb{R}^d, t > 0.$$

By (12), on account of $\|x\| = r$, $\frac{\partial k}{\partial r} = \frac{-r}{2t^{\frac{d}{2}+1}} \exp\left(-\frac{r^2}{4t} \right)$, we have

$$\Delta k(x,t) = \left(\frac{r^2}{4t^{\frac{d}{2}+2}} - \frac{d}{2t^{\frac{d}{2}+1}} \right) \exp\left(-\frac{r^2}{4t} \right)$$

and $\dfrac{\partial k(x,t)}{\partial t} = \left(\dfrac{r^2}{4t^{\frac{d}{2}+2}} - \dfrac{d}{2t^{\frac{d}{2}+1}} \right) \exp\left(-\dfrac{r^2}{4t} \right),$

and therefore

$$\frac{\partial k(x,t)}{\partial t} - \Delta k(x,t) = 0, \text{ for } x \in \mathbb{R}^d, t > 0.$$

The functions $k(x,t)$ are called fundamental solutions of the heat equation. We have, for $x \neq 0$, $\lim\limits_{t \searrow 0} k(x,t) = 0$ and $\lim\limits_{t \searrow 0} k(0,t) = \infty$, and for all x,

$\lim\limits_{t \to \infty} k(x,t) = 0.$

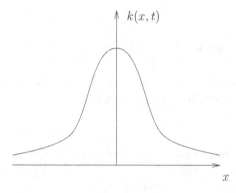

For every fixed $t > 0$, the curve $k(x,t)$ is bell-shaped, and it is the steeper the smaller t is; in the illustration, $n = 1$.

Physically, the function $k(x,t)$ describes the process of heat conduction in the space $\mathbb{R}^d$ as a function of time t, where the initial values are concentrated in a point and form an infinitely hot heat source there. This is an idealised situation, but we shall see later that a general initial state can be represented by a superposition (i.e. integration) of such idealised initial states. Moreover, the bell shaped curve described by $k(x,t)$ (for t fixed) also plays a role in probability theory. Namely, it gives the socalled Gaussian distribution.

Finally, we want to present a solution of the wave equation. Let $c > 0$, $v \in \mathbb{R}^d, w = c\|v\|$, and

$$\sigma(x,t) := g(\langle v, x \rangle - wt)$$
$$\text{with } g \in C^2(\mathbb{R}, \mathbb{R}).$$

We have

$$\frac{1}{c^2} \frac{\partial^2}{\partial t^2} \sigma(x,t) - \Delta\sigma(x,t) = \frac{w^2}{c^2} g'' - \|v\|^2 g'' = 0.$$

Therefore $\sigma(x,t)$ is a solution of the wave equation.

The Laplace equation, the heat equation and the wave equation are prototypes of a certain class of partial differential equations ("partial" means that the unknown function is a function of several variables, in contrast to ordinary differential equations), which we want to define briefly.

Definition 9.16 A linear differential operator of second order is a continuous linear map of $C^2(\Omega, \mathbb{R})$ into $C^0(\Omega, \mathbb{R})$ of the following form:

$$f(x) \mapsto Lf(x) = \sum_{i,j=1}^{d} a_{ij}(x)\frac{\partial^2 f(x)}{\partial x^i \partial x^j} + \sum_{i=1}^{d} b_i(x)\frac{\partial f(x)}{\partial x^i} + c(x)f(x) + d(x), \quad (13)$$

where $a_{ij}(x), b_i(x), c(x), d(x) \in C^0(\Omega, \mathbb{R})$.

If $(a_{ij}(x))_{i,j=1,\ldots,d}$ (for all $x \in \Omega$) is a positive definite matrix, then the differential equation

$$Lf(x) = 0$$

is called elliptic.

A differential equation of the form

$$\frac{\partial}{\partial t}f(x,t) - Lf(x,t) = 0$$

($f : \Omega \times \mathbb{R}_+ \to \mathbb{R}$ with corresponding differentiability properties) with elliptic L (as in (13)) is called parabolic, and an equation of the form

$$\frac{\partial^2}{\partial t^2}f(x,t) - Lf(x,t) = 0 \quad (f \in C^2(\Omega \times \mathbb{R}_+, \mathbb{R})),$$

with again an elliptic L, is called hyperbolic.

These are the types of differential equations of second order which are of principal interest in mathematics and physics. Elliptic equations describe states of equilibrium, parabolic and hyperbolic ones diffusion and vibration processess, respectively. For the Laplace operator

$$a_{ij}(x) = \delta_{ij} \left(= \begin{cases} 1 & \text{for } i = j \\ 0 & \text{for } i \neq j \end{cases} \right) \quad \text{for all } x.$$

The matrix $(a_{ij}(x))$ therefore is the identity matrix and hence positive definite.

B. Vector Valued Functions

As before, let Ω be an open subset of $\mathbb{R}^d$. We consider the map

$$f : \Omega \to \mathbb{R}^m$$

and write $f = (f^1, \ldots, f^m)$.

If f is differentiable at $x \in \Omega$, the differential $Df(x)$ is a linear map $\mathbb{R}^d \to \mathbb{R}^m$, and indeed $Df(x)$ is given by the matrix

$$\left(\frac{\partial f^j(x)}{\partial x^i} \right)_{i=1,\ldots,d, j=1,\ldots,m}$$

(compare example 5 at the beginning and (10) of the previous chapter). This matrix is called the Jacobi matrix of f. Furthermore in case $d = m$, the Jacobian or functional determinant of f at x is defined as

$$\det \left(\frac{\partial f^j(x)}{\partial x^i} \right);$$

in other words, it is the determinant of the Jacobi matrix.

If $g : \Omega' \to \mathbb{R}^l$, Ω' open in $\mathbb{R}^m$, is differentiable at the point $f(x) \in \Omega'$, the chain rule gives

$$D(g \circ f)(x) = (Dg)(f(x)) \circ Df(x),$$

or, in matrix notation

$$\frac{\partial (g \circ f)^k(x)}{\partial x^j} = \sum_{j=1}^{m} \frac{\partial g^k(f(x))}{\partial y^j} \cdot \frac{\partial f^j(x)}{\partial x^i} \quad (k = 1, \ldots, l, i = 1, \ldots, d).$$

In particular in the case $d = m = l$ we have

$$\det \left(\frac{\partial (g \circ f)^k(x)}{\partial x^i} \right) = \det \left(\frac{\partial g^k(f(x))}{\partial y^j} \right) \det \left(\frac{\partial f^j(x)}{\partial x^i} \right).$$

Definition 9.17 A function $f : \Omega \to \mathbb{R}^d$ (so $m = d$) is, in many contexts, also called a vector field, as to any point $x \in \Omega \subset \mathbb{R}^d$ a vector $f(x)$ in $\mathbb{R}^d$ is assigned by f. If f is partially differentiable, the divergence of f is defined as

$$\operatorname{div} f(x) := \sum_{i=1}^{d} \frac{\partial f^i(x)}{\partial x^i}.$$

Theorem 9.18 *If $\varphi : \Omega \to \mathbb{R}$ is partially differentiable, then*

$$\frac{\partial}{\partial x^i}(\varphi f^i) = \varphi \frac{\partial f^i}{\partial x^i} + \frac{\partial \varphi}{\partial x^i} f^i$$

and by summation over i,

$$\text{div}\,(\varphi f) = \varphi \,\text{div}\, f + \langle\, \text{grad}\,\varphi, f\,\rangle.$$

If $\varphi : \Omega \to \mathbb{R}$ is twice partially differentiable, then $\text{grad}\,\varphi = \nabla\varphi = \left(\frac{\partial\varphi}{\partial x^1}, \ldots, \frac{\partial\varphi}{\partial x^n}\right)$ is a partially differentiable vector field and we have

$$\text{div}\,(\text{grad}\,\varphi) = \sum_{i=1}^{n} \frac{\partial^2\varphi}{(\partial x^i)^2} = \Delta\varphi.$$

Definition 9.19 In the special case $d = 3$ one defines for a partially differentiable vector field $f : \Omega \to \mathbb{R}^3$ yet another vector field $\text{rot}\, f$ (the rotation of f) by

$$\text{rot}\, f := \left(\frac{\partial f^3}{\partial x^2} - \frac{\partial f^2}{\partial x^3}, \frac{\partial f^1}{\partial x^3} - \frac{\partial f^3}{\partial x^1}, \frac{\partial f^2}{\partial x^1} - \frac{\partial f^1}{\partial x^2}\right).$$

Corollary 9.20 *If $\varphi : \Omega \to \mathbb{R}$ is twice differentiable, then*

$$\text{rot}\,\text{grad}\,\varphi = 0.$$

Proof. The proof follows from direct calculations. The first component of rot grad φ is e.g.

$$\frac{\partial}{\partial x^2}\left(\frac{\partial\varphi}{\partial x^3}\right) - \frac{\partial}{\partial x^3}\left(\frac{\partial\varphi}{\partial x^2}\right) = 0,$$

by corollary 8.14. $\qquad\qquad\square$

If $f : \Omega \to \mathbb{R}^3 (\Omega \subset \mathbb{R}^3)$ is a twice differentiable vector field, then

$$\text{div}\,\text{rot}\, f = 0.$$

The proof is again a direct calculation.

Exercises for § 9

1)   We define $f : \mathbb{R}^2 \to \mathbb{R}$ by

$$f(\xi, \eta) := \begin{cases} \frac{\xi\eta}{r}\sin(\frac{1}{r}) & \text{for } (\xi, \eta) \neq (0,0) \\ 0 & \text{for } (\xi, \eta) = (0,0), \end{cases}$$

where $r(\xi, \eta) := (\xi^2 + \eta^2)^{\frac{1}{2}}$.

Where do the partial derivatives $\frac{\partial f}{\partial \xi}, \frac{\partial f}{\partial \eta}$ exist? For which $(a, b) \in \mathbb{R}^2$ are the maps $\xi \mapsto \frac{\partial f}{\partial \xi}(\xi, b), \eta \mapsto \frac{\partial f}{\partial \eta}(\xi, b), \eta \mapsto \frac{\partial f}{\partial \eta}(a, \eta)$ continuous?

f is not differentiable at $(0,0)$. Is this compatible with the results demonstrated in §9?

2) Determine the local minima and maxima of $f : \mathbb{R}^2 \to \mathbb{R}$,

a) $f(\xi, \eta) := (4\xi^2 + \eta^2)e^{-\xi^2 - 4\eta^2}$,

b) $f(\xi, \eta) := \xi^2 - \eta^2 + 1$,

c) $f(\xi, \eta) := \xi^3 + \eta^3 - \xi - \eta$.

3) Show that $f(x, y) = (y - x^2)(y - 2x^2)$ does not have a local minimum at $0 \in \mathbb{R}^2$. However, the restriction of f to any straight line through 0 does have a local minimum at 0.

4) Let $\Omega \subset \mathbb{R}^d$ be open, $f : \Omega \to \mathbb{R}$ continuously differentiable. Show that the gradient of f is orthogonal to the level set $N_f(c) := \{x \in \Omega : f(x) = c\}$ $(c \in \mathbb{R})$ in the following sense:
If $\gamma : (-\varepsilon, \varepsilon) \to \mathbb{R}^d (\varepsilon > 0)$ is continuously differentiable with $\gamma(0) = x_0 \in \Omega, \gamma(t) \subset N_f(c)$ for $-\varepsilon < t < \varepsilon$, then

$$\langle \gamma'(0), \text{ grad } f(x_0) \rangle = 0.$$

5) Let $f : \mathbb{R}^d \to \mathbb{R}$ be a continuous function with continuous partial derivatives. Show that f is homogeneous of degree α (i.e. $f(tx) = t^\alpha f(x)$ for all $x \in \mathbb{R}^d, t > 0$, precisely if Euler's formula

$$\sum_{i=1}^{d} x^i \frac{\partial f}{\partial x^i}(x) = \alpha f(x)$$

holds for all $x \in \mathbb{R}^d$.

6) Compute $\nabla \log \log r$ and $\Delta \log \log r$, for $r = \left(\Sigma(x^i)^2\right)^{\frac{1}{2}} \neq 0$ $(x = (x^1, \ldots, x^d) \in \mathbb{R}^d)$.

7) Let $\Omega \subset \mathbb{R}^2$ be open, $0 \notin \Omega$, $f \in C^2(\Omega)$ a solution of $\Delta f = 0$ in Ω. Then $g(x) := f\left(\frac{x}{\|x\|^2}\right)$ is a solution of $\Delta g = 0$ in $\Omega^* := \{x : \frac{x}{\|x\|^2} \in \Omega\}$. Similarly, if $\Omega \subset \mathbb{R}^3$ instead, $g(x) = \frac{1}{\|x\|} f\left(\frac{x}{\|x\|}\right)$ yields a solution of $\Delta g = 0$.

8) Determine all harmonic polynomials in $\mathbb{R}^2$ of degree ≤ 3.

9) Let $\Omega := \{(\xi, \eta, \zeta) \in \mathbb{R}^3 : \xi^2 + \eta^2 < 1\}$. Let $u \in C^3(\Omega)$ be harmonic, i.e. $\Delta u = 0$ in Ω. Suppose that u is rotationally symmetric about the ζ-axis (express this condition through a formula!), and suppose that

$$f(\xi) := u(0, 0, \xi)$$

is given for $\xi \in \mathbb{R}$. We wish to compute u on Ω from f by Taylor expansion w.r.t. ξ and η, where we consider ξ as a parameter. Thus, we write

$$u(\xi, \eta, \xi) = a_0(\xi) + a_1(\xi)\xi + a_2(\xi)\eta + a_{11}(\xi)\xi^2 + 2a_{12}(\xi)\xi\eta + a_{22}(\xi)\eta^2$$

+ higher order terms.

Use the equation $\Delta u = 0$ to determine the functions $a_0(\xi), \ldots, a_{22}(\xi)$ in terms of f.

10. The Implicit Function Theorem. Applications

The Banach fixed point theorem is used to derive the implicit function theorem. Corollaries are the inverse function theorem and the Lagrange multiplier rules for extrema with side conditions.

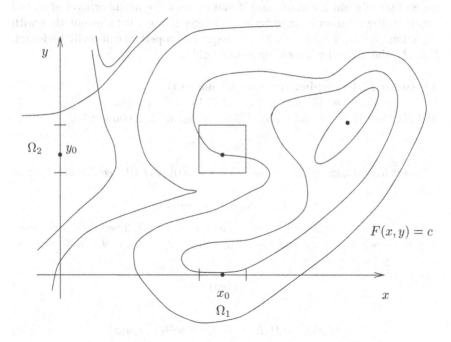

We shall first explain the setting of the problem. Let a function $F(x, y)$ be represented by its contour lines. We have already illustrated a few examples in the previous chapter. Practical examples occur in maps, where $F(x, y)$ gives the height above the sea level, or in weather charts, where $F(x, y)$ describes the air pressure. Here the contours are the socalled isobars – lines of equal pressure. Now the following question can be posed: Given $c \in \mathbb{R}$ and (x_0, y_0) such that $F(x_0, y_0) = c$, can a function $x \mapsto y = g(x)$ be found, at least locally, i.e. in a neighborhood of (x_0, y_0), that satisfies the equation

$$F(x, g(x)) = c. \qquad (1)$$

Here, g should be as continuous as possible and even differentiable, provided F is assumed to be itself differentiable. Besides, in this neighborhood, for every x, $g(x)$ should be the unique solution of (1). We thus want to represent the contours locally as functions of x, in graphical representations as graphs over the x-axis.

By illustrations one can convince oneself quickly that this is not always possible. The contours should, first of all, be non-degenerate, i.e., they should not be single points, nor should they have corners or self intersections. Also, such a solution in x can certainly not function where the contour has a vertical tangent, at least not with a differentiable g, and not at all if the contour has a recurring point. Of course, at such points, one could try to find a solution in y instead of x. Finally, the restriction to a small neighborhood is important. If one chooses too big a neighborhood of x_0, there exists in general no solution anymore, and if one chooses the neighborhood of y_0 too big, then the solution is, in general, no longer unique. It turns out that with these limitations, a solution with the required properties can really be found. Namely, the following important result holds.

Theorem 10.1 (Implicit function theorem)

Let V_1, V_2, W be Banach spaces, $\Omega \subset V_1 \times V_2$ an open set, $(x_0, y_0) \in \Omega$ and $F : \Omega \to W$ be continuously differentiable in Ω. Assume that

$$F(x_0, y_0) = 0$$

(without loss of generality, we have taken $c = 0$). Let (the continuous) linear map

$$D_2 F(x_0, y_0) : V_2 \to W$$

be invertible, and let its inverse be likewise continuous. Then there exist open neighborhoods Ω_1 of x_0 and Ω_2 of y_0, $\Omega_1 \times \Omega_2 \subset \Omega$, and a differentiable function $g : \Omega_1 \to \Omega_2$ such that

$$F(x, g(x)) = 0 \qquad (2)$$

and

$$Dg(x) = -(D_2 F(x, g(x)))^{-1} \circ D_1 F(x, g(x)) \qquad (3)$$

for all $x \in \Omega_1$.

Furthermore, for every $x \in \Omega_1, g(x)$ is the only solution of (2) in Ω_2.

Proof. For abbreviation, we set $L_0 := D_2 F(x_0, y_0)$. The equation $F(x, y) = 0$ is the equivalent to

$$y = y - L_0^{-1} F(x, y) =: G(x, y).$$

Thus, we have transformed our problem to that of finding, for every x, a fixed point of the map $y \mapsto G(x, y)$. The advantage of this reformulation is that we can use the Banach fixed point theorem. For recollection we formulate this once more as

Lemma 10.2 *Let A be a closed subspace of a Banach space B and let $T :$ $A \to A$ satisfy*

$$\|Ty_1 - Ty_2\| \leq q\|y_1 - y_2\| \text{ for all } y_1, y_2 \in A \tag{4}$$

with a fixed $q < 1$.

Then there exists a unique $y \in A$ such that

$$Ty = y. \tag{5}$$

Furthermore, if we have a family $T(x)$, where all the $T(x)$ fulfil (4) for a fixed q independent of x (x is allowed to vary in an open subset of a Banach space B_0), then the solution $y = y(x)$ of

$$T(x)y = y \tag{6}$$

depends continuously on x.

As we have previously not proved continuous dependence on x, we give a proof again.

Proof. We set, recursively, for every $x \in \Omega$

$$y_n := T(x)y_{n-1}, \tag{7}$$

where $y_0 \in A$ is chosen arbitrarily.

So

$$y_n = \sum_{\nu=1}^{n}(y_\nu - y_{\nu-1}) + y_0 = \sum_{\nu=1}^{n}(T(x)^{\nu-1}y_1 - T(x)^{\nu-1}y_0) + y_0. \tag{8}$$

Now

$$\sum_{\nu=1}^{n}\|T(x)^{\nu-1}y_1 - T(x)^{\nu-1}y_0\| \leq \sum_{\nu=1}^{n}q^{\nu-1}\|y_1 - y_0\| \leq \frac{1}{1-q}\|y_1 - y_0\|$$

and therefore the series (8) converges absolutely and uniformly, and the limit function $y(x) = \lim\limits_{n\to\infty} y_n$ is continuous. We also have

$$y(x) = \lim_{n\to\infty} T(x)y_{n-1} = T(x)(\lim_{n\to\infty} y_{n-1}) = T(x)y(x),$$

and as A is closed, $y(x) \in A$, and consequently $y(x)$ solves (6). The uniqueness follows again from (4): If y_1, y_2 are two solutions of $y = Ty$, then

$$\|y_1 - y_2\| = \|Ty_1 - Ty_2\| \leq q\|y_1 - y_2\|$$

and therefore as $q < 1$, we have $y_1 = y_2$.   □

With that, the proof of lemma 10.2 is complete and we can proceed with the proof of the theorem. First we obtain $(L_0^{-1} \circ L_0 = Id)$

$$G(x, y_1) - G(x, y_2) = L_0^{-1}(D_2 F(x_0, y_0)(y_1 - y_2) - (F(x, y_1) - F(x, y_2))).$$

As F is differentiable at (x_0, y_0) and L_0^{-1} is continuous, there exist such $\delta_1 > 0, \eta > 0$ that for $\|x - x_0\| \leq \delta_1, \|y_1 - y_0\| \leq \eta, \|y_2 - y_0\| \leq \eta$ (and therefore also $\|y_1 - y_2\| \leq 2\eta$) we have

$$\|G(x, y_1) - G(x, y_2)\| \leq \frac{1}{2}\|y_1 - y_2\| \tag{9}$$

(instead of $\frac{1}{2}$ we could work here with any other $q < 1$).

Besides, there exists such a $\delta_2 > 0$ that for $\|x - x_0\| \leq \delta_2$

$$\|G(x, y_0) - G(x_0, y_0)\| < \frac{\eta}{2}. \tag{10}$$

If then $\|y - y_0\| \leq \eta$, we have, using $G(x_0, y_0) = y_0$

$$\begin{aligned}
\|G(x, y) - y_0\| &= \|G(x, y) - G(x_0, y_0)\| \\
&\leq \|G(x, y) - G(x, y_0)\| \\
&\quad + \|G(x, y_0) - G(x_0, y_0)\| \\
&\leq \frac{1}{2}\|y - y_0\| + \frac{\eta}{2} \\
&\leq \eta.
\end{aligned}$$

by (9), (10), for every x with $\|x - x_0\| \leq \delta := \min(\delta_1, \delta_2)$.

Therefore, $G(x, y)$ maps the closed ball $B(y_0, \eta)$ onto itself (and similarly also the open ball $U(y_0, \eta)$). Therefore lemma 10.2 can be applied to the function $y \mapsto G(x, y)$, and for every x such that $\|x - x_0\| \leq \delta$ there exists therefore a unique $y = y(x)$ with $\|y - y_0\| \leq \eta$ and $y = G(x, y)$, so $F(x, y) = 0$, and y depends continuously on x. We set $\Omega_1 := \{x : \|x - x_0\| < \delta\}$, $\Omega_2 := \{y : \|y - y_0\| < \eta\}$; without restriction, let $\Omega_1 \times \Omega_2 \subset \Omega$. We denote such a y with $g(x)$, and it just remains to show the differentiability of g. Let $(x_1, y_1) \in \Omega_1 \times \Omega_2, y_1 = g(x_1)$. Since F is differentiable at (x_1, y_1) we obtain, setting $K = D_1 F(x_1, y_1)$ and $L = D_2 F(x_1, y_1)$,

$$F(x, y) = K(x - x_1) + L(y - y_1) + \varphi(x, y) \tag{11}$$

for $(x, y) \in \Omega$ with

$$\lim_{(x,y) \to (x_1, y_1)} \frac{\varphi(x, y)}{\|(x - x_1, y - y_1)\|} = 0. \tag{12}$$

Since F is continuously differentiable, we may choose δ and η so small that our operator L satisfies the assumption of Lemma 7.21 and therefore has a continuous inverse L^{-1}.

As $F(x, g(x)) = 0$ holds for $x \in \Omega_1$, it follows that

$$g(x) = -L^{-1}K(x - x_1) + y_1 - L^{-1}\varphi(x, g(x)). \tag{13}$$

On account of (12) there exist $\rho_1, \rho_2 > 0$ with the property that for

$$\|x - x_1\| \le \rho_1, \|y - y_1\| \le \rho_2$$
$$\|\varphi(x, y)\| \le \frac{1}{2\|L^{-1}\|}(\|x - x_1\| + \|y - y_1\|),$$

so

$$\|\varphi(x, g(x))\| \le \frac{1}{2\|L^{-1}\|}\|x - x_1\| + \|g(x) - g(x_1)\| \tag{14}$$

also holds. From (13) and (14) it follows that

$$\|g(x) - g(x_1)\| \le \|L^{-1}K\| \, \|x - x_1\| + \frac{1}{2}\|x - x_1\| + \frac{1}{2}\|g(x) - g(x_1)\|,$$

so

$$\|g(x) - g(x_1)\| \le c\|x - x_1\| \text{ with } c := 2\|L^{-1}K\| + 1. \tag{15}$$

Setting $\psi(x) := -L^{-1}\varphi(x, g(x))$ it follows from (13) that

$$g(x) - g(x_1) = -L^{-1}K(x - x_1) + \psi(x), \tag{16}$$

and we have

$$\lim_{x \to x_1} \frac{\psi(x)}{\|x - x_1\|} = 0 \tag{17}$$

using $\|\psi(x)\| \le \|L^{-1}\| \, \|\varphi(x, g(x))\|$ and $\lim_{x \to x_1} \frac{\varphi(x, g(x))}{\|x - x_1\|} = 0$ by (12) and (15).

(16) and (17) mean that $g(x)$ is differentiable at x_1 with

$$Dg(x_1) = -L^{-1}K = -(D_2F(x_1, y_1))^{-1}D_1F(x_1, y_1).$$

This is equivalent to (3). $\qquad\qquad\square$

A direct consequence of the implicit function theorem is the

Theorem 10.3 (Inverse function theorem) *Let V and W be Banach spaces, Ω open in V, $f : \Omega \to W$ continuously differentiable and $y_0 \in \Omega$. Let $Df(y_0)$ be invertible and the inverse $(Df(y_0))^{-1}$ likewise continuous. There exists an open neighborhood $\Omega' \subset \Omega$ of y_0 which is mapped bijectively under the function f onto an open neighborhood Ω'' of $x_0 = f(y_0)$, and the inverse function $g = f^{-1} : \Omega'' \to \Omega'$ is differentiable with*

$$Dg(x_0) = (Df(y_0))^{-1}. \tag{18}$$

Proof. We consider $F(x, y) = f(y) - x$. By assumption $D_2F(x_0, y_0) = Df(y_0)$ is invertible, with continuous inverse. Therefore by the implicit function theorem there exists an open neighborhood Ω'' of x_0, as well as a differentiable function $g : \Omega'' \to V$ such that $g(\Omega'') \subset \Omega_2$ for a neighborhood Ω_2 of y_0, $F(x, g(x)) = 0$, so $f(g(x)) = x$ for $x \in \Omega''$, and $g(x_0) = y_0$. In what follows, we restrict f to $g(\Omega'')$, without changing the notations. On account of $f(g(x)) = x$, the function g is injective on Ω''; therefore g establishes a bijection of Ω'' onto $g(\Omega'')$. Furthermore $g(\Omega'') = f^{-1}(\Omega'')$ is open, as f is continuous. We set accordingly $\Omega' = g(\Omega'')$. Then f maps Ω' bijectively onto Ω''. Finally (18) follows from (3) as well as from the relation $f(g(x)) = x$, so by the chain rule,

$$Df(g(x_0)) \circ Dg(x_0) = \mathrm{Id}.$$

$\square$

We shall now consider further explicit consequences in the finite dimensional case.

Corollary 10.4 *Let Ω be an open set in $\mathbb{R}^d$, $x_0 \in \Omega$, $\varphi : \Omega \to \mathbb{R}^k$ continuously differentiable, $\varphi(x_0) = y_0$.*

(i)   *If $d \leq k$ and $D\varphi(x_0)$ is of maximal rank $(= d)$, there exist open neighborhoods Ω' of y_0 and Ω'' of x_0 and a differentiable function*

$$g : \Omega' \to \mathbb{R}^k$$

with the property that for all $x \in \Omega''$

$$g \circ \varphi(x) = i(x),$$

$i : \mathbb{R}^d \to \mathbb{R}^k$, $d \leq k$, being the canonical injection $i(x^1, \dots x^d) = (x^1, \dots, x^d, 0, \dots, 0)$.

(ii)   *If $d \geq k$ and $D\varphi(x_0)$ is of maximal rank $(= k)$, there exists an open neighborhood $\tilde{\Omega}$ of x_0 as well as a differentiable map*

$$h : \tilde{\Omega} \to \Omega$$

with $h(x_0) = x_0$ and $\varphi \circ h(x) = \pi(x)$,
$\pi : \mathbb{R}^d \to \mathbb{R}^k$ (for $d \geq k$) being the canonical projection $\pi(x^1, \dots, x^d) = (x^1, \dots, x^k)$.

Proof.

(i)   By assumption, we have rank $(D_j\varphi^i)_{\substack{i=1,\dots,k \\ j=1,\dots,d}} = d$ at x_0. By changing notations, if necessary, we can assume that $\det(D_j\varphi^i)_{i,j=1,\dots,d} \neq 0$ at x_0.
We define $f : \Omega \times \mathbb{R}^{k-d} \to \mathbb{R}^k$ by

$$f(x^1, \ldots, x^k) = \varphi(x^1, \ldots, x^d) + (0, \ldots, 0, x^{d+1}, \ldots, x^k).$$

Now

$$\det(D_j f^i)_{i,j=1,\ldots,k} = \det(D_j \varphi^i)_{i,j=1,\ldots,d} \neq 0$$

at $\{x_0\} \times \{0\}$.

By the implicit function theorem, there exists locally a differentiable inverse g of f such that

$$i(x) = g \circ f(i(x)) = g \circ \varphi(x).$$

(ii)   As before, we assume, without loss of generality, that $\det(D_j \varphi^i)_{i,j=1,\ldots,k} \neq 0$ at x_0.
We define $f : \Omega \to \mathbb{R}^d$ by

$$f(x^1, \ldots, x^d) = (\varphi^1(x), \ldots, \varphi^k(x), x^{k+1}, \ldots, x^d).$$

We have

$$\det(D_j f^i)_{i,j=1,\ldots,d} = \det(D_j \varphi^i)_{i,j=1,\ldots,k} \neq 0$$

at x_0.

Therefore there exists, by the inverse function theorem, an inverse function h of f with

$$\pi(x) = \pi \circ f \circ h(x) = \varphi \circ h(x).$$

$\square$

In case (i), φ thus locally looks like the inclusion

$$\Omega'' \subset \mathbb{R}^d \overset{\varphi}{\to} \Omega' \subset \mathbb{R}^k$$
$$i \searrow \quad \swarrow g$$
$$\mathbb{R}^k$$

and in case (ii) like the projection

$$\tilde{\Omega} \subset \mathbb{R}^d$$
$$h \swarrow \qquad \searrow \pi$$
$$h(\tilde{\Omega}) \subset \Omega \subset \mathbb{R}^d \underset{\varphi}{\to} \mathbb{R}^k$$

As $i = g \circ \varphi$ and $\pi = \varphi \circ h$, respectively, both diagrams can be traced in any way in the direction of the arrows. Such diagrams are called commutative diagrams.

Example. As an example, we want to consider again the familiar polar coordinates in $\mathbb{R}^2$. Let $f : \mathbb{R}_+ \times \mathbb{R} \to \mathbb{R}^2$ be given by

$$f(r, \varphi) = (r \cos \varphi, r \sin \varphi).$$

Now

$$Df(r, \varphi) = \begin{pmatrix} \frac{\partial f^1}{\partial r} & \frac{\partial f^1}{\partial \varphi} \\ \frac{\partial f^2}{\partial r} & \frac{\partial f^2}{\partial \varphi} \end{pmatrix} = \begin{pmatrix} \cos \varphi & -r \sin \varphi \\ \sin \varphi & r \cos \varphi \end{pmatrix},$$

so $\det(Df) = r > 0$ for $r > 0$ and any φ.

Consequently, $Df(r, \varphi)$ for $r > 0$ is invertible and by the inverse function theorem, f is therefore locally invertible, with inverse g and

$$Dg(x, y) = (Df(r, \varphi))^{-1} = \begin{pmatrix} \cos \varphi & \sin \varphi \\ \frac{-\sin \varphi}{r} & \frac{\cos \varphi}{r} \end{pmatrix}$$

for $(x, y) = f(r, \varphi)$.

We have $\frac{x}{r} = \cos \varphi$, $\frac{y}{r} = \sin \varphi$, $r = \sqrt{x^2 + y^2}$ and moreover $\varphi = \arctan \frac{y}{x}$, so that here the solution can be given even explicitly. In particular

$$Dg(x, y) = \begin{pmatrix} \frac{x}{\sqrt{x^2+y^2}} & \frac{y}{\sqrt{x^2+y^2}} \\ \frac{-y}{x^2+y^2} & \frac{x}{x^2+y^2} \end{pmatrix}.$$

From this example one sees that inversion is, in general, possible only locally, for we have

$$f(r, \varphi + 2\pi m) = f(r, \varphi) \text{ for all } m \in \mathbb{Z}.$$

If x, y are cartesian coordinates of a point in $\mathbb{R}^2$, then r and φ, with $x = r \cos \varphi$, $y = r \sin \varphi$, are called polar coordinates of (x, y). For $(x, y) \neq (0, 0)$ the coordinate φ is completely undetermined; but here, the above function f is also not invertible. In complex notation we have

$$x + iy = re^{i\varphi}.$$

We now want to derive the *Lagrange multiplier* rule for local extrema with constraints. For this, let Ω be an open subset of $\mathbb{R}^d$, $m < d$, and let $F : \Omega \to \mathbb{R}^m$ be given. Now, local extrema of a function $h : \Omega \to \mathbb{R}$ under the constraint $F(x) = 0$ are to be found. So, in other words, we consider the restriction of h to the set $\{x : F(x) = 0\}$ and look in this set for locally smallest or greatest values, whereby these extremal properties are to hold only in comparison with other points of the set $\{F(x) = 0\}$. For this, the following result holds.

Theorem 10.5 *Let $F : \Omega \to \mathbb{R}^m$, $h : \Omega \to \mathbb{R}$ be continuously differentiable ($\Omega \subset \mathbb{R}^d$, $m < d$), and let h have a local extremum at the point x_0 subject to the constraint $F(x) = 0$.*

Let $DF(x_0)$ have maximal rank m, i.e. assume that there exists a nonvanishing subdeterminant of m rows of

$$DF(x_0) = \begin{pmatrix} \frac{\partial F^1(x_0)}{\partial x^1} & \cdots & \frac{\partial F^1(x_0)}{\partial x^d} \\ \vdots & & \\ \frac{\partial F^m(x_0)}{\partial x^1} & \cdots & \frac{\partial F^m(x_0)}{\partial x^d} \end{pmatrix}.$$

Then there exist m real numbers $\lambda_1, \ldots, \lambda_m$ (Lagrange multipliers) which satisfy the equation

$$Dh(x_0) = \sum_{j=1}^{m} \lambda_j DF^j(x_0). \tag{19}$$

In the special case $m = 1$ this simply means the existence of a real λ with

$$\operatorname{grad} h(x_0) = \lambda \operatorname{grad} F(x_0).$$

Proof. By renumbering the coordinates $x^1, \ldots, x^d$ if necessary, we may assume that

$$\det \left(\frac{\partial F^i(x_0)}{\partial x^j} \right)_{\substack{i=1,\ldots,m \\ j=n-m+1,\ldots,d}} \neq 0, \tag{20}$$

so the subdeterminant of $DF(x_0)$ formed from the last m rows does not vanish. We write $z = (x^1, \ldots, x^{d-m})$, $y = (x^{d-m+1}, \ldots, x^d)$, and similarly z_0, y_0. The inequality (20) means that $D_2 F(z_0, y_0)$ $(x = (z, y) \in \mathbb{R}^{d-m} \times \mathbb{R}^m)$ is invertible. The implicit function theorem implies that in a neighborhood of (z_0, y_0) the condition $F(z, y) = 0$ is described by $y = g(z)$, so $F(z, g(z)) = 0$. The function $H(z) := h(z, g(z))$ then has an (unconstrained) local extremum at x_0. Therefore, by theorem 9.9

$$DH(z_0) = 0 \iff \frac{\partial H}{\partial z^i}(z_0) = 0 \text{ for } i = 1, \ldots, d - m,$$

so

$$\frac{\partial h(z_0, g(z_0))}{\partial z^i} + \sum_{j=1}^{m} \frac{\partial h(z_0, g(z_0))}{\partial y^j} \cdot \frac{\partial g^j(z_0)}{\partial z^i} = 0$$

for $i = 1, \ldots, d - m$, or in other notation,

$$D_1 h(z_0, g(z_0)) + D_2 h(z_0, g(z_0)) Dg(z_0) = 0.$$

On the other hand, by the implicit function theorem,

$$Dg(z_0) = -(D_2 F(z_0, g(z_0)))^{-1} D_1 F(z_0, g(z_0)),$$

so altogether

$$D_1 h - D_2 h (D_2 F)^{-1} D_1 F = 0,$$

always at the point $(z_0, g(z_0))$. We set

$$\Lambda := (\lambda_1, \ldots, \lambda_m) := D_2 h (D_2 F)^{-1}(z_0, g(z_0)).$$

We then have

$$D_1 h = \Lambda D_1 F \tag{21}$$

and by definition of Λ also

$$D_2 h = \Lambda D_2 F, \tag{22}$$

again evaluated at the point $(z_0, g(z_0))$. The equations (21) and (22) are equivalent to (19). $\qquad\square$

Example. As an example, we want to determine the extrema of $h : \mathbb{R}^3 \to \mathbb{R}$, $h(x) = x^1 + x^2 + x^3$ subject to the condition $F(x) = 0$ with $F(x) = (x^1)^2 + (x^2)^2 + (x^3)^2 - 1$. At an extreme point we must have $\mathrm{grad}\, h(x_0) = \lambda\, \mathrm{grad}\, F(x_0)$, so here

$$1 = 2\lambda x_0^i \text{ for } i = 1, 2, 3, \tag{23}$$

and in addition $F(x_0) = 0$, so

$$(x_0^1)^2 + (x_0^2)^2 + (x_0^3)^2 = 1. \tag{24}$$

These are four equations for the three components of x_0 and λ, so for altogether four unknowns. Equations (23) give $x_0^1 = x_0^2 = x_0^3$, and by (24)

$$x_0 = \pm\left(\frac{1}{\sqrt{3}}, \frac{1}{\sqrt{3}}, \frac{1}{\sqrt{3}}\right). \tag{25}$$

On the other hand, h must assume its maximum and minimum on the compact set
$\{F(x) = 0\}$, therefore at both the points determined by (25). The plus sign obviously goes with the maximum and the minus with the minimum in (25).

Exercises for § 10

1)   Discuss the map

$$f : \mathbb{C} \to \mathbb{C}, f(z) := z^2.$$

Where is f differentiable when considered as a map of $\mathbb{R}^2$ to itself? Where is the Jacobian different from 0? Is the image of each open set open? Construct a maximal region in $\mathbb{C}$ where f is bijective and f as well as f^{-1} are continuously differentiable. Study the images under f of parallels to the real and the imaginary axis, as well as of rays starting at 0 and of circles with center at 0.

2)   Define $f : \mathbb{R}^2 \to \mathbb{R}^2$ by

$$f^1(\xi, \eta) := \xi$$

$$f^2(\xi, \eta) := \eta - \xi^2 \quad \text{for } \xi^2 \leq \eta$$

$$f^2(\xi, \eta) := \frac{\eta^2 - \xi^2 \eta}{\xi^2} \quad \text{for } 0 \leq \eta \leq \xi^2, (\xi, \eta) \neq (0, 0)$$

$$f^2(\xi, \eta) := -f^2(\xi, -\eta) \quad \text{for } \eta \leq 0.$$

Show that f is differentiable everywhere, $Df(0,0) = \text{Id}$, but Df is not continuous at $(0,0)$. Every neighborhood of $(0,0)$ contains points $x_1 \neq x_2$ with $f(x_1) = f(x_2)$. Is this compatible with the inverse function theorem?

3)   Let $B := \{x \in \mathbb{R}^d : \|x\| < 1\}, A := \{x \in \mathbb{R}^d, \|x\| > 1\}$. Consider $f : B \setminus \{0\} \to A$, $f(x) = \frac{x}{\|x\|^2}$. Compute Df. Is f bijective? Supply a geometric construction for $f(x)$ if $d = 2$ or 3.

4)   **(Spatial polar coordinates)** Let $\mathbb{R}^d_+ := \{(r, t^2, \ldots, t^d) \in \mathbb{R}^d : r \geq 0\}$ $(d \geq 3)$. We define a map $F_d : \mathbb{R}^d_+ \to \mathbb{R}^d$ by the following equations:

$$x^1 := r \prod_{i=2}^{d} \cos t^i, \, x^j := r \sin t^j \prod_{i=j+1}^{d} \cos t^i \quad \text{for } j = 2, \ldots, d-1$$

$$x^d := r \sin t^d.$$

Show that F_d is surjective. Compute its Jacobian. Where is the Jacobian $\neq 0$? What are the images under F_d of the following sets?

$$Q_R := \{(r, t^2, \ldots, t^d) : 0 \leq r \leq R, -\pi < t^2 \leq \pi, -\frac{\pi}{2} \leq t^j \leq \frac{\pi}{2}$$
$$\text{for } j = 3, \ldots, d\}$$

$$H_R := \{(r, t^2, \ldots, t^d) : r = R, -\pi < t^2 \leq \pi, -\frac{\pi}{2} \leq t^j \leq \frac{\pi}{2}$$
$$\text{for } j = 3, \ldots, d\}.$$

For $x \in \mathbb{R}^d$ with $x = F_d(r, t^2, \ldots, t^d)$, $(r, t^2, \ldots, t^d)$ are called polar coordinates of x. r is uniquely determined by x (give a formula!). Which convention can be employed to determine $t^2, \ldots, t^d$ uniquely for $x \neq 0$? For $d = 3$, interpret t^2 and t^3 as angles (draw a picture!).

11. Curves in $\mathbb{R}^d$. Systems of ODEs

First, some elementary properties, like rectifiability or arc length parametrization, of curves in Euclidean space are treated. Next, curves that solve systems of ODEs are considered. Higher order ODEs are reduced to such systems.

Definition 11.1 A curve in $\mathbb{R}^d$ is a continuous map $\gamma : I \to \mathbb{R}^d$, where I is an interval in $\mathbb{R}$ consisting of more than one point. The interval I can be proper or improper.

Examples.

1) A straight line $\gamma : \mathbb{R} \to \mathbb{R}^d, \gamma(t) = x_0 + vt \; (v \neq 0)$

2) A circle of radius $r > 0 : \gamma : [0, 2\pi] \to \mathbb{R}^2, \gamma(t) = (r \cos t, r \sin t)$

3) A helix: $\gamma : \mathbb{R} \to \mathbb{R}^3, \gamma(t) = (r \cos t, r \sin t, \alpha t) \; (r > 0, \alpha \neq 0)$

4) Archimedian spiral: $\gamma : \mathbb{R} \to \mathbb{R}^2, \gamma(t) = (\alpha t \cos t, \alpha t \sin t), \; \alpha > 0$

5) Logarithmic spiral: $\gamma : \mathbb{R} \to \mathbb{R}^2, \gamma(t) = (\alpha e^{\lambda t} \cos t, \alpha e^{\lambda t} \sin t), \; \alpha > 0, \lambda > 0$

6) The graph of a continuous function $f : I \to \mathbb{R} : \gamma : I \to \mathbb{R}^2, \gamma(t) = (t, f(t))$.

Definition 11.2 Let $\gamma : I \to \mathbb{R}^d$ be a differentiable curve. For $t \in I$ we call

$$\dot{\gamma}(t) := D\gamma(t)$$

the tangent vector of γ at t. $\dot{\gamma}(t)$ is the limit of secants: $\dot{\gamma}(t) = \lim\limits_{\substack{h \to 0 \\ h \neq 0}} \frac{\gamma(t+h) - \gamma(t)}{h}$.

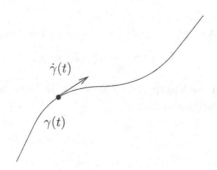

One can interpret a curve in $\mathbb{R}^n$ also kinematically as the orbit of a point mass which is situated at position $\gamma(t)$ at time t. $\dot\gamma(t)$ then is the velocity vector at time t.

The trace of a continuously differentiable curve could still have corners, as the example of Neil's parabola shows:

$$\gamma : \mathbb{R} \to \mathbb{R}^2, \gamma(t) = (t^2, t^3), \text{ so } \gamma(\mathbb{R}) = \{(\xi, \eta) \in \mathbb{R}^2 : \eta = \pm\xi^{\frac{3}{2}}\}.$$

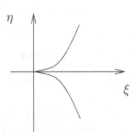

In order to exclude this, we make the following definition

Definition 11.3 Let $\gamma : I \to \mathbb{R}^d$ be a continuously differentiable curve. It is called regular if $\dot\gamma(t) \neq 0$ for all $t \in I$, and singular if $\dot\gamma(t) = 0$ for some $t \in I$.

We now want to define the length of a curve. Let $\gamma : [a, b] \to \mathbb{R}^d$ be a curve; a partition Z of $[a, b]$ consists of points $t_0 = a < t_1 < \ldots < t_k = b \, (k \in \mathbb{N})$. We set

$$L(\gamma, Z) := \sum_{i=1}^{k} \|\gamma(t_i) - \gamma(t_{i-1})\|,$$

so $L(\gamma, Z)$ is the length of the polygon joining the points $\gamma(t_0), \gamma(t_1), \ldots, \gamma(t_k)$.

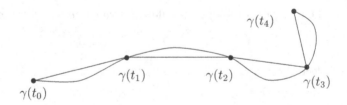

Definition 11.4 A curve $\gamma : [a, b] \to \mathbb{R}^d$ is called rectifiable if

$$L(\gamma) := \sup_{Z} L(\gamma, Z) < \infty.$$

$L(\gamma)$ is then called the length of γ. For $\alpha, \beta \in [a, b]$ we denote the restriction of γ to $[\alpha, \beta]$ by $\gamma_{|[\alpha,\beta]}$, so $\gamma_{|[\alpha,\beta]} : [\alpha, \beta] \to \mathbb{R}^d$ with $\gamma_{|[\alpha,\beta]}(t) = \gamma(t)$ for $t \in [\alpha, \beta]$.

Lemma 11.5 *Let $\gamma : [a,b] \to \mathbb{R}^d$ be a curve, $c \in (a,b)$. The curve γ is rectifiable precisely if $\gamma_{|[a,c]}$ and $\gamma_{|[c,b]}$ are rectifiable and then*

$$L(\gamma) = L(\gamma_{|[a,c]}) + L(\gamma_{|[c,b]}).$$

Proof. Let Z be partition of $[a,b]$ with $t_i < c < t_{i+1}$ for an index i. The points $t_0, t_1, \ldots, t_i, c, t_{i+1}, \ldots t_k$ likewise define a partition Z' of $[a,b]$ and we have

$$L(\gamma, Z) \leq L(\gamma, Z').$$

We set further

$$\gamma_1 = \gamma_{|[a,c]}, \gamma_2 = \gamma_{|[c,b]}.$$

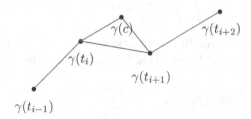

The partition Z_1' defined by $t_0, \ldots, t_i, c$, is a partition of $[a,c]$ and similarly Z_2', given by $c, t_{i+1}, \ldots t_k$ is one of $[c,b]$. It follows that

$$L(\gamma, Z) \leq L(\gamma, Z') = L(\gamma_1, Z_1') + L(\gamma_2, Z_2') \leq L(\gamma). \tag{1}$$

Conversely, arbitrary partitions Z_1, Z_2 of $[a,c]$ and $[c,b]$, respectively, define a partition $Z = Z_1 \cup Z_2$ of $[a,b]$, and we have

$$L(\gamma_1, Z_1) + L(\gamma_2, Z_2) = L(\gamma, Z) \leq L(\gamma). \tag{2}$$

From (1) and (2) the assertion follows easily.   □

Definition 11.6 For a rectifiable curve $\gamma : [a,b] \to \mathbb{R}^d$ we define its arc length function $s : [a,b] \to \mathbb{R}_+$, by

$$s(t) := L(\gamma_{|[a,t]}).$$

By lemma 11.5, this definition is possible and $s(t)$ is monotonically increasing.

Theorem 11.7 *Let $\gamma : [a,b] \to \mathbb{R}^d$ be a continuously differentiable curve. Then γ is rectifiable and*

$$L(\gamma) = \int_a^b \|\dot{\gamma}(\tau)\| d\tau,$$

and $\dot{s}(t) = \|\dot{\gamma}(t)\|$ *for all* $t \in [a, b]$.

Proof. First of all, for a partition Z of $[a, b]$ given by $t_0, \ldots, t_k$ we have

$$\|\gamma(t_i) - \gamma(t_{i-1})\| = \|\int_{t_{i-1}}^{t_i} \dot{\gamma}(\tau) d\tau\| \quad (i = 1, \ldots, k), \tag{3}$$

where we have set

$$\int_\alpha^\beta \dot{\gamma}(\tau) d\tau := (\int_\alpha^\beta \dot{\gamma}^1(\tau) d\tau, \ldots, \int_\alpha^\beta \dot{\gamma}^d(\tau) d\tau).$$

Furthermore, for an arbitrary continuous $f : [a, b] \to \mathbb{R}^d$ and $\alpha, \beta \in [a, b]$ we have

$$\|\int_\alpha^\beta f(\tau) d\tau\| \leq \int_\alpha^\beta \|f(\tau)\| d\tau. \tag{4}$$

This follows by approximating f uniformly by (vectorvalued) step functions $(s_n)_{n \in \mathbb{N}}, s_n = (s_n^1, \ldots, s_n^d)$, and using (by the triangle inequality)

$$\|\sum_{\lambda=1}^{l} s(\tau_\lambda)(\tau_\lambda - \tau_{\lambda-1})\| \leq \sum_{\lambda=1}^{l} \|s(\tau_\lambda)\|(\tau_\lambda - \tau_{\lambda-1}); \tag{5}$$

here the τ_λ are precisely the points of discontinuity of s. For $s = s_\nu$ both sides of (5) converge to the corresponding sides of (4). By means of (4) we now conclude from (3) that

$$\|\gamma(t_i) - \gamma(t_{i-1})\| \leq \int_{t_{i-1}}^{t_i} \|\dot{\gamma}(\tau)\| d\tau,$$

and from this

$$L(\gamma) \leq \int_a^b \|\dot{\gamma}(\tau)\| d\tau. \tag{6}$$

In particular, γ is rectifiable. For $t, t + h \in [a, b]$ we now have (assuming $h > 0$)

$$\|\gamma(t + h) - \gamma(t)\| \leq L(\gamma_{|[t,t+h]}) = s(t + h) - s(t)$$
$$\leq \int_t^{t+h} \|\dot{\gamma}(\tau)\| d\tau,$$

by lemma 11.5 and by (6) applied to $\gamma_{|[t,t+h]}$. It follows that

$$\|\frac{\gamma(t + h) - \gamma(t)}{h}\| \leq \frac{s(t + h) - s(t)}{h} \leq \frac{1}{h} \int_t^{t+h} \|\dot{\gamma}(\tau)\| d\tau.$$

For $h \to 0$, the left and right hand sides converge to $\|\dot{\gamma}(t)\|$ and it follows that $\dot{s}(t) = \|\dot{\gamma}(t)\|$ and from this also

$$L(\gamma) = s(b) = \int_a^b \dot{s}(\tau)d\tau = \int_a^b \|\dot{\gamma}(\tau)\|d\tau .$$

$\square$

Definition 11.8 A curve $\gamma : [a, b] \to \mathbb{R}^d$ is called piecewise continuously differentiable if there exist points $t_0 = a < t_1 < \ldots < t_k = b$ with the property that for $i = 1, \ldots, k$, $\gamma_{|[t_{i-1}, t_i]}$ is continuously differentiable.

From lemma 11.5 and theorem 11.7, we conclude

Corollary 11.9 *Every piecewise continuously differentiable curve is rectifiable and its length is the sum of the lengths of its continuously differentiable pieces.*

Definition 11.10 Let $\gamma : [a, b] \to \mathbb{R}^d$ be a curve, $[\alpha, \beta] \subset \mathbb{R}$. A parameter transformation is a bijective continuous map $\varphi : [\alpha, \beta] \to [a, b]$. It is called orientation preserving if it is strictly monotonically increasing, orientation reversing if it is strictly monotonically decreasing (as φ is bijective, one of these two cases must occur). Two curves γ_1 and γ_2 are called equivalent if $\gamma_2 = \gamma_1 \circ \varphi$ for an orientation preserving transformation, and weakly equivalent if φ is not necessarily orientation preserving.

Obviously these relations (equivalent and weakly equivalent) are equivalence relations. The transitivity follows, for example, from the fact that the composition of two parameter transformations is again a parameter transformation. For (weakly) equivalent curves the image sets are the same but they are, in general, traced differently.

Definition 11.12 An arc is a class of weakly equivalent curves, an oriented arc is a class of equivalent curves.

A curve $\gamma : I \to \mathbb{R}^d$ need not necessarily be injective; it could, for example, have double points.

Example. Let $\gamma : \mathbb{R} \to \mathbb{R}^2$ be the curve $\gamma(t) = (t^2 - 1, t^3 - t)$. We have $\gamma(1) = \gamma(-1) = (0, 0)$. So γ has a double point,

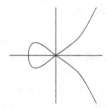

and both the tangent vectors $\gamma'(-1) = (-2,2)$ and $\gamma'(1) = (2,2)$ there have different directions.

Definition 11.13 An arc is called a Jordan arc if it is represented by an injective curve $\gamma : [a,b] \to \mathbb{R}^d$. It is called a closed Jordan arc if it is represented by a curve $\gamma : [a,b] \to \mathbb{R}^d$ which is injective on $[a,b)$ and satisfies $\gamma(a) = \gamma(b)$ (thus the initial and end points coincide).

Obviously, a curve weakly equivalent to an injective curve is again injective, so that for the Jordan property it is immaterial by which curve an arc is represented.

Lemma 11.14 Let $\gamma_1 : [a_1,b_1] \to \mathbb{R}^d$ and $\gamma_2 : [a_2,b_2] \to \mathbb{R}^d$ be weakly equivalent curves. If one is rectifiable, so is the other, and their lengths then coincide.

Proof. Let $\gamma_2 = \gamma_1 \circ \varphi$, where $\varphi : [a_2,b_2] \to [a_1,b_1]$ is bijective and continuous. Now φ induces a bijection between the partitions of $[a_2,b_2]$ and those of $[a_1,b_1]$. Namely, if $a_2 = t_0 < t_1 < \ldots t_k = b_2$ is a partition Z_2 of $[a_2,b_2]$, then $\varphi(t_0), \varphi(t_1), \ldots, \varphi(t_k)$ leads to a partition Z_1 of $[a_1,b_1]$ for increasing φ, and $\varphi(t_k), \ldots, \varphi(t_0)$ for decreasing φ. Similarly φ^{-1} carries a partition of $[a_1,b_1]$ into one such of $[a_2,b_2]$. Moreover, we obviously have $L(\gamma_2, Z_2) = L(\gamma_1, Z_1)$, for $\gamma_2(t_i) = \gamma_1(\varphi(t_i))$ for $i = 1, \ldots, k$. From this, the assertion follows directly by the definition of $L(\gamma_j), j = 1, 2$. $\square$

Lemma 11.14 allows

Definition 11.15 An arc is called rectifiable if it is representable by a rectifiable curve $\gamma : [a,b] \to \mathbb{R}^d$ and its length is then defined to be $L(\gamma)$.

Theorem 11.16 Let $\gamma : [a,b] \to \mathbb{R}^d$ be continuously differentiable, with $\dot\gamma(t) \neq 0$ for all $t \in [a,b]$. Then the length function $s(t) := L(\gamma_{|[a,t]})$ is invertible, and its inverse $t(s)$ is likewise differentiable. The curve $\tilde\gamma(s) := \gamma(t(s)), \tilde\gamma : [0, L(\gamma)] \to \mathbb{R}^d$ is equivalent to γ and satisfies

$$\|\dot{\tilde\gamma}(s)\| = 1 \ for \ all \ s, 0 \le s \le L(\gamma). \tag{7}$$

Before the proof, we make a definition.

Definition 11.17 A curve $\tilde\gamma(s)$ with $s = L(\tilde\gamma_{|[0,s]})$ is said to be parametrized by arc length and s in this case is called the arc length parameter.

Theorem 11.16 states that a continuously differentiable (and thus also a piecewise continuously differentiable curve) can always be parametrised by arc length. This assertion also holds for arbitrary rectifiable curves. The proof is, in principle, not difficult, but it is somewhat involved and we shall not present it here. The main point consists in showing that $s(t) := L(\gamma_{|[a,t]})$ is continuous. The interested reader may try this herself or himself.

Proof of Theorem 11.16: By theorem 11.7 and our assumptions we have

$$\dot{s}(t) = \|\dot{\gamma}(t)\| > 0 \text{ for all } t, \text{ and } L(\gamma) = \int_a^b \dot{s}(\tau)d\tau. \tag{8}$$

It follows that $t \mapsto s(t)$ yields a strictly increasing bijection of $[a,b]$ onto $[0, L(\gamma)]$ and the inverse map $t(s)$ is differentiable with $\dot{t}(s) = \frac{1}{\dot{s}(t)}$. It follows that for $\tilde{\gamma}(s) := \gamma(t(s))$ we have $\dot{\tilde{\gamma}}(s) = \dot{\gamma}(t(s)) \cdot \frac{1}{\dot{s}(t)}$, so $\|\dot{\tilde{\gamma}}(s)\| = 1$, by (8). $\qquad\square$

We now want to define briefly the intersection angle between two regular curves.

Definition 11.18 Let $\gamma_1 : I_1 \to \mathbb{R}^d, \gamma_2 : I_2 \to \mathbb{R}^d$ be regular curves, and $\gamma_1(t_1) = \gamma_2(t_2)$ for certain $t_1 \in I_1, t_2 \in I_2$. The angle between the oriented curves γ_1 and γ_2 at the point $\gamma_1(t_1) = \gamma_2(t_2)$ is defined to be the angle between the corresponding tangent vectors; thus

$$\theta := \arccos \frac{<\dot{\gamma}_1(t_1), \dot{\gamma}_2(t_2)>}{\|\dot{\gamma}_1(t_1)\| \cdot \|\dot{\gamma}_2(t_2)\|}.$$

The angle between two regular curves does not change when both the curves undergo a parameter transformation, so far as both the transformations are simultaneously orientation preserving or orientation reversing. On the contrary it changes to $\pi - \theta$ when only one of the transformations reverses the orientation.

Finally we cite, without proof, the visibly plausible but surprisingly difficult to prove

Jordan arc theorem: *A closed Jordan arc Γ in $\mathbb{R}^2$ partitions $\mathbb{R}^2$ into exactly two open and connected sets, that is, $\mathbb{R}^2 \backslash \Gamma = \Omega_1 \cup \Omega_2, \partial\Omega_1 = \Gamma = \partial\Omega_2, \Omega_1 \cap \Omega_2 = \emptyset, \Omega_1, \Omega_2$ open and connected. Of these two sets, only one is bounded.*

We now wish to study curves

$$f : I \to \mathbb{R}^d, \quad \text{for some interval } I \subset \mathbb{R},$$

that solve ordinary differential equations. For that purpose, let

$$\phi : I \times J \to \mathbb{R}^d$$

be a continuous function, where J is some subset of $\mathbb{R}^d$. A curve f then is a solution of the system of ordinary differential equations

$$f'(x) = \phi(x, f(x)) \tag{9}$$

if $f(x) \in J$ for all $x \in I$, and

$$f^i(\xi_2) - f^i(\xi_1) = \int_{\xi_1}^{\xi_2} \phi^i(\xi, f(\xi))d\xi \quad \text{for } i = 1, \ldots, d$$

whenever $\xi_1, \xi_2 \in I$ (here, $f(\xi) = (f^1(\xi), \ldots, f^d(\xi))$).

The proof of theorem 6.16 can be taken over verbatim to obtain a solution of (9) with given initial condition $f(x_0) = y_0$ if ϕ satisfies a Lipschitz condition as before:

Theorem 11.19 *Suppose $\phi(x, y)$ is continuous for $|x - x_0| \leq \rho$, $|y - y_0| \leq \rho$, with*

$$|\phi(x, y)| \leq M \quad \text{for all such } x, y. \tag{10}$$

Further, let ϕ satisfy the Lipschitz condition

$$|\phi(x, y_1) - \phi(x, y_2)| \leq L|y_1 - y_2|$$

whenever $|x - x_0| \leq \rho$, $|y_1 - y_0| \leq \eta$, $|y_2 - y_0| \leq \rho$, for some fixed $L < \infty$. Then there exists $h > 0$ with the property that (9) possesses a unique solution with $f(x_0) = y_0$ on $[x_0 - h, x_0 + h] \cap I$.

Such systems of ODEs arise naturally when one studies higher order ODEs. For example, for

$$u : I \to \mathbb{R},$$

we consider the second order ODE

$$u''(x) = \psi(x, u(x), u'(x)). \tag{11}$$

Here, $u \in C^2(I)$ is a solution of (11) if its second derivative $u''(x)$ coincides with $\psi(x, u(x), u'(x))$ for all $x \in I$. Of course, this can be expressed by an equivalent integral equation as in the case of first order ODEs as studied in §6. We wish to solve (11) with initial conditions

$$u(x_0) = u_0 \tag{12}$$
$$u'(x_0) = v_0$$

for $x_0 \in I$.

We shall reduce (11) to a system of two first order ODEs. We simply put

$$v(x) := u'(x), \tag{13}$$

and (11) becomes equivalent to the system

$$u'(x) = v(x) \tag{14}$$
$$v'(x) = \psi(x, u(x), v(x)).$$

With $f(x) := (u(x), v(x)), \phi(x, f(x)) := (v(x), \psi(x, u(x), v(x)))$, (14) is equivalent to the system

$$f'(x) = \phi(x, f(x)). \tag{15}$$

Likewise, with $y_0 := (u_0, v_0)$, the initial condition (12) becomes

$$f(x_0) = y_0. \tag{16}$$

We thus obtain

Corollary 11.20 *The second order ODE (11) with initial condition (12) possesses a unique solution u on some interval $[x_0 - h, x_0 + h] \cap I$ if ψ is bounded and satisfies a Lipschitz condition of the form*

$$|\psi(x, u_1, v_2) - \psi(x, y_2, v_2)| \leq L(|u_1 - u_2| + |v_1 - v_2|)$$

for $|x - x_0| \leq \rho, |u_i - u_0| + |v_i - v_0| \leq \eta$ for $i = 1, 2$.

In the same manner, ODEs of higher than second order or systems of ODEs of higher order can be reduced to systems of first order ODEs.

Exercises for § 11

1)   Compute the arc length function for the curves of examples 1) – 5) at the beginning of this paragraph.

2)   Determine all solutions $f : \mathbb{R} \to \mathbb{R}$ of the ODE

$$f'' = f.$$

Chapter IV.

The Lebesgue Integral

12. Preparations. Semicontinuous Functions

As a preparation for Lebesgue integration theory, lower and upper semicontinuous functions are studied.

Theorem 12.1 (Dini) *Let K be a compact subset of a metric space with distance function d, $f_n : K \to \mathbb{R}$ ($n \in \mathbb{N}$) be continuous functions with*

$$f_n \leq f_{n+1} \text{ for all } n \in \mathbb{N} \tag{1}$$

and for all $x \in K$ assume that

$$f(x) = \lim_{n \to \infty} f_n(x)$$

exists, and that the function $f : K \to \mathbb{R}$ is also continuous.
   Then $(f_n)_{n \in \mathbb{N}}$ converges uniformly to f on K.

Proof. Let $\varepsilon > 0$. For every $x \in K$ there exists such an $N(x) \in \mathbb{N}$ that

$$|f(x) - f_{N(x)}(x)| < \frac{\varepsilon}{2}. \tag{2}$$

As $f_{N(x)}$ and f are both continuous, there exists, moreover, such a $\delta(x) > 0$ that

$$|(f(y) - f_{N(x)}(y)) - (f(x) - f_{N(x)}(x))| < \frac{\varepsilon}{2}$$
$$\text{for all } y \in K \text{ with } d(x, y) < \delta(x). \tag{3}$$

From (1), (2) and (3) it follows that for $n \geq N(x), d(x, y) < \delta(x)$

$$|f(y) - f_n(y)| \leq |f(y) - f_{N(x)}(y)| < \varepsilon.$$

For $x \in K$ let
$$U_x := U(x, \delta(x)) = \{y : d(x, y) < \delta(x)\}.$$

Since K is compact and is clearly covered by (U_x), it is already covered by finitely many such balls, say $U_{x_1}, \ldots, U_{x_k}$. We then have

$$|f(y) - f_n(y)| < \varepsilon \text{ for all } y \in K \text{ and } n \geq \max(N(x_1), \ldots, N(x_k)).$$

From this the uniform convergence follows. $\qquad\square$

As simple and well known examples show, this theorem does not hold anymore if the limit function f is not assumed to be continuous, and the limit of a monotonically increasing sequence of continuous functions is not necessarily continuous itself. We shall now introduce a class of functions which contains the class of continuous functions and which is closed under monotonically increasing convergence.

Definition 12.2 Let X be a metric space, $x \in X$. A function $f : X \to \mathbb{R} \cup \{\infty\}$ is called lower semicontinuous at x if, for all $c \in \mathbb{R}$ with $c < f(x)$, there exists a neighborhood U of x such that for all $y \in U, c < f(y)$. For $X = \mathbb{R}$, this is expressible as follows: $\forall \, \epsilon > 0 \, \exists \, \delta > 0 \, \forall \, y$ with $|x - y| < \delta :$ $f(x) - f(y) < \epsilon$. The function f is called lower semicontinuous on X if it is lower semicontinuous at every $x \in X$.

Correspondingly, $f : X \to \mathbb{R} \cup \{-\infty\}$ is called upper semicontinuous at x if $-f$ is lower semicontinuous at x, or equivalently, if $\forall c \in \mathbb{R}$ with $c > f(x) \, \exists$ a neighborhood U of x such that $\forall y \in U : c > f(y)$.

The lower semicontinuity of $f : X \to \mathbb{R} \cup \{\infty\}$ means that for every $c \in \mathbb{R}$ the set $f^{-1}((c, \infty])$ is a neighborhood of all its points. So we have

Lemma 12.3 $f : X \to \mathbb{R} \cup \{\infty\}$ is lower semicontinuous if and only if for all $c \in \mathbb{R}$ $f^{-1}((c, \infty]) = \{x \in X : c < f(x)\}$ is open (in X).

Examples.

1) $f : X \to \mathbb{R}$ is continuous if and only if it is lower and upper semicontinuous.

2)   Characteristic functions.
      Let $A \subset X$. We define the characteristic function of A as

$$\chi_A(x) := \begin{cases} 1 & \text{for } x \in A \\ 0 & \text{for } x \in X \backslash A \ . \end{cases}$$

      Lemma 12.3 implies that A is open exactly when χ_A is lower semicontinuous and it is closed when χ_A is upper semicontinuous.

3)   If f has a relative minimum at x_0, so $f(x) \geq f(x_0)$ for all x in a neighborhood of x_0, then f is lower semicontinuous at x_0.

We also have a sequential criterion for lower semicontinuity.

Lemma 12.4 $f : X \to \mathbb{R} \cup \{\infty\}$ is lower semicontinuous at $x \in X$ if and only if for every sequence $(x_n)_{n \in \mathbb{N}} \subset X$ with $\lim_{n \to \infty} x_n = x$ we have

$$\liminf_{n \to \infty} f(x_n) \geq f(x).$$

Proof. Let f be lower semicontinuous at $x = \lim x_n$, and let $c < f(x)$. Then for all $y \in U$, U being a suitable neighborhood of x,

$$c < f(y).$$

On the other hand, there exists an $N \in \mathbb{N}$ with $x_n \in U$ for $n \geq N$, as $x_n \to x$. It follows that $f(x_n) > c$ for $n \geq N$ and as this holds for every $c < f(x)$, it follows that

$$\liminf f(x_n) \geq f(x).$$

Conversely, assume that the given sequential criterion in the lemma holds, and let $c < f(x)$.

We assume that there exists no neighborhood U of x with the property that for all $y \in U$, $c < f(y)$ holds. In particular, there exists then for every $n \in \mathbb{N}$ an $x_n \in X$ with

$$d(x, x_n) < \frac{1}{n} \text{ and } f(x_n) \leq c.$$

But then $\lim_{n \to \infty} x_n = x$ and $\liminf_{n \to \infty} f(x_n) \leq c < f(x)$, in contradiction to the assumption. $\qquad \square$

Lemma 12.5 *Let* $f, g : X \to \mathbb{R} \cup \{\infty\}$ *be lower semicontinuous. Then* $\sup(f, g)$, $\inf(f, g)$ *and* $f + g$ *are also lower semincontinuous.*

The proof is a simple exercise. $\qquad \square$

Lemma 12.6 *A lower semicontinuous function on a compact set* K *assumes its infimum there.*

Proof. Let $\mu := \inf_{y \in K} f(y)$. There exists a sequence $(x_n)_{n \in \mathbb{N}} \subset K$ with $f(x_n) \to \mu$. As K is compact, the sequence (x_n) converges, after choosing a subsequence, to an $x \in K$. By lemma 12.4 we have

$$f(x) \leq \liminf f(x_n) = \mu,$$

and on the other hand, by definition of μ, also $f(x) \geq \mu$ so altogether $f(x) = \mu$. Thus f assumes its minimum at the point x. $\qquad \square$

Remark. A lower semicontinuous function on a compact set need not assume its supremum. For example f defined on $[0, 1]$ by

$$f(x) = \begin{cases} x & \text{for } 0 \leq x < 1 \\ 0 & \text{for } x = 1 \end{cases}$$

is such a function.

The value of a lower semicontinuous function may jump down as a point x is approached, but it cannot jump up.

Definition 12.7 Let $f_\alpha : X \to \mathbb{R} \cup \{\infty\}, \alpha \in I$, be a family of functions. The upper envelope of this family

$$f := \sup_{\alpha \in I} f_\alpha$$

is defined by

$$f(x) := \sup_{\alpha \in I} f_\alpha(x).$$

Similarly, one can define the lower envelope $\inf f_\alpha$ of a family $f_\alpha : X \to \mathbb{R} \cup \{-\infty\}, \alpha \in I$.

Lemma 12.8 *Let $f_\alpha : X \to \mathbb{R} \cup \{\infty\}$ be a family of functions and let every $f_\alpha, \alpha \in I$, be lower semicontinuous at $x_0 \in X$. Then the upper envelope is also lower semicontinuous there.*

Proof. Let $c < f(x_0)$. As $f(x_0) = \sup_{\alpha \in I} f_\alpha(x_0)$, there exists a $\beta \in I$ with $c < f_\beta(x_0)$. Since f_β is lower semicontinuous, this also holds for y in a neighborhood U of $x_0 : c < f_\beta(y)$. As $f \geq f_\beta$, it follows that for $y \in U$ $c < f(y)$. Therefore f is lower semicontinuous at x_0. □

Therefore, the upper envelope of a family of continuous functions is also lower semicontinuous. Nevertheless, the upper envelope of a family of continuous functions is not necessarily continuous, as the following example shows:

$$f_n : [0,1] \to \mathbb{R}, \ f_n(x) := \begin{cases} nx & \text{for } 0 \leq x < \frac{1}{n} \\ 1 & \text{for } \frac{1}{n} \leq x \leq 1 \end{cases} \quad (n \in \mathbb{N})$$

We have then

$$\sup_{n \in \mathbb{N}} f_n(x) = \begin{cases} 0 & \text{for } x = 0 \\ 1 & \text{for } 0 < x \leq 1 \end{cases}.$$

Definition 12.9 Let X be a metric space, $f : X \to \mathbb{R}$. The support of f, in symbols $\text{supp}\, f$, is defined as the closure of $\{x \in X : f(x) \neq 0\}$. $C_c(X)$ is the space of continuous functions $f : X \to \mathbb{R}$ with compact support.

Theorem 12.10 *For $f : \mathbb{R}^d \to \mathbb{R} \cup \{\infty\}$ the following conditions (i) and (ii) are equivalent*

(i)   a) f *is lower semicontinuous.*
        b) *There exists a compact set $K \subset \mathbb{R}^d$ with $f(x) \geq 0$ for $x \in \mathbb{R}^d \backslash K$.*

(ii)   *There exists a monotonically increasing sequence of functions $(f_n)_{n \in \mathbb{N}}$ $\subset C_c(\mathbb{R}^d)$ (monotonically increasing means $f_n \leq f_{n+1}$ for all $n \in \mathbb{N}$) with $f = \lim_{n \to \infty} f_n$ (in the sense of pointwise convergence; so $f(x) = \lim_{n \to \infty} f_n(x)$ for all $x \in \mathbb{R}^d$).*

Proof. (ii) $\Rightarrow$ (i): As the sequence (f_n) is monotonically increasing, we have $\lim f_n = \sup f_n$, and since all the f_n are continuous, $f = \sup f_n$ is lower semicontinuous, by lemma 12.8. Moreover, $f \geq f_1$, and therefore $f(x) \geq 0$ for $x \in \mathbb{R}^d \backslash \operatorname{supp}(f_1)$ and $K := \operatorname{supp}(f_1)$ is assumed to be compact.

(i) $\Rightarrow$ (ii): First, we observe that it suffices to find a sequence $(g_m)_{m \in \mathbb{N}}$ of continuous functions with compact support that fulfils

$$f = \sup_{m \in \mathbb{N}} g_m; \tag{4}$$

that is, we then have a monotonically increasing sequence in $C_c(\mathbb{R}^d)$ defined by $f_n := \sup(g_1, \ldots, g_n)$ with $f = \lim f_n$. We shall now construct such a family $(g_m)_{m \in \mathbb{N}}$.

As a lower semicontinuous function on a compact set is bounded from below (e.g. by lemma 12.6), there exists a rational $m \geq 0$ with

$$f(x) > -m \text{ for } x \in \mathbb{R}^d.$$

Let

$$Q := \{(q, r, s) : q, r, s \text{ rational and } s \geq -m,$$
$$f(x) \geq s \text{ for all } x \text{ with } |x - q| < r\}.$$

The set Q is countable, and for every $j = (q, r, s) \in Q$ there exists a function $g_j \in C_c(\mathbb{R}^d)$ with the following properties:

1) $g_j(x) = s$ for $x \in U(q, \frac{r}{2})$.

2) $g_j(x) \leq s$ for $x \in U(q, r)$.

3) $g_j(x) = -m$ for $x \in K \backslash U(q, r)$.

4) $g_j(x) \leq 0$ for $x \in \mathbb{R}^d \backslash (K \cup U(q, r))$.

By construction, we have $f \geq g_j$ for all $j \in Q$, and also $f = \sup_{j \in Q} g_j$, which one sees as follows:

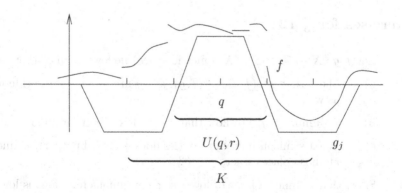

Let $x \in \mathbb{R}^d, c < f(x)$. We choose a rational s, $s \geq -m$ with $c \leq s < f(x)$. As f is lower semicontinuous, there exists a $\delta > 0$ with

$$f(y) > s \text{ for } |x - y| < \delta.$$

We choose a rational r, $0 < r \leq \frac{\delta}{2}$, as well as a $q \in \mathbb{R}^d$ with rational coordinates with $|x - q| < \frac{r}{2}$. Then $j = (q, r, s) \in Q$ and $c \leq g_j(x) < f(x)$. As we can construct such a g_j for every $c < f(x)$, we indeed have $f = \sup_{j \in Q} g_j$. As Q is countable, we have thereby constructed a family of functions which satisfies the condition (4) above.   □

Quite analogously, the upper semicontinuous functions which are nonpositive outside a compact set can be characterised as follows:

Theorem 12.11 *For $f : \mathbb{R}^d \to \mathbb{R} \cup \{-\infty\}$, the following conditions (i) and (ii) are equivalent*

(i)   a) *f is upper semicontinuous.*

   b) *$f(x) \leq 0$ outside a compact set $K \subset \mathbb{R}^d$.*

(ii)   *There exists a monotonically decreasing sequence $(f_n)_{n \in \mathbb{N}}$ of continuous functions with compact support with $f = \lim f_n$.*

Definition 12.12 Let $H_I(\mathbb{R}^d)$ be the class of functions $f : \mathbb{R}^d \to \mathbb{R} \cup \{\infty\}$ which satisfy the conditions of theorem 12.10 and $H_S(\mathbb{R}^d)$ the class of those which satisfy the conditions of theorem 12.10'.

Corollary 12.13 $H_I(\mathbb{R}^d) \cap H_S(\mathbb{R}^d) = C_c(\mathbb{R}^d)$. *A function that is the limit of a monotonically increasing, as well as of a monotonically decreasing, sequence of continuous functions with compact support is itself already continuous.*

Proof. This follows directly from theorems 12.10 and 12.11.   □

Exercises for § 12

1)   Let $f, g : X \to \mathbb{R} \cup \{\infty\}$ (X a metric space) be lower semicontinuous.

   a)   Show that $\sup(f, g)$, $\inf(f, g)$, $f + g$ are lower semicontinuous as well.

   b)   Assuming $f, g \geq 0$ show that $f \cdot g$ is lower semicontinuous, too.

   c)   What semicontinuity properties does $\frac{1}{f}$ have if properly defined at those places where $f = 0$?

2)   Show that a uniform limit of lower semicontinuous functions is lower semicontinuous itself.

3)   Let $g : \mathbb{R} \times (0, \infty) \to \mathbb{R}$ be defined by $g(x^1, x^2) = |x^1 x^2 - 1|$, $f : \mathbb{R} \to \mathbb{R}$ by $f(x^1) := \inf_{x^2 > 0} g(x^1, x^2)$. Show that f is *not* lower semicontinuous.

4)   Let $f : X \to \mathbb{R} \cup \{\infty\}$ be a function, for some metric space X. We define "regularizations" $f_{(-)}, f_{(s)}$ of f by

$$f_{(-)}(x) := \sup\{g(x) : g \leq f, g : X \to \mathbb{R} \cup \{\infty\} \quad \text{lower semicontinuous}\}$$
$$f_{(s)}(x) := \sup\{g(x) : g \leq f, g : X \to \mathbb{R} \quad \text{continuous}\}.$$

a)   Show that $f_{(-)}$ and $f_{(s)}$ are lower semicontinuous.

b)   Construct examples where $f_{(-)}$ and $f_{(s)}$ are not continuous.

c)   Does one always have $f_{(-)} = f_{(s)}$?

d)   What is the relation between $f_{(-)}, f_{(s)}$, and $\underline{f}$, defined by

$$\underline{f}(x) := \liminf_{y \to x} f(y) := \inf\{\liminf_{n \to \infty} f(y_n) : y_n \to x\}?$$

13. The Lebesgue Integral for Semicontinuous Functions. The Volume of Compact Sets.

We define the integral of semicontinuous functions, and consider properties of such integrals, like Fubini's theorem. Volumes of compact sets are defined, and certain rules, like Cavalieri's principle, for their computation are given. In particular, computations simplify in rotationally symmetric situations.

The aim of this, and of the following paragraphs, is to construct as large a class as possible of real valued functions on $\mathbb{R}^d$, for which one can define an integral in a sensible manner. The correspondence

$$f \mapsto \int_{\mathbb{R}^d} f(x)dx$$

should here be linear, monotone ($f \leq g \Rightarrow \int f \leq \int g$) and invariant under isometries of $\mathbb{R}^d$ (if A is an orthogonal matrix and $b \in \mathbb{R}^d$, then for f in our class $\int_{\mathbb{R}^d} f(Ax+b)dx = \int_{\mathbb{R}^d} f(x)dx$ should hold). Furthermore, integration should be interchangeable with general limit processes.

The construct will be made stepwise: first, we will define the integral for continuous functions with compact support, then for semicontinuous ones of the classes H_I and H_S and finally for general functions. The convergence theorems will be proved in §16.

Definition 13.1 A unit cube in $\mathbb{R}^d$ is

$$I^d := \{(x^1, \ldots, x^d) \in \mathbb{R}^d : 0 \leq x^i \leq 1 \text{ for } i = 1, \ldots, d\}$$

or, more generally, a subset $A(I^d) + b$, where $b \in \mathbb{R}^d$ and A is an orthogonal $d \times d$-matrix.

To define a cube of side length $\ell > 0$, we substitute the condition $0 \leq x^i \leq \ell$ for $i = 1, \ldots, d$.

Definition 13.2 Let $W \subset \mathbb{R}^d$ be a cube. A function $t : W \to \mathbb{R}$ is called elementary if there is a partition of W into subcubes $W_1, \ldots, W_k$ with $W = \bigcup_{i=1}^{k} W_i, \overset{\circ}{W}_i \cap \overset{\circ}{W}_j = \emptyset$ for $i \neq j$, so that t is constant on every $\overset{\circ}{W}_i, i = 1, \ldots, k$.

Let $t_{\overset{\circ}{|W_i}} = c_i$ and let W_i have side length ℓ_i. We define the integral of the elementary function t as

$$\int_W t(x)dx = \sum_{i=1}^k c_i \ell_i^d. \tag{1}$$

The right side is clearly independent of the partition of W into subcubes on which t is constant.

Now let $f \in C_c(\mathbb{R}^d)$. We choose a cube W with $\operatorname{supp} f \subset W$, as well as a sequence $(t_n)_{n\in\mathbb{N}}$ of elementary functions on W that converge uniformly to f. (The existence of such a sequence should be clear: Let ℓ be the side length of W, and for every $n \in \mathbb{N}$ we partition W in to n^d cubes $W_1, \ldots, W_{n^d}$ of side length $\frac{\ell}{n}$. For every $i = 1, \ldots, n^d$ we choose a point $x_i \in W_i$ and set $t_n(x) = f(x_i)$ for $x \in W_i$. As f is uniformly continuous on the compact cube W, the functions t_n then converge uniformly to f).

Definition 13.3 We define the integral of f to be

$$\int_{\mathbb{R}^d} f(x)dx := \lim_{n\to\infty} \int_W t_n(x)dx.$$

The existence of this limit is easy to see: For $\varepsilon > 0$ we find an $N \in \mathbb{N}$ with

$$\sup_{x\in W} |f(x) - t_n(x)| < \varepsilon \text{ for } n \geq N,$$

so also

$$\sup_{x\in W} |t_n(x) - t_m(x)| < 2\varepsilon \text{ for } n, m \geq N$$

and therefore

$$|\int_W t_n(x)dx - \int_W t_m(x)dx| < 2\varepsilon \cdot \ell^d \ (\ell = \text{side length of } W)$$

for $n, m \geq N$, so that the sequence of integrals $\int_W t_n(x)dx$ is a Cauchy sequence in $\mathbb{R}$. An analogous argument shows that $\int_{\mathbb{R}^d} f(x)dx$ is independent of the choice of the sequence $(t_n)_{n\in\mathbb{N}}$ which converges uniformly to f, as well as of the choice of the cube W which contains $\operatorname{supp} f$.

Furthermore, for $d = 1$, we recover the integral defined earlier for continuous functions. However, in contrast to the former, the main point here is to construct a definite instead of an indefinite integral.

Lemma 13.4 Let $f, g \in C_c(\mathbb{R}^d), \alpha \in \mathbb{R}$. We have

(i)
$$\int_{\mathbb{R}^d} (f(x) + g(x))dx = \int_{\mathbb{R}^d} f(x)dx + \int_{\mathbb{R}^d} g(x)dx,$$

$$\int_{\mathbb{R}^d} \alpha f(x)dx = \alpha \int_{\mathbb{R}^d} f(x)dx \ \ (linearity).$$

(ii)   If $f \le g$, then also $\int_{\mathbb{R}^d} f(x)dx \le \int_{\mathbb{R}^d} g(x)dx$.

(iii)   If $b \in \mathbb{R}^d$, A an orthogonal $d \times d-$matrix, then

$$\int_{\mathbb{R}^d} f(Ax + b)dx = \int_{\mathbb{R}^d} f(x)dx.$$

Proof. Parts (i) follows directly from the corresponding rules for the integral of elementary functions. The rule (iii) holds for elementary functions, because for an orthogonal matrix A and an elementary function t, the function $t \circ A$ is again elementary as A maps a cube onto a cube, and similarly for translation by a vector b. (ii) follows, as in case $f \le g$, we can approximate f and g by sequences (t_n) and (s_n), respectively, of elementary functions with $t_n \le s_n$. $\qquad \square$

Furthermore, we have

Lemma 13.5 *Let* $f \in C_c(\mathbb{R}^d)$ *with* $\mathrm{supp}(f)$ *in a cube of side length* ℓ. *Then*

$$|\int_{\mathbb{R}^d} f(x)dx| \le \sup |f(x)| \cdot \ell^d. \tag{2}$$

The proof again follows directly from the corresponding property for elementary functions.

Lemmas 13.4 and 13.5 mean that the correspondence

$$f \mapsto \int_{\mathbb{R}^d} f(x)dx$$

is a linear, bounded (therefore continuous) real valued functional on each Banach space $C_c(W)$ (W a cube in $\mathbb{R}^d$) that is invariant under isometries. One can show that a functional with these properties and the normalisation

$$\int_{I^d} 1dx = 1$$

is already uniquely determined.

We shall now integrate semicontinuous functions of the classes H_I and H_S.

Definition 13.6 For $f \in H_I$ we define

$$\int_{\mathbb{R}^d} f(x)dx := \sup\{\int_{\mathbb{R}^d} g(x)dx : g \in C_c(\mathbb{R}^d), g \le f\} \in \mathbb{R} \cup \{\infty\},$$

and similarly for $f \in H_S$

$$\int_{\mathbb{R}^d} f(x)dx := -\int_{\mathbb{R}^d} -f(x)dx,$$

as for $f \in H_S$, we have $-f \in H_I$.

Lemma 13.7 Let $(f_n)_{n \in \mathbb{N}} \subset H_I$ be a monotonically increasing sequence. Then $f := \sup_{n \in \mathbb{N}} f_n \in H_I$ and

$$\int_{\mathbb{R}^d} f(x)dx = \sup_{n \in \mathbb{N}} \int_{\mathbb{R}^d} f_n(x)dx = \lim_{n \to \infty} \int_{\mathbb{R}^d} f_n(x)dx. \tag{3}$$

Proof. We first consider the case where f and all f_n are in $C_c(\mathbb{R}^d)$. Then for all n we have

$$\text{supp} f_n \subset \text{supp} f_1 \cup \text{supp} f =: K,$$

on account of the assumed monotonicity. By the theorem of Dini (theorem 12.1) it follows that (f_n) even converges uniformly to f, and (2) shows that

$$|\int_{\mathbb{R}^d} (f_n(x) - f(x))dx| \le \text{const.} \cdot \sup |f_n(x) - f(x)|,$$

and this tends to zero because of uniform convergence.

Now we come to the general case. From the definition it follows directly that for all n

$$\int_{\mathbb{R}^d} f_n(x)dx \le \int_{\mathbb{R}^d} f(x)dx,$$

and it remains to show that, conversely, for every $g \in C_c(\mathbb{R}^d)$ with $g \le f$

$$\int_{\mathbb{R}^d} g(x)dx \le \sup_{n \in \mathbb{N}} \int_{\mathbb{R}^d} f_n(x)dx \tag{4}$$

holds.

By theorem 12.10, for any $n \in \mathbb{N}$ there exists a monotonically increasing sequence $(\varphi_{nm})_{m \in \mathbb{N}} \subset C_c(\mathbb{R}^d)$ with

$$f_n = \sup_{m \in \mathbb{N}} \varphi_{nm}.$$

We set

$$h_n := \sup_{p \leq n, q \leq n} \varphi_{pq}$$

$$g_n := \inf(g, h_n).$$

The sequences (h_n) and (g_n) are monotonically increasing sequences in $C_c(\mathbb{R}^d)$, and

$$f = \sup_{n \in \mathbb{N}} h_n$$

$$g = \sup_{n \in \mathbb{N}} g_n \quad (\text{by } g \leq f).$$

Since g lies also in $C_c(\mathbb{R}^d)$, it follows from the case already dealt with that

$$\int_{\mathbb{R}^d} g(x)dx = \sup_{n \in \mathbb{N}} \int_{\mathbb{R}^d} g_n(x)dx$$

$$\leq \sup_{n \in \mathbb{N}} \int_{\mathbb{R}^d} h_n(x)dx \quad \text{as } g_n \leq h_n$$

$$\leq \sup_{n \in \mathbb{N}} \int_{\mathbb{R}^d} f_n(x)dx \quad \text{as } h_n \leq f_n,$$

hence (4). Here we have repeatedly used the monotonicity of the integral (lemma 13.4). □

Lemma 13.8 *The assertions of lemma 13.4 hold for $f, g \in H_I(\mathbb{R}^d), \alpha \geq 0$.*

The proof again follows directly from the definitions and lemma 13.4. Finally, it follows from lemmas 13.8 and 13.7 that

Lemma 13.9 *Let $(f_n)_{n \in \mathbb{N}} \subset H_I$ be a sequence of non-negative functions. We have*

$$\int_{\mathbb{R}^d} \sum_{n=1}^{\infty} f_n(x)dx = \sum_{n=1}^{\infty} \int_{\mathbb{R}^d} f_n(x)dx.$$

We shall now reduce an integral in $\mathbb{R}^d$ to a d-fold iterated integral in $\mathbb{R}$:

Theorem 13.10 (Fubini) *Let $f \in H_I(\mathbb{R}^d), i_1 \ldots i_d$ a permutation of $(1, \ldots, d), 1 \leq c < d$. For $(x^{i_{c+1}}, \ldots, x^{i_d}) \in \mathbb{R}^{d-c}$, let*

$$F(x^{i_{c+1}}, \ldots, x^{i_d}) := \int_{\mathbb{R}^c} f(\xi, x^{i_{c+1}}, \ldots, x^{i_d}) d\xi \ \ with \ \xi = (x^{i_1}, \ldots, x^{i_c}).$$

Then

$$\int_{\mathbb{R}^d} f(x) dx = \int_{\mathbb{R}^{d-c}} F(\eta) d\eta \ \ with \ \eta = (x^{i_{c+1}}, \ldots, x^{i_d}).$$

(In particular, all the functions appearing are in H_I.)

One best remembers the assertion of theorem 13.10 in the following form

$$\int_{\mathbb{R}^d} f(x) dx = \int_{\mathbb{R}^{d-c}} (\int_{\mathbb{R}^c} f(\xi, \eta) d\xi) d\eta.$$

Iteratively, we can also write

$$\int_{\mathbb{R}^d} f(x) dx = \int_{\mathbb{R}} \cdots \int_{\mathbb{R}} f(x^1, \ldots, x^d) dx^1 \ldots dx^d,$$

where we may omit the brackets, as theorem 13.10 also asserts that the order of the separate integrations is immaterial.

Proof of theorem 13.10. We first consider the case $f \in C_c(\mathbb{R}^d)$. We choose an axis-parallel cube W with $\mathrm{supp} f \subset W$ and we approximate f uniformly by elementary functions t_n on W. Every t_n is a sum of terms $t_{n,i}$ with

$$t_{n,i} = \begin{cases} c_i & \text{on a subcube } W_i \\ 0 & \text{otherwise} \end{cases}.$$

Let the cube W_i have side-length ℓ_i; obviously

$$\int_{\mathbb{R}^d} t_{n,i}(x) dx = c_i \ell_i^d = c_i \ell_i^{d-c} \ell_i^c$$

$$= \int_{\mathbb{R}^{d-c}} (\int_{\mathbb{R}^c} t_{n,i}(\xi, \eta) d\xi) d\eta.$$

The assertion thus holds for $t_{n,i}$ and therefore also for $t_n = \Sigma t_{n,i}$.

If now $\sup_x |t_n(x) - f(x)| < \varepsilon$, then

$$\sup_\eta |\int_{\mathbb{R}^c} (t_n(\xi, \eta) - f(\xi, \eta)) d\xi| < \ell^c \cdot \varepsilon \ (\ell = \text{ side-length of } W),$$

and from this the assertion for f follows, if one takes into account that with $f(\xi, \eta)$ also $F(\eta) = \int_{\mathbb{R}^c} f(\xi, \eta) d\xi$ is continuous; namely we have

$$|F(\eta_1) - F(\eta_2)| \le \ell^c \sup_\xi |f(\xi, \eta_1) - f(\xi, \eta_2)|,$$

and the right hand side becomes arbitrarily small provided $|\eta_1 - \eta_2|$ is sufficiently small, because f is even uniformly continuous on its compact support.

Now let $f \in H_I(\mathbb{R}^d)$, and $(f_n)_{n\in\mathbb{N}} \subset C_c(\mathbb{R}^d)$ be a monotonically increasing sequence convergent to f. With $F_n(\eta) := \int_{\mathbb{R}^c} f_n(\xi, \eta)d\xi$, we have

$$F(\eta) = \int_{\mathbb{R}^c} f(\xi, \eta)d\xi = \lim_{n\to\infty} \int_{\mathbb{R}^c} f_n(\xi, \eta)d\xi = \lim_{n\to\infty} F_n(\eta).$$

By the previous considerations, $F_n \in C_c(\mathbb{R}^{d-c})$ and by theorem 12.10, we then have $F \in H_I(\mathbb{R}^{d-c})$. Further

$$\int_{\mathbb{R}^{d-c}} F(\eta)d\eta = \lim_{n\to\infty} \int_{\mathbb{R}^{d-c}} F_n(\eta)d\eta = \lim_{n\to\infty} \int_{\mathbb{R}^{d-c}} \int_{\mathbb{R}^c} f_n(\xi, \eta)d\xi d\eta$$

$$= \lim_{n\to\infty} \int_{\mathbb{R}^d} f_n(x)dx \text{ by what has already been proved}$$

$$= \int_{\mathbb{R}^d} f(x)dx,$$

where we have used lemma 13.7 twice. This gives the assertion.   □

Corollary 13.11 *In the notations of theorem 13.10, let $f(\xi, \eta) = \varphi(\xi) \cdot \psi(\eta)$ with $\varphi, \psi \in H_I$. Then*

$$\int_{\mathbb{R}^d} f(x)dx = \int_{\mathbb{R}^c} \varphi(\xi)d\xi \cdot \int_{\mathbb{R}^{d-c}} \psi(\eta)d\eta .$$

Lemma 13.12 *Let $f \in H_I(\mathbb{R}^d)$, $A \in GL(d, \mathbb{R})$ (so A is a $d \times d-$matrix with $\det A \neq 0$), $b \in \mathbb{R}^d$. Then the function $x \mapsto f(Ax + b)$ is also in $H_I(\mathbb{R}^d)$ and we have*

$$\int_{\mathbb{R}^d} f(Ax + b)dx = \frac{1}{|\det A|} \int_{\mathbb{R}^d} f(x)dx.$$

Proof. The first assertion is clear, and by lemma 13.8 we have

$$\int_{\mathbb{R}^d} f(Ax + b)dx = \int_{\mathbb{R}^d} f(Ax)dx.$$

Furthermore, well known results from linear algebra give the decomposition

$$A = S_1 D S_2 \tag{5}$$

where S_1 and S_2 are orthogonal matrices and D a diagonal matrix, say

$$D = \begin{pmatrix} \lambda_1 & & 0 \\ & \ddots & \\ 0 & & \lambda_d \end{pmatrix}.$$

Again, by lemma 13.7,

$$\int_{\mathbb{R}^d} f(S_1 D S_2 x) dx = \int_{\mathbb{R}^d} f(S_1 D x) dx$$

$$= \int_{\mathbb{R}} \cdots \int_{\mathbb{R}} f(S_1 D(x^1, \ldots, x^d)) dx^1 \ldots dx^d \text{ by theorem 13.10}$$

$$= \frac{1}{|\lambda_1 \ldots \lambda_d|} \int_{\mathbb{R}} \cdots \int_{\mathbb{R}} f(S_1(y^1, \ldots, y^d)) dy^1 \ldots dy^d,$$

(as for a one dimensional integral $\int_{\mathbb{R}} \varphi(\lambda x) dx = \frac{1}{|\lambda|} \int_{\mathbb{R}} \varphi(y) dy$ holds)

$$= \frac{1}{|\lambda_1 \ldots \lambda_d|} \int_{\mathbb{R}^d} f(y) dy, \text{ again by lemma 13.8,}$$

As orthogonal matrices have determinant ± 1, it follows from (5) that

$$|\det A| = |\det D| = |\lambda_1 \ldots \lambda_d|,$$

and therefore, together with the previous formulae, the assertion.   □

We want to give another proof of lemma 13.12 which does not use the decomposition (5), but rather just elementary linear algebra.

Let I be the $(d \times d)$-unit matrix. For $i \neq j, i, j \in \{1, \ldots, d\}$, and $t \in \mathbb{R}$ let $S_{ij}(t)$ be the matrix which is obtained when one adds t times the jth row of I to its ith row, and R_{ij} the matrix which results from interchanging the ith and jth rows of I. We have

$$\det S_{ij}(t) = 1 = |\det R_{ij}| \text{ for all } i \neq j, t \in \mathbb{R}.$$

Moreover,

$$\int_{\mathbb{R}^d} f(R_{ij} x) dx = \int_{\mathbb{R}^d} f(x) dx,$$

as R_{ij} is orthogonal, so it maps a cube to one with the same side-length and also

$$\int_{\mathbb{R}^d} f(S_{ij}(t) x) dx = \int_{\mathbb{R}^d} f(x^1, x^2, \ldots x^{i-1}, x^i + t x^j, x^{i+1}, \ldots, x^d) dx^1 \ldots dx^d \quad (6)$$

by the theorem 13.10 of Fubini. Further, by properties of one dimensional integrals

$$\int_{\mathbb{R}} \varphi(x^i + tx^j)dx^i = \int_{\mathbb{R}} \varphi(x^i)dx^i.$$

We therefore first perform integration over x^i in the iterated integral (6) and then the remaining integrations, which by the theorem of Fubini does not change the result, and obtain

$$\int_{\mathbb{R}^d} f(S_{ij}(t)x)dx = \int_{\mathbb{R}^d} f(x)dx.$$

We consider now the matrix A. The matrix $R_{ij}A$ arises from A by interchanging the ith and jth rows, and AR_{ij} by interchanging the ith and jth columns. So by multiplying A with the matrices R_{ij} we can bring a nonzero element to the $(1,1)$-entry. By adding suitable multiples of the new first column to the other columns, we obtain zeroes to the $(1,2), \ldots, (1,d)$-entries. This process is obtained by multiplying A with matrices $S_{ij}(t)$. Similarly, one can also bring zeroes at the entries $(2,1), (2,2), \ldots, (2,d)$. One iterates this process d-times and obtains a diagonal matrix. So we have

$$A = P_1 D P_2,$$

where P_1 and P_2 are products of matrices of the form $R_{ij}, S_{ij}(t)$ and

$$D = \begin{pmatrix} \lambda_1 & & 0 \\ & \ddots & \\ 0 & & \lambda_d \end{pmatrix}.$$

So

$$\det A = \det P_1 \det D \det P_2,$$
$$|\det P_1| = 1 = |\det P_2|, \text{ therefore}$$
$$|\lambda_1 \ldots \lambda_d| = |\det D| = |\det A| \neq 0 \text{ by assumption}.$$

Finally,

$$\int_{\mathbb{R}^d} f(P_1 D P_2 x)dx = \int_{\mathbb{R}^d} f(P_1 D x)dx$$

(by what is already proved)

$$= \int_{\mathbb{R}^d} f(P_1(\lambda_1 x^1, \ldots, \lambda_d x^d))dx^1 \ldots dx^d$$

$$= \frac{1}{|\lambda_1 \ldots \lambda_d|} \int_{\mathbb{R}^d} f(P_1 x)dx,$$

by the rules for one dimensional integrals (the theorem of Fubini has been used again)

$$= \frac{1}{|\lambda_1 \ldots \lambda_d|} \int\limits_{\mathbb{R}^d} f(x)dx,$$

again by what has already been proved.

The assertion follows on account of $|\lambda_1 \ldots \lambda_d| = |\det A|$.   □

Remark. It is very important that the preceding assertions, for example theorem 13.10 or lemma 13.12 carry over in an obvious manner to functions of the class $H_S(\mathbb{R}^d)$. We shall therefore use all these assertions in what follows without comment also for functions in $H_S(\mathbb{R}^d)$.

Notation. Let us introduce the following notation. Let A be – for the time being – a compact subset of $\mathbb{R}^d$, $f \in H_S(\mathbb{R}^d)$. Then

$$\int\limits_{A} f(x)dx := \int\limits_{\mathbb{R}^d} f(x)\chi_A(x)dx .$$

Later we shall tacitly use this notation also for more general sets and functions; for the time being, strictly speaking, we must also assume that $f \cdot \chi_A$ is likewise in H_S; for example this always holds for f positive (compare lemma 12.5).

If $d = 1$, we also write

$$\int\limits_{a}^{b} f(x)dx \quad \text{in place of} \quad \int\limits_{[a,b]} f(x)dx.$$

A quite analogous definition is possible in case A is open and bounded, $f \in H_I(\mathbb{R}^d)$ and e.g. f is again non-negative.

Now let $K \subset \mathbb{R}^d$ be a compact set. Then the characteristic function χ_K is in $H_S(\mathbb{R}^d)$ (cf. §12).

Definition 13.13 The volume of K is defined as

$$\text{Vol}(K) := \text{Vol}_d(K) := \int\limits_{\mathbb{R}^d} \chi_K(x)dx.$$

For the case $d = 1$ one says length instead of volume, and for $d = 2$, one says area.

Lemma 13.14 *Let* $1 \le c < d$, $K_1 \subset \mathbb{R}^c, K_2 \subset \mathbb{R}^{d-c}$ *compact. Then*

$$\text{Vol}_d(K_1 \times K_2) = \text{Vol}_c(K_1) \cdot \text{Vol}_{d-c}(K_2).$$

Proof. We write $x = (x^1, \ldots, x^d), \xi = (x^1, \ldots, x^c), \eta = (x^{c+1}, \ldots, x^d)$. Then

$$\chi_{K_1 \times K_2}(\xi, \eta) = \chi_{K_1}(\xi) \cdot \chi_{K_2}(\eta),$$

and the result follows from corollary 13.11. □

From this corollary, many elementary geometric formulae can be recovered. For example, the volume of a rectangular parallelepiped arises as the product of the lengths of its sides, or the volume of the cylinder $B \times [0, h] \subset \mathbb{R}^d, B \subset \mathbb{R}^{d-1}$ compact, as the product of $\mathrm{Vol}_{d-1}(B)$, that is, of its base, and of its height h.

As in the one dimensional case, one can also compute the volume of sets that are bounded from above by the graph of a continuous function.

Lemma 13.15 *Let $K \subset \mathbb{R}^d$ be compact, $f : K \to \mathbb{R}$ continuous and non-negative, $K_f := \{(x,t) \in K \times \mathbb{R} : 0 \le t \le f(x)\}$. Then*

$$\mathrm{Vol}_{d+1}(K_f) = \int_K f(x)dx$$

Proof. By means of the theorem of Fubini we obtain

$$\mathrm{Vol}_{d+1}(K_f) = \int_{\mathbb{R}^{d+1}} \chi_{K_f}(x,t)dtdx = \int_{\mathbb{R}^d} (\int_0^{f(x)} 1dt)\chi_K(x)dx = \int_{\mathbb{R}^d} f(x)\chi_K(x)dx$$

$$= \int_K f(x)dx.$$

□

A useful method for calculating volumes is the principle of Cavalieri (Cavalieri, a contemporary of Galileo, came upon corollary 13.17 infra by heuristic considerations).

Theorem 13.16 *Let $K \subset \mathbb{R}^d$ be compact. For $t \in \mathbb{R}$, let*

$$K_t := \{(x^1, \dots, x^{d-1}) \in \mathbb{R}^{d-1} : (x^1, \dots, x^{d-1}, t) \in K\}$$

be the intersection of K and the hyperplane $\{x_d = t\}$. Then

$$\mathrm{Vol}_d(K) = \int_{\mathbb{R}} \mathrm{Vol}_{d-1}(K_t)dt.$$

Corollary 13.17 *Let $K, K' \subset \mathbb{R}^d$ be compact and for all t, let $\mathrm{Vol}_{d-1}(K_t) = \mathrm{Vol}_{d-1}(K_t')$. Then also*

$$\mathrm{Vol}_d(K) = \mathrm{Vol}_d(K').$$

□

Proof of theorem 13.16. By means of the theorem of Fubini we have

$$
\mathrm{Vol}_d(K) = \int_{\mathbb{R}^d} \chi_K(x^1, \ldots, x^{d-1}, t) dx^1 \ldots dx^{d-1} dt
$$

$$
= \int_{\mathbb{R}^d} \chi_{K_t}(x^1, \ldots, x^{d-1}) dx^1 \ldots dx^{d-1} dt
$$

$$
(\text{as } \chi_K(x^1, \ldots, x^{d-1}, t) = \chi_{K_t}(x^1, \ldots, x^{d-1}))
$$

$$
= \int_{\mathbb{R}} \mathrm{Vol}_{d-1}(K_t) dt .
$$

□

Example. As an example, we shall now calculate the volume of a sphere.
Let

$$
B^d(0, r) := \{x \in \mathbb{R}^d : \|x\| \leq r\}
$$

the closed ball in $\mathbb{R}^d$ of radius r,

$$
\omega_d := \mathrm{Vol}(B^d(0, 1)).
$$

Obviously $\omega_1 = 2$. We shall now apply the Cavalieri principle: We have
$B^d(0, 1)_t = B^{d-1}(0, \sqrt{1 - t^2})$ for $|t| \leq 1$, and therefore

$$
\omega_d = \int_{-1}^{1} \mathrm{Vol}(B^{d-1}(0, \sqrt{1 - t^2})) dt \tag{7}
$$

We now need the following simple consequence of lemma 13.12

Lemma 13.18 *Let $K \subset \mathbb{R}^d$ be compact, $\lambda > 0$, $\lambda K := \{x \in \mathbb{R}^d : \frac{1}{\lambda} x \in K\}$.*
Then

$$
\mathrm{Vol}_d(\lambda K) = \lambda^d \, \mathrm{Vol}_d(K).
$$

Proof. By lemma 13.12

$$
\int_{\mathbb{R}^d} \chi_{\lambda K}(x) dx = \lambda^d \int_{\mathbb{R}^d} \chi_{\lambda K}(\lambda y) dy = \lambda^d \int_{\mathbb{R}^d} \chi_K(y) dy.
$$

□

(7) therefore yields

$$
\omega_d = \omega_{d-1} \int_{-1}^{1} (1 - t^2)^{\frac{d-1}{2}} dt .
$$

Furthermore, by the substitution $t = \cos x$,

$$s_d := \int_{-1}^{1} (1 - t^2)^{\frac{d-1}{2}} dt = \int_{0}^{\pi} \sin^d(x) dx.$$

Using integration by parts and induction over d, it easily follows that

$$s_{2k} = \pi \prod_{i=1}^{k} \frac{2i-1}{2i}, s_{2k+1} := 2 \prod_{i=1}^{k} \frac{2i}{2i+1} \quad (k \in \mathbb{N}).$$

In particular, $s_d \cdot s_{d-1} = \frac{2\pi}{d}$, so

$$\omega_d = \omega_{d-1} s_d = \omega_{d-2} s_{d-1} s_d = \frac{2\pi}{d} \omega_{d-2},$$

and therefore

$$\omega_{2k} = \frac{1}{k!} \pi^k, \omega_{2k+1} = \frac{2^{k+1}}{1 \cdot 3 \cdot \ldots \cdot (2k+1)} \pi^k.$$

In particular, we note that

$$\lim_{d \to \infty} \omega_d = 0.$$

Thus the volume of the unit ball in $\mathbb{R}^d$ tends towards zero with increasing d. From lemma 13.18 it also follows that

$$\text{Vol}_d B(0, r) = \omega_d r^d.$$

A further consequence of the principle of Cavalieri is

Corollary 13.19 (Volume of solids of revolution) *Let* $[a, b] \subset \mathbb{R}$, $f : [a, b] \to \mathbb{R}$ *continuous and positive,*

$$K := \{(x, y, t) \in \mathbb{R}^2 \times [a, b] : x^2 + y^2 \le f(t)^2\}.$$

Then

$$\text{Vol}(K) = \pi \int_{a}^{b} f(t)^2 dt.$$

Proof. For $t \in [a, b]$ we have

$$K_t = \{(x, y) \in \mathbb{R}^2 : x^2 + y^2 \le f(t)^2\}.$$

so

$$\text{Vol}(K_t) = \pi f(t)^2,$$

e.g. by the previous example, and the result follows from theorem 13.16. $\square$

Example. As a further example, we shall calculate the volume of a torus of revolution. Let $0 < \rho < R$, $B := \{(t, x, y) \in \mathbb{R}^3 : y = 0, (x - R)^2 + t^2 \le \rho^2\}$, T the solid generated by rotating B around the t-axis, so $T = \{(x, y, t) : (R - \sqrt{\rho^2 - t^2})^2 \le x^2 + y^2 \le (R + \sqrt{\rho^2 - t^2})^2\}$. As in the proof of corollary 13.19,

$$\mathrm{Vol}\,(T) = \pi \int_{-\rho}^{\rho} ((R + \sqrt{\rho^2 - t^2})^2 - (R - \sqrt{\rho^2 - t^2})^2)dt = 4\pi R \int_{-\rho}^{\rho} \sqrt{\rho^2 - t^2}\,dt$$

$$= 4\pi\rho^2 R \int_{0}^{\pi} \sin^2 x\,dx = 2\pi^2\rho^2 R.$$

We shall now integrate rotationally symmetric functions. For this we need the following simple lemmas.

Lemma 13.20 *Let $K \subset \mathbb{R}^d$ be compact, $f \in H_I(\mathbb{R}^d)$ bounded. Assume that $K = K_1 \cup K_2$ with compact sets K_1 and K_2 and $\mathrm{Vol}(K_1 \cap K_2) = 0$.*
Then

$$\int_K f(x)dx = \int_{K_1} f(x)dx + \int_{K_2} f(x)dx.$$

Proof. From the monotonicity of the integral (lemma 13.4) it follows easily that for an arbitrary compact A

$$|\int_A f(x)dx| \le \sup_{x \in A} |f(x)| \cdot \mathrm{Vol}\,A, \tag{8}$$

so

$$\int_{K_1 \cap K_2} f(x)dx = 0. \tag{9}$$

Now

$$\chi_{K_1} + \chi_{K_2} = \chi_K + \chi_{K_1 \cap K_2},$$

so

$$\int_{K_1} f(x)dx + \int_{K_2} f(x)dx = \int_K f(x)dx + \int_{K_1 \cap K_2} f(x)dx = \int_K f(x)dx\,,$$

because of (9). $\square$

More general results will be proved later. Now we can handle rotationally symmetric functions.

Theorem 13.21 Let $0 \leq R_1 < R_2$, $f : [R_1, R_2] \to \mathbb{R}$ be continuous. We then have for the d-dimensional integral

$$\int_{R_1 \leq \|x\| \leq R_2} f(\|x\|)dx = d\omega_d \int_{R_1}^{R_2} f(r)r^{d-1}dr$$

(ω_d = volume of the d-dimensional unit ball: see above).

Proof. We set, for $0 \leq \rho \leq r$, $A(\rho, r) := \{x \in \mathbb{R}^d : \rho \leq \|x\| \leq r\}$. Let $n \in \mathbb{N}, n \geq 2$,

$$r_{k,n} := R_1 + \frac{k}{n}(R_2 - R_1) \text{ for } k = 0, 1, \ldots, n$$

$$A_{k,n} := A(r_{k-1,n}, r_{k,n}) \text{ for } k = 1, \ldots, n.$$

We have

$$\text{Vol}(A(\rho, r)) = \omega_d(r^d - \rho^d),$$

the argument being the same as for calculating the volume of a ball. In particular,

$$\text{Vol}(A(r, r)) = 0 \quad \text{for all } r. \tag{10}$$

Let

$$\rho_{k,n}^{d-1} := \frac{r_{k,n}^d - r_{k-1,n}^d}{d(r_{k,n} - r_{k-1,n})}$$

$$= \frac{1}{d}\left(r_{k,n}^{d-1} + r_{k,n}^{d-2}r_{k-1,n} + \ldots + r_{k,n}r_{k-1,n}^{d-2} + r_{k-1,n}^{d-1}\right),$$

so

$$r_{k-1,n} < \rho_{k,n} < r_{k,n}. \tag{11}$$

We have, by choice of $\rho_{k,n}$,

$$\text{Vol}(A_{k,n}) = \omega_d(r_{k,n}^d - r_{k-1,n}^d)$$

$$= d\omega_d \rho_{k,n}^{d-1}(r_{k,n} - r_{k-1,n}). \tag{12}$$

Now let $t_n(\|x\|) := f(\rho_{k,n})$

$$\tau_n(\|x\|) := f(\rho_{k,n})\rho_{k,n}^{d-1} \quad \text{for } r_{k-1,n} < \|x\| < r_{k,n}$$

and continued semicontinuously for $\|x\| = r_{k,n}$. Using lemma 13.20 (using (10)) we obtain

$$\int\limits_{A(R_1,R_2)} t_n(\|x\|)dx = \sum_{k=1}^{n} f(\rho_{k,n}) \int \chi_{A_{k,n}}(x)dx \qquad (13)$$

$$= d\omega_d \sum_{k=1}^{n} f(\rho_{k,n}) \rho_{k,n}^{d-1}(r_{k,n} - r_{k-1,n}) \text{ by (12)}$$

$$= d\omega_d \int\limits_{R_1}^{R_2} \tau_n(r)dr.$$

Now, on the one hand, $t_n(\|x\|)$ converges uniformly to $f(\|x\|)$ on $A(R_1, R_2)$ as $n \to \infty$, since f is uniformly continuous on this set, and on the other hand, $\tau_n(r)$ converges uniformly to $f(r)r^{d-1}$, because of (11). Therefore letting $n \to \infty$, it follows from (13) that

$$\int\limits_{A(R_1,R_2)} f(\|x\|)dx = d\omega_d \int\limits_{R_1}^{R_2} f(r)r^{d-1}dr.$$

$\square$

Examples.
1)   For $s \in \mathbb{R}, 0 < R_1 < R_2$,

$$\int\limits_{R_1 \leq \|x\| \leq R_2} \|x\|^s dx = d\omega_d \int\limits_{R_1}^{R_2} r^{d+s-1}dr$$

$$= \begin{cases} \frac{d\omega_d}{d+s}(R_2^{d+s} - R_1^{d+s}) & \text{for } s \neq -d \\ d\omega_d(\log R_2 - \log R_1) & \text{for } s = -d \end{cases}$$

So, in particular, (using e.g. lemma 13.7)

$$\int\limits_{R_1 \leq \|x\|} \|x\|^s dx = \lim_{R_2 \to \infty} \int\limits_{R_1 \leq \|x\| \leq R_2} \|x\|^s dx < \infty \iff s < -d$$

$$\int\limits_{\|x\| \leq R_2} \|x\|^s dx = \lim_{R_1 \to 0} \int\limits_{R_1 \leq \|x\| \leq R_2} \|x\|^s dx < \infty \iff s > -d$$

2)   We want to calculate the one dimensional integral $\int\limits_{0}^{\infty} e^{-x^2} dx$.

This integral exists because e^{-x^2} decreases faster than any power of x as $|x| \to \infty$.

First of all,

$$\int\limits_{-\infty}^{\infty} e^{-x^2}\,dx = 2\int\limits_{0}^{\infty} e^{-x^2}\,dx.$$

Furthermore

$$(\int\limits_{-\infty}^{\infty} e^{-x^2}\,dx)^2 = \int\limits_{-\infty}^{\infty} e^{-x^2}\,dx \cdot \int\limits_{-\infty}^{\infty} e^{-y^2}\,dy$$

$$= \int\limits_{-\infty}^{\infty}\int\limits_{-\infty}^{\infty} e^{-(x^2+y^2)}\,dx\,dy \quad \text{by corollary 13.11}$$

$$= 2\pi \int\limits_{0}^{\infty} e^{-r^2} r\,dr \quad \text{by theorem 13.21, by taking limits}$$

$$= -\pi \int\limits_{0}^{\infty} \frac{d}{dr}(e^{-r^2})\,dr = \pi.$$

Therefore we obtain

$$\int\limits_{0}^{\infty} e^{-x^2}\,dx = \frac{1}{2}\sqrt{\pi}.$$

The passage to limits used in the above examples can be justified as follows:

As the functions f considered are all positive, the sequence

$$(f \cdot \chi_{\overset{\circ}{A}(\frac{1}{n},n)})_{n\in\mathbb{N}} \subset H_I(\mathbb{R}^d)$$

converges in a monotonically increasing manner towards f and lemma 13.7 therefore justifies the passage to the limit

$$\int\limits_{\mathbb{R}^d} f(x)\,dx = \lim_{n\to\infty} \int\limits_{\overset{\circ}{A}(\frac{1}{n},n)} f(x)\,dx.$$

(We integrate here over open sets, because their characteristic functions are lower semicontinuous; as $Vol_d(A(r,r)) = 0$ (cf. (10)), this plays no role for the value of the integral).

Exercises for § 13

1) Come up with five connected solids of $\mathbb{R}^d$ of nonzero volume, different from a rectangular parallelepiped, a pyramid, a parallelotope, a cylinder, a cone, a ball, a torus, or an ellipsoid, and compute the volume of these five solids.

2)

a) $U^d(0,1) = \{x \in \mathbb{R}^d : \|x\| < 1\}$. For which $\lambda \in \mathbb{R}$, does the following integral exist, and if it does, what is its value?

$$\int_{U^d(0,1)} \frac{dx}{(1 - \|x\|^2)^\lambda}.$$

b) For $0 < r < R, A(r,R) \subset \mathbb{R}^d$, compute

$$\int_{A(r,R)} \log \|x\| dx.$$

c) Compute

$$\int_{A(r,R)} e^{-\|x\|^2} dx.$$

14. Lebesgue Integrable Functions and Sets

The general Lebesgue integral is defined, and basic properties are derived. Here, a function is called Lebesgue integrable if approximation from above by lower semicontinuous functions leads to the same result as approximation from below by upper semicontinuous functions. Sets are called integrable when their characteristic functions are.

Definition 14.1 For $f : \mathbb{R}^d \to \mathbb{R} \cup \{\pm\infty\}$ we set

$$\int_{\mathbb{R}^d}^* f(x)dx := \inf\{\int_{\mathbb{R}^d} g(x)dx, g \in H_I(\mathbb{R}^d), g \geq f\}$$

$$\in \mathbb{R} \cup \{\pm\infty\} \quad \text{(upper integral)}$$

$$\int_{*\,\mathbb{R}^d} f(x)dx := \sup\{\int_{\mathbb{R}^d} g(x)dx : g \in H_S(\mathbb{R}^d), g \leq f\}$$

$$\in \mathbb{R} \cup \{\pm\infty\} \quad \text{(lower integral)}$$

$$(= -\int^* (-f(x))dx).$$

Remark. For every $f : \mathbb{R}^d \to \mathbb{R} \cup \{\pm\infty\}$ there exist $g_1 \in H_I, g_2 \in H_S$ with $g_2 \leq f \leq g_1$, namely $g_1 \equiv \infty$, $g_2 \equiv -\infty$. Therefore the sets on which the infimum and supremum, respectively, are taken are non-empty.

Lemma 14.2

(i)   For every $f : \mathbb{R}^d \to \mathbb{R} \cup \{\pm\infty\}$ we have

$$\int_{*\,\mathbb{R}^d} f(x)dx \leq \int_{\mathbb{R}^d}^* f(x)dx.$$

(ii)   For every $f \in H_I(\mathbb{R}^d)$ or $H_S(\mathbb{R}^d)$ we have

$$\int_{*\,\mathbb{R}^d} f(x)dx = \int_{\mathbb{R}^d}^* f(x)dx = \int_{\mathbb{R}^d} f(x)dx.$$

Proof.

(i)   It suffices to show the following: if $\varphi \in H_I, \psi \in H_S, \psi \leq \varphi$, then $\int_{\mathbb{R}^d} \psi(x)dx \leq \int_{\mathbb{R}^d} \varphi(x)dx$. Now $-\psi \in H_I$ so $\varphi - \psi \in H_I$ (lemma 12.5) and $\varphi - \psi \geq 0$, so, by lemma 13.8

$$0 \leq \int_{\mathbb{R}^d} (\varphi(x) - \psi(x))dx$$

hence

$$\int_{\mathbb{R}^d} \varphi(x)dx \geq \int_{\mathbb{R}^d} \psi(x)dx.$$

(ii)   Let $f \in H_I(\mathbb{R}^d)$. From the definitions, it follows directly that

$$\int_{\mathbb{R}^d}^* f(x)dx = \int_{\mathbb{R}^d} f(x)dx. \tag{1}$$

Now let $(g_n)_{n \in \mathbb{N}} \subset C_c(\mathbb{R}^d)$ be a monotonically increasing sequence convergent to f. Then on the one hand

$$\int_{\mathbb{R}^d} f(x)dx = \sup_{n \in \mathbb{N}} \int_{\mathbb{R}^d} g_n(x)dx \text{ (lemma 13.7).} \tag{2}$$

and on the other hand, as also $g_n \subset H_S(\mathbb{R}^d)$,

$$\sup_n \int_{\mathbb{R}^d} g_n(x)dx \leq \int_{*\mathbb{R}^d} f(x)dx \leq \int_{\mathbb{R}^d}^* f(x)dx, \tag{3}$$

by (i). The assertion follows from (1), (2) and (3).
$f \in H_S(\mathbb{R}^d)$ is treated analogously. $\square$

For later purposes, we want to write down some simple properties of the upper integral; analogous statements hold for the lower integral.

Lemma 14.3

(i)   *Let $f : \mathbb{R}^d \to \mathbb{R} \cup \{\infty\}$ be non-negative, $\lambda \geq 0$. Then*

$$\int_{\mathbb{R}^d}^* \lambda f(x)dx = \lambda \int_{\mathbb{R}^d}^* f(x)dx.$$

(ii)   *If $f \leq g$, then also*

$$\int_{\mathbb{R}^d}^* f(x)dx \leq \int_{\mathbb{R}^d}^* g(x)dx.$$

(iii)   *If A is an orthogonal $d \times d$-matrix, $b \in \mathbb{R}^d$, then for all f*

$$\int_{\mathbb{R}^d}^* f(Ax + b)dx = \int_{\mathbb{R}^d}^* f(x)dx .$$

Proof. This follows directly from the corresponding statements of lemma 13.8.
□

The following result will be used in the proof of the central convergence theorems of Lebesgue integration theory.

Theorem 14.4 *Let $f_n : \mathbb{R}^d \to \mathbb{R} \cup \{\infty\}$ be non-negative functions ($n \in \mathbb{N}$). Then*

$$\int_{\mathbb{R}^d}^* (\sum_{n=1}^{\infty} f_n(x))dx \le \sum_{n=1}^{\infty} \int_{\mathbb{R}^d}^* f_n(x)dx .$$

Proof. By definition of the upper integral, for every $\varepsilon > 0$ and $n \in \mathbb{N}$ there exists. $g_n \in H_I(\mathbb{R}^d)$ such that $g_n \ge f_n$ and

$$\int_{\mathbb{R}^d} g_n(x)dx \le \int_{\mathbb{R}^d}^* f_n(x)dx + 2^{-n}\varepsilon . \tag{4}$$

Lemmas 12.5 and 12.8 imply that

$$g := \sum_{n=1}^{\infty} g_n \in H_I(\mathbb{R}^d).$$

As $f_n \ge 0$, so is $g_n \ge 0$ and lemma 13.9 therefore implies

$$\int_{\mathbb{R}^d} g(x)dx = \sum_{n=1}^{\infty} \int_{\mathbb{R}^d} g_n(x)dx \le \sum_{n=1}^{\infty} \int_{\mathbb{R}^d}^* f_n(x)dx + \varepsilon \text{ by (4).} \tag{5}$$

But, on the other hand, $\sum_{n=1}^{\infty} f_n \le \sum_{n=1}^{\infty} g_n = g$ and therefore by lemma 14.3 (ii)

$$\int_{\mathbb{R}^d}^* (\sum_{n=1}^{\infty} f_n(x))dx \le \int_{\mathbb{R}^d} g(x)dx. \tag{6}$$

As $\varepsilon > 0$ is arbitrary, (5) and (6) yield the assertion.
□

Definition 14.5 $f : \mathbb{R}^d \to \mathbb{R} \cup \{\pm\infty\}$ is said to be (Lebesgue)-integrable if

$$-\infty < \int_{\mathbb{R}^d *} f(x)dx = \int_{\mathbb{R}^d}^* f(x)dx < \infty.$$

The common value of the upper and lower integral is then called the (Lebesgue) integral of f, and it is denoted by $\int\limits_{\mathbb{R}^d} f(x)dx$.

Using lemma 14.2 (ii), we observe

Lemma 14.6 $f \in H_I(\mathbb{R}^d)$ *is integrable if and only if*

$$\int f(x)dx < \infty.$$

Here the integral is to be understood in the sense of §13 and both concepts of integration – from §13 and the current chapter – coincide. Analogously, we have: $f \in H_S(\mathbb{R}^d)$ *is integrable* $\iff \int\limits_{\mathbb{R}^d} f(x)dx > -\infty.$

In particular, every $f \in C_c(\mathbb{R}^d)$ *is integrable.* □

Remark. In contrast to the terminology of §13, the integral $\int\limits_{\mathbb{R}^d} f(x)dx$ of an arbitrary function f is automatically finite, if it is at all defined.

For certain reasons, which we do not wish to go into precisely here, but which the reader equipped with the concept of measurable functions introduced below can perhaps consider himself, it is not meaningful to take $+\infty$ or $-\infty$ as the value of the integral of an arbitrary f whose upper and lower integrals are both simultaneously $+\infty$ or $-\infty$.

The following lemma is another simple reformulation of the definitions.

Lemma 14.7 $f : \mathbb{R}^d \to \mathbb{R} \cup \{\pm\infty\}$ *is integrable*
$\iff \forall \varepsilon > 0 \exists g \in H_S, h \in H_I : g \le f \le h, \int\limits_{\mathbb{R}^d} h(x)dx - \int\limits_{\mathbb{R}^d} g(x)dx < \varepsilon$
$\iff \exists$ *a monotonically increasing sequence* $(g_n)_{n\in\mathbb{N}} \subset H_S$, *a monotonically decreasing sequence* $(h_n)_{n\in\mathbb{N}} \subset H_I$:

$$g_n \le f \le h_n,$$

$$-\infty < \lim_{n\to\infty} \int\limits_{\mathbb{R}^d} h_n(x)dx = \lim_{n\to\infty} \int\limits_{\mathbb{R}^d} g_n(x)dx < \infty,$$

and the common value of both integrals is $\int\limits_{\mathbb{R}^d} f(x)dx.$

Remark. For the first criterion, $\int\limits_{\mathbb{R}^d} h(x)dx$ and $\int\limits_{\mathbb{R}^d} g(x)dx$, and with it also the upper and lower integrals of f, must be finite, and the upper and lower integrals must then agree.

The second criterion, apart from the requirement of monotonicity, is equivalent to the first one. Now if one has arbitrary sequences $(g'_n)_{n\in\mathbb{N}} \subset H_S$,

$(h'_n)_{n \in \mathbb{N}} \subset H_I$, which satisfy the other conditions of the criterion, then so do the monotonic sequences

$$g_n := \sup_{1 \le i \le n} \{g'_i\}, h_n := \inf_{1 \le i \le n} \{h'_i\}.$$

Theorem 14.8 *Let* $f : \mathbb{R}^d \to \mathbb{R} \cup \{\pm\infty\}$

(i)   *f is integrable* $\iff \forall \, \varepsilon > 0 \; \exists \, \varphi \in C_c(\mathbb{R}^d) : \int_{\mathbb{R}^d}^* |f(x) - \varphi(x)| dx < \varepsilon$

(ii)   *If* $(\varphi_n)_{n \in \mathbb{N}} \subset C_c(\mathbb{R}^d)$ *satisfies* $\lim_{n \to \infty} \int_{\mathbb{R}^d}^* |f(x) - \varphi_n(x)| dx = 0$, *then*

$\int_{\mathbb{R}^d} f(x) dx = \lim_{n \to \infty} \int_{\mathbb{R}^d} \varphi_n(x) dx.$

Proof.

(i)   " $\Longrightarrow$ " : By lemma 14.7 there exist $g \in H_S, h \in H_I$ with $g \le f \le h$ and $\int_{\mathbb{R}^d} h(x) dx - \int_{\mathbb{R}^d} g(x) dx < \frac{\varepsilon}{2}$. On the other hand, there exists a $\varphi \in C_c(\mathbb{R}^d)$ with $\varphi \le h$ and

$$\int_{\mathbb{R}^d} h(x) dx - \int_{\mathbb{R}^d} \varphi(x) dx < \frac{\varepsilon}{2},$$

by definition of the integral for $h \in H_I$.
Now

$$|f - \varphi| \le |h - \varphi| + |h - f| \le |h - \varphi| + |h - g| = (h - \varphi) + (h - g)$$

and therefore, by lemma 14.3 (ii),

$$\int_{\mathbb{R}^d}^* |f(x) - \varphi(x)| dx \le \int_{\mathbb{R}^d}^* (h(x) - \varphi(x)) dx + \int_{\mathbb{R}^d}^* (h(x) - g(x)) dx < \varepsilon.$$

" $\Longleftarrow$ " : If $\varphi \in C_c(\mathbb{R}^d)$ satisfies $\int_{\mathbb{R}^d}^* |f(x) - \varphi(x)| dx < \varepsilon$, then there exists an $h \in H_I$ with $|f - \varphi| \le h$ and $\int_{\mathbb{R}^d} h(x) dx < 2\varepsilon$ (by definition of the $\mathbb{R}^d$ upper integral). We then have

$$-h + \varphi \le f \le h + \varphi$$

(as φ is finite, both the sums are defined everywhere) $-h + \varphi \in H_S, h + \varphi \in H_I$ (compare §12) and

$$\int_{\mathbb{R}^d} (h(x) + \varphi(x)) dx - \int_{\mathbb{R}^d} (-h(x) + \varphi(x)) dx = 2 \int_{\mathbb{R}^d} h(x) dx < 4\varepsilon,$$

and the first criterion in lemma 14.7 is fulfilled; f is therefore integrable.

(ii)   From (i), it follows first that f is integrable. Let $\varepsilon > 0$, and $n \in \mathbb{N}$ such that

$$\int_{\mathbb{R}^d}^{*} |f(x) - \varphi_n(x)| dx < \varepsilon.$$

Then there exists an $h_n \in H_I(\mathbb{R}^d)$ with

$$|f - \varphi_n| \leq h_n, \int_{\mathbb{R}^d} h_n(x) dx < 2\varepsilon,$$

so again $-h_n + \varphi_n \leq f \leq h_n + \varphi_n$ and

$$\int_{\mathbb{R}^d} (-h_n(x) + \varphi_n(x)) dx \leq \int_{\mathbb{R}^d} f(x) dx \leq \int_{\mathbb{R}^d} (h_n(x) + \varphi_n(x)) dx$$

and thus

$$|\int_{\mathbb{R}^d} f(x) dx - \int_{\mathbb{R}^d} \varphi_n(x) dx| \leq \int_{\mathbb{R}^d} h_n(x) dx < 2\varepsilon.$$

The result follows.   □

Theorem 14.9 *If f is integrable, so are $|f|$, $f^+ := \sup(f, 0)$, $f^- := -\inf(f, 0)$ and*

$$|\int_{\mathbb{R}^d} f(x) dx| \leq \int_{\mathbb{R}^d} |f(x)| dx. \tag{7}$$

Conversely, if f^+ and f^- are integrable, so is f.

If f_1 and f_2 are integrable, then $\sup(f_1, f_2)$ and $\inf(f_1, f_2)$ are also integrable.

Proof. We use the criterion in theorem 14.8 (i). If $\varphi \in C_c(\mathbb{R}^d)$ satisfies $\int_{\mathbb{R}^d}^{*} |f(x) - \varphi(x)| dx < \varepsilon$, it follows, because of $||f| - |\varphi|| \leq |f - \varphi|$, that

$$\int_{\mathbb{R}^d}^{*} ||f(x)| - |\varphi(x)|| dx < \varepsilon,$$

hence by theorem 14.8 (i) the integrability of $|f|$. The inequality (7) then follows, on account of $-|f| \leq f \leq |f|$, from the monotonicity of the upper integral and therefore also of the integral for integrable functions (lemma 14.3 (ii)). The integrability of $f^{\pm}$ likewise follows, so from that of f, on account of $|f^{\pm} - \varphi^{\pm}| \leq |f - \varphi|$. Conversely, if f^+ and f^- are integrable, then there exist, again by theorem 14.8 (i), $\varphi, \psi \in C_c(\mathbb{R}^d)$ with

$$\int_{\mathbb{R}^d}^{*} |f^+(x) - \varphi(x)| dx < \frac{\varepsilon}{2},$$

$$\int_{\mathbb{R}^d}^{*} |f^-(x) - \psi(x)| dx < \frac{\varepsilon}{2}.$$

Since $f = f^+ - f^-$ and $\varphi - \psi \in C_c(\mathbb{R}^d)$, we have

$$\int_{\mathbb{R}^d}^{*} |f(x) - (\varphi(x) - \psi(x))| dx < \varepsilon,$$

and therefore f is integrable, as usual by theorem 14.8 (i).

For the last assertion, we use the criterion in lemma 14.7. Thus there exist $g_1, g_2 \in H_S, h_1, h_2 \in H_I$ with $g_1 \leq f_1 \leq h_1, g_2 \leq f_2 \leq h_2$ and

$$\int_{\mathbb{R}^d} h_i(x) dx - \int_{\mathbb{R}^d} g_i(x) dx < \frac{\varepsilon}{2} \quad \text{for} \quad i = 1, 2.$$

We then have $\sup(g_1, g_2) \leq \sup(f_1, f_2) \leq \sup(h_1, h_2)$ and on account of $\sup(h_1, h_2) - \sup(g_1, g_2) \leq h_1 + h_2 - g_1 - g_2$,

$$\int_{\mathbb{R}^d} \sup(h_1, h_2)(x) dx - \int_{\mathbb{R}^d} \sup(g_1, g_2)(x) dx < \varepsilon$$

and $\sup(f_1, f_2)$ is therefore integrable by lemma 14.7, and analogously, $\inf(f_1, f_2)$ is integrable. $\qquad\square$

Lemma 14.10 *Let f be integrable, $A \in GL(d, \mathbb{R}), b \in \mathbb{R}^d$. Then $x \mapsto f(Ax + b)$ is also integrable and*

$$\int_{\mathbb{R}^d} f(Ax + b) dx = \frac{1}{|\det A|} \int_{\mathbb{R}^d} f(x) dx.$$

Proof. The proof follows from lemma 13.11. $\qquad\square$

Definition 14.11 $\mathcal{L}_1(\mathbb{R}^d) := \{f : \mathbb{R}^d \to \mathbb{R} : f \text{ integrable }\}.$

Here, we do not allow $\pm\infty$ as function values, so that e.g. we can add functions from $\mathcal{L}_1(\mathbb{R}^d)$ pointwise: Below, when considering null sets, we shall see that this does not impose any restrictions for integration theory.

Theorem 14.12 $\mathcal{L}_1(\mathbb{R}^d)$ *is a vector space and $\int_{\mathbb{R}^d} : \mathcal{L}_1(\mathbb{R}^d) \to \mathbb{R}$ is linear and monotone*

$$(f \le g \Longrightarrow \int_{\mathbb{R}^d} f(x)dx \le \int_{\mathbb{R}^d} g(x)dx).$$

Proof. That $\mathcal{L}_1(\mathbb{R}^d)$ is a vector space follows from the preceding remarks, the linearity of the integral follows as usual from theorem 14.8, and monotonicity follows from the monotonicity of the upper integral (lemma 14.3 (ii)).   □

Definition 14.13 $A \subset \mathbb{R}^d$ is called integrable $\iff \chi_A$ is integrable. In case A is integrable, we define the Lebesgue measure or volume of A to be

$$\text{Vol}\,(A) := \text{Vol}\,_d(A) := \int_{\mathbb{R}^d} \chi_A(x)dx\,.$$

For an arbitrary subset A of $\mathbb{R}^d$ one also defines the outer and inner measure, respectively, of A as

$$\mu^*(A) := \int_{\mathbb{R}^d}^{*} \chi_A(x)dx \text{ and } \mu_*(A) := \int_{*\,\mathbb{R}^d} \chi_A(x)dx,$$

and A is thus integrable if and only if $\mu^*(A)$ and $\mu_*(A)$ agree and are finite.

Theorem 14.14 *If A is open or closed, then A is integrable precisely if $\mu^*(A)$ is finite. In particular, compact as well as bounded open sets are integrable.*

Proof. If A is open or closed, then χ_A is, respectively, lower or upper semi-continuous (compare §12) and for semicontinuous functions, the upper and lower integrals are equal by lemma 14.1 (ii). Furthermore, every bounded set A is contained in a cube and therefore, because of the monotonicity of the upper integral,

$$\mu^*(A) \le \mu^*(W) < \infty.$$

□

Theorem 14.15 *If A and B are integrable, then so are $A \cup B, A \cap B$, and $A\backslash B$, and*

$$\text{Vol}\,(A \cup B) + \text{Vol}\,(A \cap B) = \text{Vol}\,A + \text{Vol}\,B,$$
$$\text{Vol}\,(A\backslash B) = \text{Vol}\,(A) - \text{Vol}\,(A \cap B).$$

Proof. For the characteristic functions, we have

$$\chi_{A \cap B} = \inf(\chi_A, \chi_B)$$
$$\chi_{A \cup B} = \chi_A + \chi_B - \chi_{A \cap B}$$
$$\chi_{A\backslash B} = \chi_A - \chi_{A \cap B}.$$

The integrability of $A \cap B$ therefore follows from theorem 14.9, and that of $A \cup B$ and $A \backslash B$ together with the corresponding volume formulae then follows from theorem 14.12. $\qquad\qquad\qquad\qquad\qquad\qquad\qquad\qquad\qquad\qquad\qquad$ $\square$

Definition 14.16 Let $A \subset \mathbb{R}^d$, $f : A \to \mathbb{R} \cup \{\pm\infty\}$. f is called integrable on A provided that the function $\tilde{f} : \mathbb{R}^d \to \mathbb{R} \cup \{\pm\infty\}$, defined by

$$\tilde{f}(x) := \begin{cases} f(x) & \text{for } x \in A \\ 0 & \text{for } x \in \mathbb{R}^d \backslash A \end{cases},$$

is integrable. In this case, we set

$$\int_A f(x)dx := \int_{\mathbb{R}^d} \tilde{f}(x)dx.$$

Exercises for § 14

1)

a) Let $W \subset \mathbb{R}^d$ be a cube, $f : W \to \mathbb{R}$ nonnegative and integrable. Assume that for some $x_0 \in W$, f is continuous at x_0 with $f(x_0) > 0$. Concluse

$$\int_W f(x)dx > 0.$$

b) Let $W \subset \mathbb{R}^d$ be a cube, $f : W \to \mathbb{R}$ continuous and nonnegative, $\int_W f(x)dx = 0$. Conclude

$$f \equiv 0 \quad \text{on } W.$$

c) Extend a) and b) to sets more general than cubes.

2) Compute the following integrals:

a) $K := [0,2] \times [3,4]$, $\int_K (2x + 3y)dxdy$.

b) $K := [1,2] \times [1,2]$, $\int_K e^{x+y}dxdy$.

c) $K := [0,1] \times [0,1]$, $\int_K (xy + y^2)dxdy$.

d) $K := [0, \frac{\pi}{2}] \times [0, \frac{\pi}{2}]$, $\int_K \sin(x + y)dxdy$.

e) $K := [1,2] \times [2,3] \times [0,2]$, $\int_K \frac{2z}{(x+y)^2}dxdydz$.

f) $K := [0,1] \times [0,1] \times [0,1]$, $\int\limits_{K} \frac{x^2 z^3}{1+y^2} dx\,dy\,dz$.

g) $K := \{(x,y) \in \mathbb{R}^2 : x^2 + y^2 \leq 1\}, n, m \in \mathbb{N}$, $\int\limits_{K} x^n y^m dx\,dy$.

3)   Let V be a finite dimensional vector space (over $\mathbb{R}$). Develop an analogue of the concept of the Lebesgue integral for maps $f : \mathbb{R}^d \to V$. State criteria for integrability. In the special case $V = \mathbb{C}$, show

$$\left| \int\limits_{\mathbb{R}^d} f(x)dx \right| \leq \int\limits_{\mathbb{R}^d} |f(x)|dx.$$

4)   Let $f : [r_1, R_1] \times [r_2, R_2]$ be bounded. Assume that for any fixed $x \in [r_1, R_1]$,
$f(x, \cdot) \to \mathbb{R}$ is integrable on $[r_2, R_2]$, and for any fixed $y \in [r_2, R_2]$, $f(\cdot, y)$ is continuous on $[r_1, R_1]$. Conclude that the function

$$F(x) := \int\limits_{r_2}^{R_2} f(x,y)dy$$

is continuous on $[r_1, R_1]$.

5)   Let $f : \mathbb{R}^d \to \mathbb{R}$ be integrable. For $x \in \mathbb{R}^d, n \in \mathbb{N}$, put

$$f_n(x) := \begin{cases} f(x) & \text{if } -n \leq f(x) \leq n \\ n & \text{if } f(x) > n \\ -n & \text{if } f(x) < -n \end{cases}.$$

Show that f_n is integrable, that $|f_n - f|$ converges to 0 pointwise, and that

$$\int\limits_{\mathbb{R}^d} f(x)dx = \lim_{n \to \infty} \int\limits_{\mathbb{R}^d} f_n(x)dx.$$

6)   **(Riemann integral)** Let $T(\mathbb{R}^d)$ be the vector space of elementary functions. Thus, each $t \in T(\mathbb{R}^d)$ is defined on some cube $W \subset \mathbb{R}^d$, and constant on suitable subcubes, i.e.

$$W = \bigcup_{i=1}^{n} W_i, \quad W_i \text{ cubes with } \overset{\circ}{W}_i \cap \overset{\circ}{W}_j = \emptyset \text{ for } i \neq j,$$

$$t_{|\overset{\circ}{W}_i} = c_i.$$

Denote the length of the edges of W_i by l_i. Show that t is integrable, with

$$\int\limits_{\mathbb{R}^d} t(x)dx = \sum_{i=1}^{n} c_i l_i^d.$$

$f : \mathbb{R}^d \to \mathbb{R}$ is called Riemann integrable if for any $\varepsilon > 0$, we may find $\varphi, \psi \in T(\mathbb{R}^d)$ with

$$\psi \leq f \leq \varphi$$

$$\int\limits_{\mathbb{R}^d} \varphi(x)dx - \int\limits_{\mathbb{R}^d} \psi(x)dx < \varepsilon.$$

Show that any Riemann integrable function is Lebesgue integrable as well, with

$$\int\limits_{\mathbb{R}^d} f(x)dx = \inf\{\int\limits_{\mathbb{R}^d} \varphi(x)dx : \varphi \in T(\mathbb{R}^d), \varphi \geq f\}$$

$$= \sup\{\int\limits_{\mathbb{R}^d} \psi(x)dx : \psi \in T(\mathbb{R}^d), \psi \leq f\}.$$

7)   In the terminology of the preceding exercise, show that the following function $f : [0,1] \to \mathbb{R}$ is not Riemann integrable:

$$f(x) := \begin{cases} 1 & \text{if } x \text{ is rational} \\ 0 & \text{if } x \text{ is irrational} \end{cases}.$$

8)   Let f, g be integrable, with f bounded. Show that fg is integrable. Given an example to show that the boundedness of f is necessary.

9)   Let $f : \mathbb{R}^d \to \mathbb{R}$ be bounded and satisfy

$$|f(x)| \leq \frac{c}{\|x\|^{d+\lambda}} \quad \text{for some constant } c \text{ and some } \lambda > 0.$$

Show that f is integrable.

15. Null Functions and Null Sets. The Theorem of Fubini

Null functions, i.e. those whose integral is 0 on every set, and null sets, i.e. those whose volume vanishes, are negligible for purposes of integration theory. In particular, countable sets, or lower dimensional subsets of Euclidean spaces are null sets. The general theorem of Fubini saying that in multiple integrals the order of integration is irrelevant is shown.

We begin with an example. Let $f : \mathbb{R} \to \mathbb{R}$ be defined as follows:

$$f(x) := \begin{cases} 1, & \text{if } x \text{ rational} \\ 0, & \text{if } x \text{ irrational} \end{cases}$$

We claim that f is integrable and $\int_{\mathbb{R}} f(x)dx = 0$.

For the proof, let $n \mapsto x_n$ be a bijection of $\mathbb{N}$ onto the set of rational numbers. (We recall that the rational numbers are countable.) Further, let $\varepsilon > 0$.

For $n \in \mathbb{N}$, let

$$U_n := \{x \in \mathbb{R} : |x - x_n| < \varepsilon \cdot 2^{-n}\}$$

and

$$U := \bigcup_{n \in \mathbb{N}} U_n.$$

By construction, U contains all the rational numbers. Therefore

$$f \le \chi_U \le \sum_{n=1}^{\infty} \chi_{U_n} \tag{1}$$

Furthermore, by theorem 14.4,

$$\int_{\mathbb{R}}^{*} \sum_{n=1}^{\infty} \chi_{U_n}(x)dx \le \sum_{n=1}^{\infty} \int^{*} \chi_{U_n}(x)dx = \sum_{n=1}^{\infty} \varepsilon \cdot 2^{1-n} = 4\varepsilon,$$

and therefore by (1)

$$\int_{\mathbb{R}}^{*} f(x)dx \le 4\varepsilon. \tag{2}$$

On the other hand, as f is nonnegative,

$$\int_{\mathbb{R}}^{*} f(x)dx \geq 0. \tag{3}$$

Since $\varepsilon > 0$ can be chosen arbitrarily small, it follows from lemma 14.2 (i) and inequalities (2) and (3), that f is integrable and

$$\int_{\mathbb{R}} f(x)dx = 0,$$

as asserted.

We shall now investigate systematically the phenomenon which has appeared in this example and develop the concepts required for it.

Definition 15.1 $f : \mathbb{R}^d \to \mathbb{R} \cup \{\pm\infty\}$ is called a null function if $\int_{\mathbb{R}^d}^{*} |f(x)|dx = 0$. $A \subset \mathbb{R}^d$ is called a null set if χ_A is a null function.

The following lemma sums up a few obvious remarks.

Lemma 15.2 Let f be a null function, $\lambda \in \mathbb{R}, |g| \leq |f|$. Then λf and g are also null functions. Every subset of a null set is again a null set. □

The next lemma explains our example above.

Lemma 15.3 Let $(f_n)_{n \in \mathbb{N}}$ be a sequence of nonnegative null functions. Then $\sum_{n=1}^{\infty} f_n$ is also a null function. If $(N_m)_{m \in \mathbb{N}}$ is a sequence of null sets in $\mathbb{R}^d$, then $N := \bigcup_{m \in \mathbb{N}} N_m$ is also a null set.

Proof. The first statement follows directly from theorem 14.4, and the second follows from the first on account of $\chi_N \leq \sum_{m=1}^{\infty} \chi_{N_m}$. □

Corollary 15.4
(i)   Every countable subset of $\mathbb{R}^d$ is a null set.

(ii)   Every hyperplane in $\mathbb{R}^d$ is a null set.

Proof.
(i)   First, every one point set in $\mathbb{R}^d$ is a null set, as it is contained in a cube of arbitrarily small volume. Lemma 15.3 implies then that any countable as a countable union of one-point sets is also a null set.

(ii)   Let L be a hyperplane in $\mathbb{R}^d$, $x_0 \in L$, $x_0 = (x_0^1, \ldots x_0^d)$, $W(L, n) := \{x = (x^1, \ldots x^d) \in L : \sup_{1 \leq i \leq d} |x_0^i - x^i| \leq n\}$. One computes, e.g. as follows, that $\operatorname{Vol}_d W(L, n) = 0$. As Lebesgue measure

is invariant under isometries of $\mathbb{R}^d$, we may assume, without loss of generality, that $L = \{x = (x^1, \ldots, x^d) : x^d = 0\}$.
Then for $\varepsilon > 0$

$$Q(L, n, \varepsilon) := \{x = (x^1, \ldots, x^d) : (x^1, \ldots, x^{d-1}, 0) \in W(L, n), |x^d| \le \varepsilon\}.$$

$Q(L, n, \varepsilon)$ is a rectangular parallelepiped and

$$\text{Vol }_d Q(L, n, \varepsilon) = (2n)^{d-1} \cdot \varepsilon \quad \text{(compare lemma 13.14)}.$$

So, because $w(L, n) \subset Q(L, n, \varepsilon)$, we see that for every $\varepsilon > 0$

$$\text{Vol}_d w(L, n) \le (2n)^{d-1} 2\varepsilon,$$

thus

$$\text{Vol}_d w(L, n) = 0.$$

Since $L = \underset{n \in \mathbb{N}}{\cup}\, w(L, n)$ the result follows again from lemma 15.3.   □

The following result should already be known to the reader, but it is given here a new proof.

Corollary 15.5 $\mathbb{R}$ *is uncountable.*

Proof. Were $\mathbb{R}$ countable, so would $[0, 1]$ be and corollary 15.4 (i) would then give the contradiction

$$1 = \int_0^1 dx = \text{ vol } [0, 1] = 0.$$

□

Example. We shall now construct an uncountable null set in $\mathbb{R}$, the so-called Cantor set.
From $[0, 1]$ we remove the middle third $(\frac{1}{3}, \frac{2}{3})$, and obtain the set

$$S_1 := [0, \frac{1}{3}] \cup [\frac{2}{3}, 1].$$

From S_1 we remove again the middle thirds of both the subintervals, that is, $(\frac{1}{9}, \frac{2}{9})$ and $(\frac{7}{9}, \frac{8}{9})$ and obtain

$$S_2 := [0, \frac{1}{9}] \cup [\frac{2}{9}, \frac{3}{9}] \cup [\frac{6}{9}, \frac{7}{9}] \cup [\frac{8}{9}, 1].$$

This process is iterated: thus we remove at the nth-step the open middle third of all the subintervals of the set S_{n-1}. One obtains in this manner a set S_n, which consists of 2^n disjoint compact intervals of length 3^{-n}. Finally, we set

$$S = \bigcap_{n \in \mathbb{N}} S_n.$$

Obviously, for all $n \geq 2$

$$S \subset S_n \subset S_{n-1}$$

and

$$\text{Vol } S_n = (\frac{2}{3})^n,$$

so in particular $\lim_{n \to \infty} \text{Vol } S_n = 0$. Therefore, S is a null set. In order to show that S is uncountable, we derive a contradiction to the assumption that there is a bijection

$$\mathbb{N} \to S$$

$$m \mapsto x_m$$

For this, we associate to every $x \in S$ a sequence, which assumes only the values 0 and 1. We set

$$\lambda_1(x) = \begin{cases} 0 & \text{if } x \in [0, \frac{1}{3}] \\ 1 & \text{if } x \in [\frac{2}{3}, 1] \end{cases}$$

and iteratively for $n \in \mathbb{N}$, $\lambda_n(x) = 0$ in case x, in the passage from S_{n-1} to S_n, lies in the first subinterval of the interval of S_{n-1} which contains x, and $\lambda_n(x) = 1$, in case x lies in the third subinterval. This defines the sequence $(\lambda_n(x))_{n \in \mathbb{N}}$. Conversely, every sequence $(\lambda_n)_{n \in \mathbb{N}}$ which assumes only the values 0 and 1 defines in this manner a point in S, and different sequences define different points in S.

Now let $m \mapsto x_m$ be a counting of S.

We define $x \in S$ by

$$\lambda_m(x) = \begin{cases} 0 & \text{if } \lambda_m(x_m) = 1 \\ 1 & \text{if } \lambda_m(x_m) = 0 \end{cases}.$$

Then for no $m \in \mathbb{N}$, x can be of the form x_m, for we always have

$$\lambda_m(x_m) \neq \lambda_m(x).$$

So x could not have been contained in the counting and consequently the existence of a counting of S is contradicted.

Definition 15.6 We say that a property $E(x)$ holds for almost all $x \in \mathbb{R}^d$, or it holds almost everywhere if

$$\{x \in \mathbb{R}^d : E(x) \text{ does not hold}\}$$

is a null set.

Lemma 15.7 $f : \mathbb{R}^d \to \mathbb{R} \cup \{\pm\infty\}$ *is a null function precisely when* $f = 0$ *almost everywhere (i.e. when* $\{x \in \mathbb{R}^d : f(x) \neq 0\}$ *is a null set).*

Proof. Let $N := \{x \in \mathbb{R}^d : |f(x)| > 0\}$.

Now

$$|f| \leq \sup_{m \in \mathbb{N}} m\chi_N \quad \text{and} \quad \chi_N \leq \sup_{m \in \mathbb{N}} m|f|.$$

Thus if N is a null set, then

$$\int_{\mathbb{R}^d}^* |f(x)|dx \leq \int_{\mathbb{R}^d}^* \sup_{m \in \mathbb{N}} m\chi_N(x)dx \leq \int_{\mathbb{R}^d}^* \sum_{m=1}^{\infty} m\chi_N(x)dx = 0,$$

by theorem 14.4, so f is a null function, and conversely if f is a null function then

$$\int_{\mathbb{R}^d}^* \chi_N(x)dx \leq \int_{\mathbb{R}^d}^* \sup_{m \in \mathbb{N}} m|f(x)|dx \leq \int_{\mathbb{R}^d}^* \sum_{m=1}^{\infty} m|f(x)|dx = 0,$$

again by theorem 14.4, whence N is a null set. □

Lemma 15.8 *Let* $f : \mathbb{R}^d \to \mathbb{R} \cup \{\pm\infty\}$ *satisfy* $\int_{\mathbb{R}^d}^* |f(x)|dx < \infty$ *(e.g. let f be integrable). Then f is almost everywhere finite.*

Proof. We have to show that $N := \{x \in \mathbb{R}^d : f(x) = \pm\infty\}$ is a null set. For every $\varepsilon > 0$ we have

$$\chi_N \leq \varepsilon|f|,$$

consequently

$$\int_{\mathbb{R}^d}^* \chi_N(x)dx \leq \varepsilon \int_{\mathbb{R}^d}^* |f(x)|dx.$$

As this holds for arbitrary positive ε, $\int_{\mathbb{R}^d}^* \chi_N(x)dx = 0$, so N is a null set. □

Lemma 15.9 *Let* $f, g : \mathbb{R}^d \to \mathbb{R} \cup \{\pm\infty\}$ *with* $f = g$ *almost everywhere. If f is integrable then so is g, and their integrals are then equal.*

Proof. Let $N := \{x \in \mathbb{R}^d : f(x) \neq g(x)\}$. Let $(\varphi_n)_{n \in \mathbb{N}}$ be a sequence in $C_c(\mathbb{R}^d)$ with

$$\lim_{n \to \infty} \int_{\mathbb{R}^d}^* |f(x) - \varphi_n(x)|dx = 0. \tag{4}$$

Now $|g - \varphi_n| \leq |f - \varphi_n| + \sup_{m \in \mathbb{N}} m\chi_N$ and therefore as in the proof of lemma 15.7

$$\int_{\mathbb{R}^d}^* |g(x) - \varphi_n(x)|dx \leq \int_{\mathbb{R}^d}^* |f(x) - \varphi_n(x)|dx$$

and therefore by (4)

$$\lim \int_{\mathbb{R}^d}^{*} |g(x) - \varphi_n(x)| dx = 0.$$

Theorem 14.8 now gives the integrability of g and we have

$$\int_{\mathbb{R}^d} g(x) dx = \lim_{n \to \infty} \int_{\mathbb{R}^d} \varphi_n(x) dx = \int_{\mathbb{R}^d} f(x) dx.$$

$\square$

From lemmas 15.8 and 15.5 it follows that one can substitute an integrable function f by a function which coincides with f almost everywhere and which is finite almost everywhere, without changing the value of the integral of f. This justifies the restriction we made in the definition of the space $\mathcal{L}_1(\mathbb{R}^d)$: see definition 14.11.

For later purposes we introduce the following terminology.

Definition. Two functions $f, g : A \to \mathbb{R} \cup \{\pm\infty\}$ are called equivalent when they are equal almost everywhere on A (that is, when $N := \{x \in A : f(x) \neq g(x)\}$ is a null set).

Here, we can even call f and g equivalent when one or both of them are defined only almost everywhere on A, that is, when one is dealing with functions

$$f : A \backslash N_1 \to \mathbb{R} \cup \{\pm\infty\}, g : A \backslash N_2 \to \mathbb{R} \cup \{\pm\infty\}$$

for which N_1, N_2 and $\{x \in A \backslash (N_1 \cup N_2) : f(x) \neq g(x)\}$ are null sets. Lemma 15.9 then states that together with f, every function which is equivalent to f is integrable and has the same integral.

Definition 15.10 For $A \subset \mathbb{R}^d$ let $L^1(A)$ be the set of equivalence classes of functions integrable on A (as said earlier, we need only consider $f : A \to \mathbb{R}$). For $f \in L^1(A)$ we set

$$\|f\|_{L^1(A)} := \int_A |f(x)| dx$$

where the integral is to be understood as the integral of any function in the equivalence class of f.

Theorem 15.11 $L^1(A)$ is a vector space and $\| \cdot \|_{L^1(A)}$ is a norm on $L^1(A)$.

Proof. If f_1 and g_1 as well as f_2 und g_2 are equivalent and $\lambda \in \mathbb{R}$, then so are $f_1 + f_2$ and $g_1 + g_2$ as well as λf_1 and λg_1. Therefore the set of equivalence classes $L^1(A)$ forms a vector space.

It follows from the elementary properties of the integral established in §14 that $\| \cdot \|_{L^1(A)}$ satisfies the triangle inequality and that one can take out scalars with their absolute value. In order for $\| \cdot \|_{L^1(A)}$ to be a norm, it remains only to show its positive definiteness. The latter means

$$\|f\|_{L^1(A)} \iff f = 0 \text{ almost everywhere on } A$$

(i.e. f lies in the equivalence class of the function which is identically zero). But this follows at once from lemma 15.7.   □

We shall show later that $L^1(A)$ is even a Banach space.
We now come to the general form of the theorem of Fubini.

Theorem 15.12 *Let* $f : \mathbb{R}^c \times \mathbb{R}^d \to \mathbb{R} \cup \{\pm\infty\}$ *be integrable, let* $x = (\xi, \eta) \in \mathbb{R}^c \times \mathbb{R}^d$. *Then there exists a null set* $N \subset \mathbb{R}^d$ *with the property that for every* $\eta \in \mathbb{R}^d \backslash N$ *the function*

$$\mathbb{R}^c \to \mathbb{R} \cup \{\pm\infty\}$$
$$\xi \mapsto f(\xi, \eta)$$

is integrable. For $\eta \in \mathbb{R}^d \backslash N$ *we set*

$$f_1(\eta) := \int_{\mathbb{R}^c} f(\xi, \eta) d\xi$$

(and for $\eta \in N$ *we define* $f_1(\eta)$ *arbitrarily, e.g.* $= 0$*). Then* f_1 *is integrable and*

$$\int_{\mathbb{R}^d} f_1(\eta) d\eta = \int_{\mathbb{R}^{c+d}} f(x) dx.$$

(One remembers the assertion – theorem of Fubini – most easily in the following form
$(x = (\xi, \eta))$

$$\int_{\mathbb{R}^{c+d}} f(x) dx = \int_{\mathbb{R}^d} (\int_{\mathbb{R}^c} f(\xi, \eta) d\xi) d\eta$$
$$= \int_{\mathbb{R}^c} (\int_{\mathbb{R}^d} f(\xi, \eta) d\eta) d\xi,$$

for an integrable function f; the last equality follows from the symmetry in the roles of ξ and η or from the fact that $(\xi, \eta) \mapsto (\eta, \xi)$ is an orthogonal transformation and the Lebesgue integral is invariant under orthogonal coordinate transformations).

Proof. As f is integrable, there exist for every $\varepsilon > 0$ functions $g \in H_S(\mathbb{R}^{c+d})$, $h \in H_I(\mathbb{R}^{c+d})$ with $g \le f \le h$ and

$$\int_{\mathbb{R}^{c+d}} h(x) dx - \int_{\mathbb{R}^{c+d}} g(x) dx < \varepsilon, \text{ by lemma 14.7.} \tag{5}$$

Let

$$f_{1*}(\eta) = \int_{\mathbb{R}^c}^* f(\xi, \eta)d\xi, \ f_1^*(\eta) = \int_{\mathbb{R}^c}^* f(\xi, \eta)d\xi$$

and further

$$g_1(\eta) := \int_{\mathbb{R}^c} g(\xi, \eta)d\xi, \ h_1(\eta) := \int_{\mathbb{R}^c} h(\xi, \eta)d\xi.$$

By Fubini's theorem for semicontinuous functions (theorem 13.9), $g_1 \in H_S(\mathbb{R}^d)$, $h_1 \in H_I(\mathbb{R}^d)$ as well as

$$\int_{\mathbb{R}^d} g_1(\eta)d\eta = \int_{\mathbb{R}^{c+d}} g(\xi, \eta)d\xi d\eta,$$

$$\int_{\mathbb{R}^d} h_1(\eta)d\eta = \int_{\mathbb{R}^{c+d}} h(\xi, \eta)d\eta. \tag{6}$$

As $g \leq f \leq h$ it follows that

$$g_1(\eta) \leq f_{1*}(\eta) \leq f_1^*(\eta) \leq h_1(\eta) \tag{7}$$

by lemma 14.2 (i).

From (5) and (6) it follows that

$$\int_{\mathbb{R}^d} h_1(\eta)d\eta - \int_{\mathbb{R}^d} g_1(\eta)d\eta < \varepsilon. \tag{8}$$

Again, by lemma 14.7 it follows (by means of (7), (8)) that both $f_{1*}(\eta)$ and $f_1^*(\eta)$ are integrable and (by means of (5), (6), (7)) that both the integrals coincide with one another and with $\int_{\mathbb{R}^{c+d}} f(x)dx$:

$$\int_{\mathbb{R}^d} f_{1*}(\eta)d\eta = \int_{\mathbb{R}^d} f_1^*(\eta)d\eta = \int_{\mathbb{R}^{c+d}} f(x)dx. \tag{9}$$

As $f_{1*} \leq f_1^*$ it follows from lemma 15.8 and (9) that f_{1*} and f_1^* are equal almost everywhere, and that both are finite almost everywhere.

Thus for almost all η, the integrals $\int_{\mathbb{R}^c}^* f(\xi, \eta)d\xi$ and $\int_{\mathbb{R}^c}^* f(\xi, \eta)d\xi$ agree and are finite. Thus for a null set N and every $\eta \in \mathbb{R}^d \backslash N$ the function

$$\xi \mapsto f(\xi, \eta)$$

is integrable over $\mathbb{R}^c$, and

$$f_1(\eta)(= \int_{\mathbb{R}^c} f(\xi, \eta)d\xi) = f_{1*}(\eta) = f_1^*(\eta).$$

The assertion then follows from (9) and lemma 15.9.   □

Exercises for § 15

1)   Let $(f_n)_{n\in\mathbb{N}}$ be a bounded sequence of null functions. Then $\sup\limits_{n\in\mathbb{N}}\{f_n(x)\}$ is a null function, too.

2)

a)   Let $Q := \{(x,y) \in \mathbb{R}^2 : 0 < x,y < 1\}$ be the unit square,

$$f(x,y) := \frac{x^2 - y^2}{(x^2 + y^2)^2}.$$

Show that the following integrals both exist, but

$$\int\limits_0^1 \left(\int\limits_0^1 f(x,y)dx\right)dy \neq \int\limits_0^1 \left(\int\limits_0^1 f(x,y)dy\right)dx.$$

Is this compatible with Fubini's theorem?

b)   In a similar vein, consider $f : Q \to \mathbb{R}$ defined by

$$f(x,y) := \begin{cases} y^{-2} & \text{for } 0 < x < y < 1 \\ -x^{-2} & \text{for } 0 < y < x < 1 \\ 0 & \text{otherwise,} \end{cases}$$

and derive the same conclusion as in a).

16. The Convergence Theorems of Lebesgue Integration Theory

We discuss the fundamental convergence theorems of Fatou, B. Levi, and Lebesgue, saying that under certain assumptions, the integral of a limit of a sequence of functions equals the limit of the integrals. Instructive examples show the necessity of those assumptions. As an application, results that justify the derivation under the integral sign w.r.t. a parameter are given.

In this paragraph, we shall consider the following question: Let $(f_n)_{n \in \mathbb{N}}$ be a sequence of integrable functions. The functions f_n converge, in a sense still to be made precise, to a function f, say pointwise or almost everywhere. Under which assumptions is then f integrable, and when is $\int f(x)dx = \lim_{n \to \infty} \int f_n(x)dx$?

To have an idea of the problems involved, we shall first consider some examples.

Examples.

1)   Let $f_n : \mathbb{R}^d \to \mathbb{R}$ be defined thus:

$$f_n(x) := \begin{cases} 1 & \text{if } \|x\| < n \\ 0 & \text{if } \|x\| \geq n \end{cases}$$

f_n is in $H_I(\mathbb{R}^d)$ for all n and therefore integrable. However $\lim_{n \to \infty} f_n(x) = 1$ for all $x \in \mathbb{R}^d$, and $\lim f_n$ is not integrable.

2)   We now show by an example that even when $\lim f_n$ is integrable, the relationship $\int \lim_{n \to \infty} f_n(x)dx = \lim_{n \to \infty} \int f_n(x)dx$ may not hold necessarily. For this, let $f_n : [0, 1] \to \mathbb{R}$ be defined as follows:

$$f_n(x) := \begin{cases} n & \text{for } 0 \leq x \leq \frac{1}{n} \\ 0 & \text{for } \frac{1}{n} < x \leq 1 \end{cases}$$

Again, all the f_n are integrable, and for every n

$$\int_{[0,1]} f_n(x)dx = 1.$$

On the other hand, $\lim_{n \to \infty} f_n = 0$ almost everywhere (namely, not for $x = 0$ only), so

$$\int\limits_{0}^{1} \lim f_n(x)dx = 0.$$

We first prove the *monotone convergence theorem of Beppo Levi*.

Theorem 16.1 *Let $f_n : \mathbb{R}^d \to \mathbb{R} \cup \{\pm\infty\}$ form a monotonically increasing sequence (so*
$f_n \le f_{n+1}$ for all n) of integrable functions. If

$$\lim_{n \to \infty} \int\limits_{\mathbb{R}^d} f_n(x)dx < \infty$$

(the limit exists in $\mathbb{R} \cup \{\infty\}$ on account of the monotonicity of (f_n)), then

$$f := \lim_{n \to \infty} f_n$$

is also integrable (by monotonicity of f_n, the above limit exists for every
$x \in \mathbb{R}^d$ as an element of $\mathbb{R} \cup \{\pm\infty\}$), and

$$\int\limits_{\mathbb{R}^d} f(x)dx = \lim_{n \to \infty} \int\limits_{\mathbb{R}^d} f_n(x)dx.$$

Proof. Let

$$N_n := \{x \in \mathbb{R}^d : f_n(x) = \pm\infty\},$$
$$N = \bigcup_{n \in \mathbb{N}} N_n.$$

By lemma 15.8, every N_n is a null set and so is N, by lemma 15.2.

Now, because of monotonicity,

$$\int\limits_{\mathbb{R}^d}^{*} |f(x) - f_m(x)|dx = \int\limits_{\mathbb{R}^d}^{*} \sum_{n=m}^{\infty} (f_{n+1}(x) - f_n(x))dx$$

$$\le \sum_{n=m}^{\infty} \int\limits_{\mathbb{R}^d}^{*} (f_{n+1}(x) - f_n(x))dx$$

$$= \lim_{n \to \infty} \int\limits_{\mathbb{R}^d} (f_{n+1}(x) - f_m(x))dx$$

using integrability of the f_n

$$= \lim_{n \to \infty} \int\limits_{\mathbb{R}^d} f_n(x)dx - \int\limits_{\mathbb{R}^d} f_m(x)dx.$$

Now for every $\varepsilon > 0$ we can choose $m \in \mathbb{N}$ so large that

$$|\lim_{n\to\infty} \int f_n(x)dx - \int f_m(x)dx| < \frac{\varepsilon}{2} \tag{1}$$

and thus obtain from the preceding inequality also

$$\int_{\mathbb{R}^d}^{*} |f(x) - f_m(x)|dx < \frac{\varepsilon}{2}. \tag{2}$$

As f_m is integrable, there exists moreover, by theorem 14.8, a $\varphi \in C_c(\mathbb{R}^d)$ with

$$\int_{\mathbb{R}^d}^{*} |f_m(x) - \varphi(x)| \, dx < \frac{\varepsilon}{2}. \tag{3}$$

By (2) and (3) it follows that

$$\int_{\mathbb{R}^d}^{*} |f(x) - \varphi(x)|dx < \varepsilon$$

and from (1)

$$|\lim_{n\to\infty} \int_{\mathbb{R}^d} f_n(x)dx - \int_{\mathbb{R}^d} \varphi(x)dx| < \varepsilon$$

holds also.

Theorem 14.8 now implies that f is integrable and

$$\int_{\mathbb{R}^d} f(x)dx = \lim_{n\to\infty} \int_{\mathbb{R}^d} f_n(x)dx.$$

$\square$

An analogous result holds, of course, for monotonically decreasing sequences. In the example 1 above, the limit of the integrals is infinite, whereas in example 2 the convergence is not monotone.

A simple reformulation of theorem 16.1 is

Corollary 16.2 (B. Levi) *Let $f_n : \mathbb{R}^d \to \mathbb{R} \cup \{\pm\infty\}$ be non-negative integrable functions. If*

$$\sum_{n=1}^{\infty} \int_{\mathbb{R}^d} f_n(x)dx < \infty,$$

then $\sum\limits_{n=1}^{\infty} f_n$ is also integrable, and

$$\int_{\mathbb{R}^d} \sum_{n=1}^{\infty} f_n(x)dx = \sum_{n=1}^{\infty} \int_{\mathbb{R}^d} f_n(x)dx.$$

Proof. One applies theorem 16.1 to the sequence of partial sums.   □

We now prove

Lemma 16.3 *Let $f_n : \mathbb{R}^d \to \mathbb{R} \cup \{\pm\infty\}$ be a sequence of integrable functions. Assume that there exists an integrable function $F : \mathbb{R}^d \to \mathbb{R} \cup \{\pm\infty\}$ with*

$$f_n \geq F \text{ for all } n \in \mathbb{N}.$$

Then $\inf\limits_{n \in \mathbb{N}} f_n$ *is also integrable.*

Similarly, $\sup\limits_{n \in \mathbb{N}} f_n$ *is integrable, provided there is an integrable $G : \mathbb{R}^d \to \mathbb{R} \cup \{\pm\infty\}$ with*

$$f_n \leq G \text{ for all } n.$$

Proof. We set

$$g_n := \inf_{i \leq n} f_i$$

and thus obtain a monotonically decreasing sequence of integrable functions, so $g_{n+1} \leq g_n$ for all n and

$$g_n \geq F,$$

and therefore also

$$\int_{\mathbb{R}^d} g_n(x)dx \geq \int_{\mathbb{R}^d} F(x)dx > -\infty.$$

By the monotone convergence theorem, we have the integrability of $f = \lim\limits_{n \to \infty} g_n$. The second statement is proved analogously by considering

$$h_n := \sup_{i \leq n} f_i.$$

□

The next result is usually referred to as *Fatou's lemma.*

Theorem 16.4 *Let $f_n : \mathbb{R}^d \to \mathbb{R} \cup \{\pm\infty\}$ be a sequence of integrable functions. Assume that there is an integrable function $F : \mathbb{R}^d \to \mathbb{R} \cup \{\pm\infty\}$ with*

$$f_n \geq F \text{ for all } n.$$

Furthermore, let

$$\int_{\mathbb{R}^d} f_n(x)dx \leq K < \infty \text{ for all } n.$$

Then $\liminf\limits_{n \to \infty} f_n$ *is integrable, and*

$$\int_{\mathbb{R}^d} \liminf_{n\to\infty} f_n(x)dx \leq \liminf_{n\to\infty} \int_{\mathbb{R}^d} f_n(x)dx.$$

Proof. We set $g_n := \inf_{i \geq n} f_i$. By lemma 16.3, all the g_n are integrable. Moreover, we have

$$g_n \leq g_{n+1}, g_n \leq f_m \text{ for all } n \text{ and } m \geq n.$$

Therefore

$$\int_{\mathbb{R}^d} g_n(x)dx \leq K$$

also holds for all n.

Thus by the monotone convergence theorem

$$\liminf_{n\to\infty} f_n = \lim_{n\to\infty} g_n$$

is integrable and

$$\int_{\mathbb{R}^d} \liminf_{n\to\infty} f_n(x)dx = \lim_{n\to\infty} \int_{\mathbb{R}^d} g_n(x)dx$$

$$\leq \liminf_{n\to\infty} \int_{\mathbb{R}^d} f_n(x)dx.$$

$\square$

In the example 1 above, the integrals are again not uniformly bounded; the corresponding statement with lim sup fails due to lack of an integrable upper bound. Example 2 shows that in theorem 16.4, in general, one cannot expect equality.

We prove now the *dominated convergence theorem of H. Lebesgue.*

Theorem 16.5 *Let $f_n : \mathbb{R}^d \to \mathbb{R}\cup\{\pm\infty\}$ be a sequence of integrable functions which converge on $\mathbb{R}^d$ pointwise almost everywhere to a function $f : \mathbb{R}^d \to \mathbb{R} \cup \{\pm\infty\}$. Moreover, assume that there is an integrable function $G : \mathbb{R}^d \to \mathbb{R} \cup \{\infty\}$ with*

$$|f_n| \leq G \text{ for all } n \in \mathbb{N}.$$

Then f is integrable and

$$\int_{\mathbb{R}^d} f(x)dx = \lim_{n\to\infty} \int_{\mathbb{R}^d} f_n(x)dx \tag{4}$$

Proof. In as much as we can change, if necessary, the functions f_n and f on a null set, we may assume that all the functions have finite values (notice that

f can be infinite at most on a null set, where G is so, compare lemma 15.8) and

$$f(x) = \lim_{n \to \infty} f_n(x) \text{ for all } x \in \mathbb{R}^d.$$

The assumptions of theorem 16.4 are fulfilled with $K = \int_{\mathbb{R}^d} G(x)dx$ and therefore f is integrable (notice that $f = \lim f_n = \liminf f_n$) and

$$\int_{\mathbb{R}^d} f(x)dx \leq \liminf_{n \to \infty} \int_{\mathbb{R}^d} f_n(x)dx.$$

Analogously, one shows that

$$\int_{\mathbb{R}^d} f(x)dx \geq \limsup_{n \to \infty} \int_{\mathbb{R}^d} f_n(x)dx,$$

and from these inequalities, the result follows directly.   □

In the examples 1 and 2 the functions f_n are not bounded in absolute value by a fixed integrable function, so that theorem 16.5 is not applicable there. However, the following holds

Corollary 16.6 *Let $f_n : \mathbb{R}^d \to \mathbb{R} \cup \{\pm\infty\}$ be integrable functions. Further, assume that there is an integrable function $F : \mathbb{R}^d \to \mathbb{R} \cup \{\pm\infty\}$ with*

$$|\lim_{n \to \infty} f_n| \leq F. \tag{5}$$

Then $\lim_{n \to \infty} f_n$ is integrable.

Proof. We set
$$h_n := \sup\{\inf(f_n, F), -F\}.$$

Then $\lim_{n \to \infty} h_n = \lim_{n \to \infty} f_n$, on account of (5), and the sequence $(h_n)_{n \in \mathbb{N}}$ satisfies the assumptions of theorem 16.5. This gives the assertion.   □

Corollary 16.7 *Let $A_1 \subset A_2 \subset A_3 \subset \ldots$ be subsets of $\mathbb{R}^d$,*

$$A := \bigcup_{n \in \mathbb{N}} A_n.$$

For $f : A \to \mathbb{R} \cup \{\pm\infty\}$, let $f_{|A_n}$ be integrable over A_n for every n; and assume that

$$\lim_{n \to \infty} \int_{A_n} |f(x)|dx < \infty.$$

Then f is integrable over A and

$$\int_A f(x)dx = \lim_{n\to\infty} \int_{A_n} f(x)dx.$$

Proof. In order to connect with the previous terminology, we define $f(x) = 0$ for $x \in \mathbb{R}^d \backslash A$ and thereby obtain, without changing notations, a function $f : \mathbb{R}^d \to \mathbb{R} \cup \{\pm\infty\}$.

Then we consider the functions

$$f_n = f\chi_{A_n}.$$

First of all, $|f_n|$ converges monotonically to $|f|$ which is integrable. But then, the assumptions of theorem 16.5 hold with $G = |f|$, and this then yields the assertion. □

Corollary 16.8 *Let $\Omega \subset \mathbb{R}^d$ be open, $f : \Omega \to \mathbb{R} \cup \{\pm\infty\}$ be integrable and $\varepsilon > 0$. Then there exists an open bounded set Ω' with $\overline{\Omega'} \subset \Omega$ and*

$$|\int_\Omega f(x)dx - \int_{\Omega'} f(x)dx| < \varepsilon.$$

Proof. First, for every open bounded set $\Omega' \subset \Omega$, $f_{|\Omega'}$ is integrable. To see this, by theorem 14.9 we can assume, without loss of generality, that $f \geq 0$. Then $f_{|\Omega'} = \lim_{n\to\infty} (\inf(f, n\chi_{\Omega'}))$, which is integrable, e.g. by theorem 16.5 (notice that the characteristic function $\chi_{\Omega'}$ is integrable as Ω' is open and bounded). We set $\Omega_n := \{x \in \Omega : \|x\| < n, \text{dist}(x, \partial\Omega) > \frac{1}{n}\}$. Then Ω_n is open and bounded and $\Omega = \bigcup_{n\in\mathbb{N}} \Omega_n$. By corollary 16.7

$$\int_\Omega f(x)dx = \lim_{n\to\infty} \int_{\Omega_n} f(x)dx.$$

Thus one can take Ω' to be Ω_n for n sufficiently large (depending on ε). □

Corollary 16.9 *Let $\Omega \subset \mathbb{R}^d$ be open, $f : \Omega \to \mathbb{R} \cup \{\pm\infty\}$ be integrable, and $\varepsilon > 0$. Then there exists $\delta > 0$ such that whenever $\Omega_0 \subset \Omega$ satisfies $Vol(\Omega_0) < \delta$, then*

$$|\int_{\Omega_0} f(x)dx| < \varepsilon. \tag{6}$$

Proof. Otherwise, there exist $\varepsilon_0 > 0$ and a sequence $(\Omega_n)_{n\in\mathbb{N}}$ of subsets of Ω with

$$\lim_{n\to\infty} Vol(\Omega_n) = 0 \tag{7}$$

and

$$\left| \int_{\Omega_n} f(x)dx \right| > \varepsilon_0 \quad \text{for all } n. \tag{8}$$

We let $f_n := f\chi_{\Omega \setminus \Omega_n}$. Then $|f_n| \leq |f|$ for all n, and, moreover, f_n converges to f pointwise almost everywhere, because of (7). Therefore, by the dominated convergence theorem 16.5,

$$\lim_{n \to \infty} \int_{\Omega} f_n(x)dx = \int_{\Omega} f(x)dx.$$

However,

$$\int_{\Omega} f(x)dx = \int_{\Omega \setminus \Omega_n} f(x)dx + \int_{\Omega_n} f(x)dx$$

$$= \int_{\Omega} f_n(x)dx + \int_{\Omega_n} f(x)dx,$$

and therefore, by (8),

$$\left| \int_{\Omega} f(x)dx - \int_{\Omega} f_n(x)dx \right| > \varepsilon_0,$$

a contradiction.

$\qquad\qquad\qquad\qquad\qquad\qquad\qquad\qquad\qquad\qquad\qquad\qquad\qquad$ $\square$

As a direct application of the dominated convergence theorem we now treat parameter dependent integrals.

Theorem 16.10 *Let $U \subset \mathbb{R}^d, y_0 \in U, f : \mathbb{R}^c \times U \to \mathbb{R} \cup \{\pm\infty\}$. Assume that*

a)   *for every fixed $y \in U$ $x \mapsto f(x, y)$ is integrable*

b)   *for almost all $x \in \mathbb{R}^c$ $y \mapsto f(x, y)$ is continuous at y_0*

c)   *there exists an integrable function $F : \mathbb{R}^c \to \mathbb{R} \cup \{\infty\}$ with the property that for every $y \in U$,*
$$|f(x, y)| \leq F(x).$$

   holds almost everywhere on $\mathbb{R}^c$.
   Then the function
$$g(y) := \int_{\mathbb{R}^c} f(x, y)dx$$

is continuous at the point y_0.

Proof. We have to show that for any sequence $(y_n)_{n \in \mathbb{N}} \subset U$ with $y_n \to y_0$,

$$g(y_n) \to g(y_0) \tag{9}$$

holds. We set

$$f_n(x) := f(x, y_n) \text{ for } n \in \mathbb{N}, f_0(x) := f(x, y_0).$$

By b), for almost all $x \in \mathbb{R}^c$ we have

$$f_0(x) = \lim_{n \to \infty} f_n(x).$$

By a) and c), the assumptions of theorem 16.5 are fulfilled and thus

$$\lim_{n \to \infty} g(y_n) = \lim_{n \to \infty} \int_{\mathbb{R}^c} f_n(x) dx$$

$$= \int_{\mathbb{R}^c} f_0(x) dx = g(y_0).$$

$\square$

Theorem 16.11 (Differentiation under the integral sign) *Let $I \subset \mathbb{R}$ be an open interval, $f : \mathbb{R}^c \times I \to \mathbb{R} \cup \{\pm\infty\}$. Assume that*

a)   *for every $t \in I$ the function $x \mapsto f(x, t)$ is integrable*

b)   *for almost all $x \in \mathbb{R}^c$, $t \mapsto f(x, t)$ is finite and is differentiable on I with respect to t*

c)   *there exists an integrable function $F : \mathbb{R}^c \to \mathbb{R} \cup \{\infty\}$ with the property that for every $t \in I$*

$$|\frac{\partial f}{\partial t}(x, t)| \leq F(x)$$

   holds for almost all $x \in \mathbb{R}^c$.
   Then the function

$$g(t) := \int_{\mathbb{R}^c} f(x, t) dx$$

is differentiable on I and

$$g'(t) = \int_{\mathbb{R}^c} \frac{\partial f}{\partial t}(x, t) dx.$$

Proof. We have to show that for any sequence $(h_n)_{n \in \mathbb{N}} \subset \mathbb{R} \setminus \{0\}$, $h_n \to 0$

$$\lim_{n \to \infty} \int_{\mathbb{R}^c} \frac{1}{h_n}(f(x, t + h_n) - f(x, t)) dx$$

$$= \int_{\mathbb{R}^c} \frac{\partial f}{\partial t}(x, t) dx \tag{10}$$

holds, as the left side of this equation is the differential quotient of $g(t)$.

We set

$$f_n(x) = \frac{1}{h_n}(f(x, t + h_n) - f(x, t)),$$

$$f_0(x) = \frac{\partial f}{\partial t}(x, t). \tag{11}$$

By the mean value theorem for almost all $x \in \mathbb{R}^c$, there exists, on account of (b), a $\theta_n = \theta_n(x)$ with $-h_n \leq \theta_n \leq h_n$ such that

$$f_n(x) = \frac{\partial f}{\partial t}(x, t + \theta_n).$$

By a), f_n is integrable and by c) $|f_n(x)| \leq F(x)$ holds. Therefore it follows from theorem 16.5 that

$$\lim_{n \to \infty} \int_{\mathbb{R}^c} f_n(x)dx = \int_{\mathbb{R}^c} f_0(x)dx$$

and by our notations (11), this is equivalent to (10).   □

Exercises for § 16

1)   Let $f : \mathbb{R} \to \mathbb{R}$ be integrable, $a, b \in \mathbb{R}, a < b$. Show

$$\lim_{t \to 0} \int_a^b |f(x + t) - f(x)|dx = 0.$$

Conclude $\lim_{t \to 0} \int_a^b f(x + t)dx = \int_a^b f(x)dx.$

2)

a)   **(Gamma function)** Let $\lambda > 0$. Show that

$$\Gamma(\lambda) := \int_0^\infty e^{-x} x^{\lambda - 1} dx$$

exists.

b)   For $n \in \mathbb{N}, \frac{|x|}{n} < 1$, show

$$(1 - \frac{x}{n})^n \leq (1 - \frac{x}{n + 1})^{n+1} \quad \text{(cf. §0)}.$$

c)   For $x > 0$, show

$$\frac{x-1}{x} \leq \int_1^x \frac{1}{t} dt \leq x - 1.$$

d)   Use a) – c) to show $\lim\limits_{n\to\infty} \int_0^n (1 - \frac{x}{n})^n x^{\lambda-1} dx = \Gamma(\lambda)$.

e)   For $\lambda > 1$, show $\int_0^\infty \frac{e^{-x}}{1-e^{-x}} x^{\lambda-1} dx = \Gamma(\lambda) \sum\limits_{n=1}^\infty \frac{1}{n^\lambda}$.

   (Hint: $\frac{e^{-x}}{1-e^{-x}} = \sum\limits_{n=1}^\infty e^{-nx}$.)

3)   Let $(f_n)_{n\in\mathbb{N}}$ be a monotonically decreasing sequence of nonnegative integrable functions with

$$\lim_{n\to\infty} \int_{\mathbb{R}^d} f_n(x) dx = 0.$$

Show that (f_n) converges to 0 almost everywhere. Give an example to show that the convergence need not take place everywhere.

4)   For $x > 0$, consider

$$f(x) := \int_0^\infty e^{-xt} dt = \frac{1}{x}.$$

Show that one may differentiate infinitely often w.r.t. x under the integral, and derive the formula

$$\int_0^\infty t^n e^{-t} dt = n!$$

without integration by parts.

5)   Let $f_n : \mathbb{R}^d \to \mathbb{R} \cup \{\pm\infty\}$ be a sequence of integrable functions for which $\sum\limits_{n=1}^\infty |f_n|$ is integrable as well. Show that

$$\int_{\mathbb{R}^d} \sum_{n=1}^\infty f_n(x) dx = \sum_{n=1}^\infty \int_{\mathbb{R}^d} f_n(x) dx.$$

17. Measurable Functions and Sets. Jensen's Inequality. The Theorem of Egorov

We introduce the general notion of a measurable function and a measurable set. Measurable functions are characterized as pointwise limits of finite valued functions. Jensen's inequality for the integration of convex functions and Egorov's theorem saying that an almost everywhere converging sequence of functions also converges almost uniformly, i.e. uniformly except on a set of arbitrarily small measure, are derived. We conclude this § with an introduction to the general theory of measures.

In the definition of integrability of a function f we required that f must have a finite integral. So not all continuous functions, e.g. the non-zero constants, are integrable on $\mathbb{R}^d$. We are now going to introduce a larger class of functions which includes the integrable as well as the continuous functions on $\mathbb{R}^d$.

Definition 17.1 Let f, g, h be functions defined on $\mathbb{R}^d$ with $g \leq h$. We define the medium function $\mathrm{med}(f, g, h)$ by:

$$\mathrm{med}\,(g, f, h) := \inf\{\sup(f, g), h\}.$$

This expression arises in that one cuts f from above by h and from below by g.

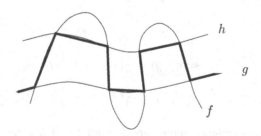

Definition 17.2 $f : \mathbb{R}^d \to \mathbb{R} \cup \{\pm\infty\}$ is called measurable if for every compact cube $W \subset \mathbb{R}^d$ and every $\mu > 0$ the function $\mathrm{med}\,(-\mu\chi_W, f, \mu\chi_W)$ is integrable.

Theorem 17.3

(i)   *All continuous and all integrable functions are measurable.*

(ii)   *If f and g are measurable then so are $|f|, f^+, f^-, \alpha f + \beta g$ $(\alpha, \beta \in \mathbb{R})$, $\sup(f, g)$ and $\inf(f, g)$.*

(iii)   *If $(f_n)_{n \in \mathbb{N}}$ is a sequence of measurable functions which converges almost everywhere to a function f, then f is also measurable.*

Proof.

(ii)   Let W be a compact cube, $\mu > 0$. We have

$$\mathrm{med}\,(-\mu\chi_W, |f|, \mu\chi_W) = |\,\mathrm{med}\,(-\mu\chi_W, f, \mu\chi_W)|.$$

If f is measurable, the right hand side is integrable (theorem 14.9) and therefore $|f|$ is measurable.
We now set, for $n \in \mathbb{N}$,

$$f_n := \mathrm{med}\,(-n\mu\chi_W, f, n\mu\chi_W)$$

and define g_n analogously. Then f_n and g_n are integrable, as f and g are measurable. For $x \in \mathbb{R}^d \backslash W$ we have $f_n(x) = 0 = g_n(x)$ and for $x \in W$, $\lim_{n \to \infty} f_n(x) = f(x)$, $\lim_{n \to \infty} g_n(x) = g(x)$ holds and therefore also

$$\lim \mathrm{med}\,(-\mu\chi_W, \alpha f_n + \beta g_n, \mu\chi_W) = \mathrm{med}\,(-\mu\chi_W, \alpha f + \beta g, \mu\chi_W).$$

$\mathrm{med}\,(-\mu\chi_W, \alpha f_n + \beta g_n, \mu\chi_W)$ is integrable (by theorem 14.9) and

$$|\,\mathrm{med}\,(-\mu\chi_W, \alpha f_n + \beta g_n, \mu\chi_W)| \leq \mu\chi_W.$$

Therefore, by the dominated convergence theorem, $\mathrm{med}\,(-\mu\chi_W, \alpha f + \beta g, \mu\chi_W)$ is integrable and hence $\alpha f + \beta g$ is measurable.
The measurability of $f^+, f^-, \sup(f, g), \inf(f, g)$ follows from the relations

$$f^+ = \frac{1}{2}(|f| + f)$$

$$f^- = \frac{1}{2}(|f| - f)$$

$$\sup(f, g) = \frac{1}{2}(f + g) + \frac{1}{2}|f - g|$$

$$\inf(f, g) = \frac{1}{2}(f + g) - \frac{1}{2}|f - g|.$$

(i)   If f is continuous, then so are f^+ and f^-. The functions $f^+\chi_W$ and $f^-\chi_W$ then are in the class $H_S(\mathbb{R}^d)$ and thereby integrable. Therefore

$$\mathrm{med}\,(-\mu\chi_W, f^+, \mu\chi_W)$$

is integrable and f^+ is thus measurable, and similarly also f^-. By (ii), $f = f^+ - f^-$ is then also measurable.
If f is integrable, so is $\mathrm{med}\,(-\mu\chi_W, f, \mu\chi_W)$, by theorem 14.9, and so f is measurable.

(iii)   Since the functions f_n are measurable, the functions med $(-\mu\chi_W, f_n,$
$\mu\chi_W)$ are integrable and converge, by assumption, almost everywhere
to med $(-\mu\chi_W, f, \mu\chi_W)$. Moreover, $|\,\text{med}\,(-\mu\chi_W, f_n, \mu\chi_W)| \le \mu\chi_W$
and the dominated convergence theorem gives the integrability of med
$(-\mu\,\chi_W, f, \mu\chi_W)$ and thereby the measurability of f.   □

Lemma 17.4 $f : \mathbb{R}^d \to \mathbb{R} \cup \{\pm\infty\}$ *is measurable precisely when for every*
non-negative integrable function $g : \mathbb{R}^d \to \mathbb{R} \cup \{\pm\infty\}$,

$$\text{med}\,(-g, f, g)$$

is integrable.

Proof. " $\Longleftarrow$ " : obvious

"$\Longrightarrow$ " : We set
$$W_n = \{x = (x^1, \ldots x^d) \in \mathbb{R}^d : |x^i| \le n \text{ for } i = 1, \ldots d\},$$
$$f_n = \text{med}\,(-n\chi_{W_n}, f, n\chi_{W_n}) \quad (n \in \mathbb{N}).$$

Then, by assumption, f_n is integrable. Furthermore, med $(-g, f_n, g)$ is also
integrable (theorem 14.9), bounded in absolute value by g and converges to
med $(-g, f, g)$. By the dominated convergence theorem, med $(-g, f, g)$ is then
integrable.   □

Theorem 17.5 *If* f *is measurable and* $|f| \le g$ *for an integrable function* g,
then f *is integrable.*

Proof. By assumption, $f = \text{med}\,(-g, f, g)$ and the result follows from lemma
17.4.   □

Corollary 17.6 *If* f *is measurable and* $|f|$ *integrable, then* f *is integrable.*
                □

We note yet another consequence of the statements of §15.

Lemma 17.7 *If* f *is measurable and* $f = g$ *almost everywhere, then* g *is also*
measurable.

Definition 17.8 A subset $A \subset \mathbb{R}^d$ is called measurable if its characteristic
function is measurable. In case A is measurable but not integrable, we set
$\text{Vol}\,(A) = \infty$.

Similar to theorem 14.15 one proves, this time with the help of theorem
17.3, that for measurable sets A and B, $A \cup B$, $A \cap B$ and $A \backslash B$ are also
measurable. However, using the dominated convergence theorem, we obtain
a yet stronger statement

Theorem 17.9

(i)   If $A_1 \subset A_2 \subset \ldots \subset \mathbb{R}^d$ are measurable, then $\overset{\infty}{\underset{n=1}{\cup}} A_n$ is measurable

and $\mathrm{Vol}\left(\overset{\infty}{\underset{n=1}{\cup}} A_n\right) = \underset{n\to\infty}{\lim} \mathrm{Vol}(A_n)$.

(ii)   Let $\{B_n\}_{n\in\mathbb{N}}$ be a sequence of measurable sets. Then $\overset{\infty}{\underset{n=1}{\cup}} B_n$ is mea-

surable and $\mathrm{Vol}\left(\overset{\infty}{\underset{n=1}{\cup}} B_n\right) \leq \overset{\infty}{\underset{n=1}{\sum}} \mathrm{Vol}(B_n)$. If the sets B_n are disjoint,

then $\mathrm{Vol}\left(\overset{\infty}{\underset{n=1}{\cup}} B_n\right) = \overset{\infty}{\underset{n=1}{\sum}} \mathrm{Vol}(B_n)$.

(iii)   If $C_1 \supset C_2 \subset \ldots$ are measurable with $\mathrm{Vol}(C_1) < \infty$, then $\mathrm{Vol}\left(\overset{\infty}{\underset{n=1}{\cap}} C_n\right)$

$= \underset{n\to\infty}{\lim} \mathrm{Vol}(C_n)$.

Proof. (i) follows by applying the dominated convergence theorem to the functions χ_{A_n}, in case $\underset{n\in\mathbb{N}}{\sup} \mathrm{Vol}(A_n) < \infty$. In the other case, $\mathrm{Vol}\left(\overset{\infty}{\underset{n=1}{\cup}} A_n\right) = \infty$.

(ii) follows from (i) applied to $A_n = \overset{n}{\underset{j=1}{\cup}} B_j$, using the formula

$$\mathrm{Vol}(A \cup B) + \mathrm{Vol}(A \cap B) = \mathrm{Vol}\, A + \mathrm{Vol}\, B$$

for measurable sets A and B (theorem 14.15).

(iii) follows from (i) applied to $A_n = C_1 \backslash C_n$. Namely, $C_1 = \overset{\infty}{\underset{n=1}{\cap}} C_n \cup \overset{\infty}{\underset{n=1}{\cup}} A_n$, hence

$$\mathrm{Vol}(C_1) = \mathrm{Vol}\left(\overset{\infty}{\underset{n=1}{\cap}} C_n\right) + \mathrm{Vol}\left(\overset{\infty}{\underset{n=1}{\cup}} A_n\right), \quad \text{again by theorem 14.15}$$

$$= \mathrm{Vol}\left(\overset{\infty}{\underset{n=1}{\cap}} C_n\right) + \mathrm{Vol}(C_1) - \underset{n\to\infty}{\lim} \mathrm{Vol}(C_n)$$

by (i) and theorem 14.15.

$\square$

Definition 17.10 $\varphi : \mathbb{R}^d \to \mathbb{R}$ is said to be a simple function if there exist disjoint measurable sets $B_1, \ldots, B_k \subset \mathbb{R}^d$ and $c_1, \ldots, c_k \in \mathbb{R}$ such that

$$\varphi = \sum_{j=1}^{k} c_j \chi_{B_j}.$$

Theorem 17.11 *The following statements are equivalent for a function*

$f : \mathbb{R}^d \to \mathbb{R} \cup \{\pm\infty\}$.

(i) f is measurable.

(ii)   For every $c \in \mathbb{R}$, the set $A_c := \{x \in \mathbb{R}^d : f(x) \geq c\}$ is measurable.

(iii) f is the pointwise limit of simple functions (the convergence thus occurs everywhere and not just almost everywhere).

Proof. (iii) $\Rightarrow$ (i): this follows directly from theorem 17.3 (iii).

(i) $\Rightarrow$ (ii): For $c \in \mathbb{R}$, we put

$$f_n(x) := n \min(f(x), c) - n \min(f(x), c - \frac{1}{n}).$$

If $f(x) \geq c$, then $f_n(x) = 1$, and if $f(x) < c$, then for $n \geq N$ (where N depends on x) $f_n(x) = 0$. Hence

$$\lim_{n \to \infty} f_n = \chi_{A_c},$$

and by theorem 17.3 (iii) χ_{A_c}, and thus A_c, is measurable.

(ii) $\Rightarrow$ (iii): For $n \in \mathbb{N}$ and $j = 1, \ldots, n2^{n+1}$ we set

$$c_{n,j} := -n + (j-1)2^{-n}$$
$$B_{n,j} := \{c_{n,j} \leq f(x) < c_{n,j+1}\},$$
$$\varphi_n := \sum_{j=1}^{n \cdot 2^{n+1}} c_{n,j} \chi_{B_{n,j}}$$

Now $B_{n,j} = A_{c_{n,j}} \setminus A_{c_{n,j+1}}$ is, by assumption, measurable and therefore φ_n is simple. Moreover, φ_n converges pointwise to f. $\square$

Remark. If f is measurable and bounded, say $|f| \leq K$, then for $n \geq K$

$$|f - \varphi_n| \leq 2^{-n}$$

with the functions φ_n of the previous proof. In this case, f is even the uniform limit of simple functions. In this formulation, the basic idea of Lebesgue integration theory does not consist in dividing the domain of a function uniformly as in Riemann integration theory and expressing the function as a limit of step functions, but rather to divide the range uniformly and to consider limits of simple functions. One notes also that the functions φ_n have been constructed so that in case f is bounded from below, say $f \geq -K$,

$$\varphi_n \leq \varphi_{n+1} \leq f$$

holds for all $n \geq K$. In this case, f is even the limit of monotonically increasing simple functions.

Corollary 17.12 *Let $f_1, \ldots f_k : \mathbb{R}^d \to \mathbb{R}$ be measurable, $Q : \mathbb{R}^k \to \mathbb{R}$ continuous. Then the function h defined by $h(x) = Q(f_1(x), \ldots, f_k(x))$ is also measurable.*

Proof. By theorem 17.11, the functions $f_1, \ldots, f_k$ are limits of simple functions $\varphi_{1,n}, \ldots, \varphi_{k,n}$. Then $\eta_n(x) := Q(\varphi_{1,n}(x), \ldots, \varphi_{k,n}(x))$ defines likewise a simple function. As Q is continuous, η_n converges pointwise to h, and theorem 17.11 yields the assertion.   □

Corollary 17.13 *If f and g are measurable then so is $f \cdot g$.*

Proof. One sets $Q(x^1, x^2) = x^1 \cdot x^2$ in corollary 17.12.   □

We now wish to derive *Jensen's inequality* concerning convex functions. We recall that a function
$$K : \mathbb{R} \to \mathbb{R}$$
is convex if for all $x, y \in \mathbb{R}, 0 \leq t \leq 1$,

$$K(tx + (1-t)y) \leq tK(x) + (1-t)K(y). \tag{1}$$

Inductively, one verifies that a convec function satisfies

$$K\left(\sum_{i=1}^{n} t_i x_i\right) \leq \sum_{i=1}^{n} t_k K(x_i) \tag{2}$$

whenever $x_i \in \mathbb{R}, 0 \leq t_i \leq 1$ for $i = 1, \ldots, n$ and $\sum_{i=1}^{n} t_i = 1$.

Theorem 17.14 (Jensen's inequality) *Let $B \subset \mathbb{R}^d$ be bounded and measurable, $f : B \to \mathbb{R}$ be integrable, $K : \mathbb{R} \to \mathbb{R}$ be convex. Suppose that $K \circ f : B \to \mathbb{R}$ is also integrable. Then*

$$K\left(\frac{1}{\text{Vol}(B)} \int_B f(x)dx\right) \leq \frac{1}{\text{Vol}(B)} \int_B K(f(x))dx. \tag{3}$$

Proof. We first consider the case where f is simple, i.e.

$$f = \sum_{i=1}^{n} c_i \xi_{B_i},$$

where the $B_i, i = 1, \ldots, n$ are disjoint measurable sets with

$$\bigcup_{i=1}^{n} B_i = B. \tag{4}$$

In that case

$$K\left(\frac{1}{\mathrm{Vol}\,(B)}\int_B \sum_{i=1}^n c_i\chi_{B_i}\right) = K\left(\sum_{i=1}^n \frac{\mathrm{Vol}\,(B_i)}{\mathrm{Vol}\,(B)}c_i\right)$$

$$\leq \sum_{i=1}^n \frac{\mathrm{Vol}\,(B_i)}{\mathrm{Vol}\,(B)}K(c_i)\quad\text{by (2), since}$$

$$\sum_{i=1}^n \frac{\mathrm{Vol}\,(B_i)}{\mathrm{Vol}\,(B)} = 1\quad\text{because of (4)}$$

$$= \frac{1}{\mathrm{Vol}\,(B)}\sum_{i=1}^d \int_B K(c_i)\chi_{B_i}$$

which is the required inequality.

We now wish to treat the case of a general integrable f through approximating f by simple functions as described in theorem 17.11. If K is bounded on $f(B)$, we may then use the dominated convergence theorem 16.5 to obtain the inequality for f from the corresponding ones for the approximating simple functions. In the general case, we consider

$$A_n := \{x \in B : |f(x)| \leq n\}.$$

Since $|f|$ is integrable together with f (see theorem 14.9), and since

$$\int_{B\setminus A_n} |f(x)|dx \geq n\,\mathrm{Vol}\,(B\setminus A_n),$$

we conclude that

$$\lim_{n\to\infty}\,\mathrm{Vol}\,(B\setminus A_n) = 0, \tag{5}$$

hence

$$\lim_{n\to\infty}\,\mathrm{Vol}\,(A_n) = \mathrm{Vol}\,(B). \tag{6}$$

By what we have already shown, Jensen's inequality holds for A_n, i.e.

$$K\left(\frac{1}{\mathrm{Vol}\,(A_n)}\int_{A_n} f(x)dx\right) \leq \frac{1}{\mathrm{Vol}\,(A_n)}\int_{A_n} K(f(x))dx.$$

Using (5) and (6), applying corollary 16.7 and noting that convex functions are continuous (see also lemma 22.5 below for a proof in a more general context), letting $n \to \infty$ then yields (3).   □

We next show Egorov's theorem

Theorem 17.15 *Let $A \subset \mathbb{R}^d$ be measurable, $\mathrm{Vol}\,(A) < \infty$, and suppose the sequence $(f_n)_{n\in\mathbb{N}}$ of measurable functions converges to the measurable* *function f almost everywhere on A. Then for every $\varepsilon > 0$, there exists a measurable $B \subset A$ satisfying*

$$\text{Vol } (A \backslash B) < \varepsilon$$

with the property that f_n converges to f uniformly on B. (One says that $(f_n)_{n \in \mathbb{N}}$ converges to f almost uniformly on A.)

Proof. We put, for $m, n \in \mathbb{N}$,

$$C_{m,n} := \bigcup_{\nu=n}^{\infty} \{x \in A : |f_\nu(x) - f(x)| \geq 2^{-m}\}.$$

Theorem 17.9 (iii) implies for every m that

$$\lim_{n \to \infty} \text{Vol } (C_{m,n}) = \text{Vol } \left(\bigcap_{n=1}^{\infty} C_{m,n} \right) = 0,$$

since f_n converges to f almost everywhere on A.

Therefore, for each $m \in \mathbb{N}$, we may find $N(m) \in \mathbb{N}$ with

$$\text{Vol } (C_{m,N(m)}) < \varepsilon 2^{-m}.$$

Thus, $B := A \backslash \bigcup_{m=1}^{\infty} C_{m,N(m)}$ satisfies

$$\text{Vol } (A \backslash B) < \varepsilon \quad \text{by theorem 17.9 (ii).}$$

Also, by construction, for $x \in B$, we have

$$|f_\nu(x) - f(x)| < 2^{-m} \quad \text{for all } \nu \geq N(m)$$

which implies the uniform convergence on B. $\qquad\qquad\square$

In this book, we are exclusively working with the Lebesgue measure. In many fields of mathematics, however, also other measures occur and need to be considered. We therefore now give a brief sketch of abstract measure theory. This theory in fact can be presented as a natural abstraction of the theory of the Lebesgue measure. For the axiomatic approach, one needs a collection of sets with certain properties:

Definition 17.16 A nonempty collection Σ of subsets of a set M is called a σ-algebra if:

(i)   Whenever $A \in \Sigma$, then also $M \backslash A \in \Sigma$.

(ii)   Whenever $(A_n)_{n \in \mathbb{N}} \subset \Sigma$, then also $\bigcup_{n=1}^{\infty} A_n \in \Sigma$.

One easily observes that for a σ-algebra on M, necessarily $\emptyset \in \Sigma$ and $M \in \Sigma$, and whenever $(A_n)_{n\in\mathbb{N}} \subset \Sigma$, then also $\bigcap_{n=1}^{\infty} A_n \in \Sigma$.

Definition 17.17 Let M be a set with a σ-algebra Σ.
A function $\mu : \Sigma \to \mathbb{R}^+ \cup \{\infty\}$, i.e.,

$$0 \le \mu(A) \le \infty \quad \text{for every } A \in \Sigma,$$

with

$$\mu(\emptyset) = 0,$$

is called a measure on (M, Σ) if
whenever the sets $A_n \in \Sigma, n \in \mathbb{N}$, are pairwise disjoint, then

$$\mu\left(\bigcup_{n=1}^{\infty} A_n\right) = \sum_{n=1}^{\infty} \mu(A_n)$$

(countable additivity).
A measure μ with

$$\mu(M) = 1$$

is called a probability measure.

The essence of this definition is that the properties derived for the Lebesgue measure in theorems 14.15 and 17.9 now become the constitutive axioms for a measure. It is easy to verify that not only (ii), but also (i) and (iii) of theorem 17.9 hold for a measure μ and sets $A_n, B_n, C_n \in \Sigma$ for $n \in \mathbb{N}$. In the proof, one uses the formula

$$\mu(A_1 \cup A_2) + \mu(A_1 \cap A_2) = \mu(A_1) + \mu(A_2) \quad \text{for all } A_1, A_2 \in \Sigma$$

and the monotonicity

$$\mu(A_1) \le \mu(A_2)$$

whenever $A_1 \subset A_2, A_1, A_2 \in \Sigma$. For example, the latter inequality follows from $A_2 \backslash A_1 = M \backslash (A_1 \cup M \backslash A_2) \in \Sigma$ (by (i) and (ii) of definition 17.16), and then

$$\mu(A_2) = \mu(A_1) + \mu(A_2 \backslash A_1) \quad \text{by the additivity of } \mu$$
$$\ge \mu(A_1) \qquad\qquad\qquad \text{by the nonnegativity of } \mu.$$

Definition 17.18 A triple (M, Σ, μ) consisting of a set M, a σ-algebra Σ on M, and a measure $\mu : \Sigma \to \mathbb{R}^+ \cup \{\infty\}$ is called a measure space. The elements of Σ then are called measurable.

Definition 17.19 Let (M, Σ, μ) be a measure space.
We say that a property holds μ-everywhere if it holds on a set $A \in \Sigma$ with $\mu(M \backslash A) = 0$, i.e., outside a set of measure 0.

Definition 17.20 Let (M, Σ, μ) be a measure space. A function $f : M \to \mathbb{R} \cup \{\pm\infty\}$ is called measurable if for every $c \in \mathbb{R}$, the set

$$A_c := \{x \in M : f(x) \geq c\} \in \Sigma.$$

Lemma 17.21 *Let* $(f_n)_{n\in\mathbb{N}}$ *be a sequence of measurable functions on a measure space* (M, Σ, μ). *Then*

$$\liminf_{n\to\infty} f_n \quad and \quad \limsup_{n\to\infty} f_n$$

are measurable. In particular, when it exists,

$$\lim_{n\to\infty} f_n$$

is measurable.

Proof. A function f is measurable precisely if the sets

$$B_c := \{x \in M : f(x) < c\}$$

are measurable, because $B_c = M \backslash A_c$.

We now have, for any sequence $(g_n)_{n\in\mathbb{N}}$ of measurable functions,

$$\{x \in M : \inf_{n\in\mathbb{N}} g_n(x) < c\} = \bigcup_{n\in\mathbb{N}} \{x \in M : g_n(x) < c\},$$

and since the latter sets are measurable, so then is the former.

Therefore, $\inf_{n\in\mathbb{N}} g_n$ is also a measurable function.

By the same reasoning, $\sup_{n\in\mathbb{N}} g_n$ is measurable.

Then

$$\liminf_{n\to\infty} f_n(x) = \sup_m \inf_{n\geq m} f_n(x)$$

and

$$\limsup_{n\to\infty} f_n(x) = \inf_m \sup_{n\geq m} f_n(x)$$

are also measurable, if the f_n are.

$\square$

Equipped with lemma 17.21, we may now extend all the results derived in the present § about functions that are measurable for the Lebesgue measure to measurable functions for an arbitrary measure μ. In particular, measurable functions can be characterized as pointwise limits of simple functions as in theorem 17.11, and Egorov's theorem 17.15 holds in our general context.

Exercises for § 17

1) Let $f : \mathbb{R}^d \to \mathbb{R}$ be measurable. Put

$$g(x) := \begin{cases} \frac{1}{f(x)} & \text{if } f(x) \neq 0 \\ 0 & \text{if } f(x) = 0. \end{cases}$$

Prove that g is measurable as well,
a)   by showing that the sets $A_c := \{x \in \mathbb{R}^d : g(x) \geq c\}$ are measurable for all $c \in \mathbb{R}$.
b)   by showing that g is the pointwise limit of simple functions.

2) Let $f : \mathbb{R}^d \to \mathbb{R}$ be nonnegative and measurable. Then f is the pointwise limit of a monotonically increasing sequence $(\varphi_n)_{n \in \mathbb{N}}$ of simple functions. Show that f is integrable if and only if the sequence $\int_{\mathbb{R}^d} \varphi_n(x)dx$ converges to some finite value. Moreover, in that case

$$\int_{\mathbb{R}^d} f(x)dx = \lim_{n \to \infty} \int_{\mathbb{R}^d} \varphi_n(x)dx.$$

3) Let $A \subset \mathbb{R}^d$ be integrable, $f : A \to \mathbb{R} \cup \{\pm\infty\}$ be measurable and finite almost everywhere. Show that for every $\varepsilon > 0$ there exists a bounded measurable function g with

$$\text{Vol } \{x \in A : g(x) \neq f(x)\} < \varepsilon.$$

4) Let $A \subset \mathbb{R}^d$ be measurable and bounded.
a)   Let $f : A \to \mathbb{R}$ be nonnegative and integrable. Show

$$\lim_{n \to \infty} n \text{ Vol } \{x : f(x) > n\} = 0.$$

b)   Does the vanishing of this limit conversely imply the integrability of a nonnegative measurable function?
c)   Show that a measurable function $f : A \to \mathbb{R}$ is integrable if and only if

$$\sum_{n=1}^{\infty} \text{Vol } \{x : |f(x)| \geq n\} < \infty.$$

5) Let $A \subset \mathbb{R}^d$ be measurable, $f_n, f : A \to \mathbb{R}$ be measurable $(n \in \mathbb{N})$. We say that $(f_n)_{n \in \mathbb{N}}$ converges to f almost uniformly if for every $\varepsilon > 0$ there exists $B \subset A$ with Vol $(B) < \varepsilon$ such that f_n converges to f uniformly on $A \backslash B$. Show that if f_n converges to f almost uniformly, then f_n converges to f pointwise almost everywhere, i.e. except on a null set.
(This is of course the converse of Egorov's theorem.)

18. The Transformation Formula

The general transformation formula for multiple integrals is derived. Transformation from Euclidean to polar coordinates is discussed in detail.

Definition 18.1 Let $U, V \subset \mathbb{R}^d$ be open. A bijective map $\Phi : U \to V$ is called a homeomorphism if Φ and Φ^{-1} are continuous, and a diffeomorphism if Φ and Φ^{-1} are continuously differentiable.

The aim of this paragraph is to prove the following transformation formula for multiple integrals.

Theorem 18.2 Let $U, V \subset \mathbb{R}^d$ be open, $\Phi : U \to V$ a diffeomorphism. A function $f : V \to \mathbb{R} \cup \{\pm\infty\}$ is integrable precisely when the function $(f \circ \Phi)|\det D\Phi|$ is integrable over U, and in this case one has

$$\int_{\Phi(U)} f(y)dy = \int_U f(\Phi(x))|\det D\Phi(x)|dx.$$

We first prove

Lemma 18.3 Let $U, V \subset \mathbb{R}^d$ be open, $\Phi : U \to V$ a homeomorphism. Let Φ be differentiable at $x_0 \in U$. Then for every sequence $(W_n)_{n\in\mathbb{N}}$ of open or closed cubes in U which contain x_0 and whose side length tends to 0 as $n \to \infty$, one has

$$\lim_{n\to\infty} \frac{\mathrm{Vol}\,(\Phi(W_n))}{\mathrm{Vol}\,(W_n)} = |\det D\Phi(x_0)|. \tag{1}$$

Proof. We remark first that the assertion holds when Φ is an affine linear transformation, as follows from lemma 13.12. We now set

$$A := D\Phi(x_0).$$

As in the proof of lemma 13.12, there exist matrices P_1, P_2 with $|\det P_1| = 1 = |\det P_2|$ and a diagonal matrix D such that

$$A = P_1 D P_2.$$

Using the fact that the assertion of the lemma holds for linear transformations, as already remarked at the beginning, and that, if necessary, we can replace Φ by $P_1^{-1}\Phi P_2^{-1}$, we may assume that $D\Phi(x_0)$ is a diagonal matrix

$$D = \begin{pmatrix} \lambda_1 & & 0 \\ & \ddots & \\ 0 & & \lambda_d \end{pmatrix}.$$

Namely, one has for any measurable set A and every linear transformation P with $|\det P| = 1$, $\mathrm{Vol}\,(PA) = \mathrm{Vol}\,(A)$.

To simplify the notations, we can also assume that

$$x_0 = 0 = \Phi(x_0).$$

By definition of the differential, there exists for any $\varepsilon > 0$ a $\delta > 0$ such that

$$\|\Phi(x) - \Phi(x_0) - D\Phi(x_0)(x - x_0)\| \leq \varepsilon\|x - x_0\|,$$

if $\|x - x_0\| < \delta$, thus by our simplifications

$$\|\Phi(x) - D \cdot x\| \leq \varepsilon\|x\| \tag{2}$$

for $\|x\| < \delta$.

We now distinguish two cases

1) $|\det D| = |\det D\Phi(x_0)| = 0$. Then at least one of the diagonal elements λ_i vanishes and DW_n lies in the hyperplane $\{x^i = 0\}$. For $\varepsilon > 0$, choose $\delta > 0$ as in (2) and let the side length ℓ_n of W_n satisfy $\ell_n < \delta$, which holds, by assumption, for sufficiently large n. On account of (2), $\Phi(W_n)$ lies in a parallelepiped with side lengths

$$(|\lambda_i| + 2\varepsilon)\ell_n.$$

As at least one of the λ_i's vanishes, it follows, with $L := \max\limits_{i=1,\dots d} |\lambda_i|$, that

$$\frac{\mathrm{Vol}\,(\Phi(W_n))}{\mathrm{Vol}\,(W_n)} \leq (L + 2\varepsilon)^{d-1} 2\varepsilon,$$

thus

$$\lim_{n\to\infty} \frac{\mathrm{Vol}\,(\Phi(W_n))}{\mathrm{Vol}\,(W_n)} = 0.$$

2) $|\det D| = |\det D\Phi(x_0)| \neq 0$. By considering the map $D^{-1} \circ \Phi$ instead of Φ, which has the differential $D^{-1} \cdot D\Phi(x_0) = \mathrm{id}$ at x_0, we can assume that

$$D = \mathrm{id}.$$

With these simplifications we obtain

$$\|\Phi(x) - x\| \leq \varepsilon\|x\| \tag{3}$$

for $\|x\| < \delta$.

Let $\varepsilon < \frac{1}{4}$ and let the side length ℓ_n of W_n satisfy again

$$\ell_n < \delta.$$

Let W_n^- and W_n^+ be cubes concentric with W_n of side lengths $(1 - 3\varepsilon)\ell_n$ and $(1 + 3\varepsilon)\ell_n$, respectively. For n sufficiently large, W_n^+ is also contained in U.

By (3) one has

$$\Phi(W_n) \subset W_n^+$$

and as Φ is a homeomorphism, $\Phi(\partial W_n) = \partial\Phi(W_n)$ also holds. By (3) one has

$$\Phi(\partial W_n) \cap W_n^- = \emptyset,$$

thus

$$\partial\Phi(W_n) \cap W_n^- = \emptyset. \tag{4}$$

On the other hand, on account of (3)

$$\Phi(W_n) \cap W_n^- \neq \emptyset, \tag{5}$$

for Φ moves the centre of W_n at most by $\frac{1}{2}\varepsilon\ell_n$, thus by less than $\frac{1}{2}(1 - 3\varepsilon)\ell_n$, since $\varepsilon < \frac{1}{4}$. As W_n^- is connected and satisfies (4) and (5), it follows that

$$W_n^- \subset \Phi(W_n).$$

Thus, altogether

$$W_n^- \subset \Phi(W_n) \subset W_n^+,$$

and therefore

$$(1 - 3\varepsilon)^d \leq \frac{\mathrm{Vol}\,(\Phi(W_n))}{\mathrm{Vol}\,(W_n)} \leq (1 + 3\varepsilon)^d,$$

so far as $\ell_n < \delta$. It follows that

$$\lim_{n \to \infty} \frac{\mathrm{Vol}\,(\Phi(W_n))}{\mathrm{Vol}\,(W_n)} = 1.$$

$\square$

Definition 18.4 We say that Φ has a measure derivative $\Delta_\Phi(x_0)$ at the point x_0, if for any sequence of open or closed cubes W_n containing x_0 and whose side lengths tend to zero,

$$\lim_{n \to \infty} \frac{\mathrm{Vol}\,(\Phi(W_n))}{\mathrm{Vol}\,(W_n)} = \Delta_\Phi(x_0).$$

Lemma 18.5 *Let Φ have measure derivative $\Delta_\Phi(x)$ at every point of a closed cube W with*

$$\Delta_\Phi(x) \leq K. \tag{5}$$

Then for every cube $W' \subset W$ one has

$$\frac{\mathrm{Vol}\,(\Phi(W'))}{\mathrm{Vol}\,(W')} \leq K. \tag{6}$$

Proof. Assume that

$$\frac{\mathrm{Vol}\,(\Phi(W'))}{\mathrm{Vol}\,(W')} \geq K_1 > K. \tag{7}$$

We may assume that W' is closed, as $\partial W'$ is a null set (e.g. by corollary 15.4) and so $\mathrm{Vol}\,(\overline{W}') = \mathrm{Vol}(W')$, and $\mathrm{Vol}\,(\Phi(\overline{W}')) \geq \mathrm{Vol}\,(\Phi(W'))$ and thereby (7) continues to hold by passage to the closure. We subdivide W' into 2^d subcubes of the same size, with disjoint interiors. At least one of these, say W_1, satisfies

$$\frac{\mathrm{Vol}\,(\Phi(W_1))}{\mathrm{Vol}\,(W_1)} \geq K_1.$$

By continued subdivisions we obtain a sequence of cubes $(W_n)_{n \in \mathbb{N}}$ with

$$\frac{\mathrm{Vol}\,(\Phi(W_n))}{\mathrm{Vol}\,(W_n)} \geq K_1. \tag{8}$$

Let x_0 be the limit of the sequence of middle points of the W_n. x_0 is contained in all W_n. From (8) it follows that

$$\Delta_\Phi(x_0) \geq K_1,$$

in contradiction to the assumption (5). Therefore (7) could not hold.   □

Lemma 18.6 *Let $U \subset \mathbb{R}^d$ be open, $\Phi : U \to \mathbb{R}^d$ injective and continuous with measure derivative $\Delta_\Phi(x)$ in U. Let Δ_Φ be bounded on every compact subset of U. Then for every (open or closed) cube W with $\overline{W} \subset U$ one has*

$$\mathrm{Vol}\,(\Phi(W)) = \int_W \Delta_\Phi(x)dx. \tag{9}$$

Proof. Divide W into subcubes $W_1, \ldots, W_k$; we shall here demand that

$$W = \bigcup_{i=1}^k W_i, \; W_i \cap W_j = \emptyset \text{ for } i \neq j,$$

and therefore we can choose W_i neither open nor closed, rather we must attach the common sides with exactly one of the cubes under consideration. But as the previous statements hold for open as well as for closed cubes, they hold also for cubes W' which are neither open nor closed, on account of $\mathring{W}' \subset W' \subset \overline{W}'$ and as $\mathrm{Vol}\,(\mathring{W}') = \mathrm{Vol}\,(\overline{W}')$. For such a subdivision we now set

$$\varphi := \sum_{i=1}^{k} \frac{\mathrm{Vol}\,(\Phi(W_i))}{\mathrm{Vol}\,(W_i)} \chi_{W_i}.$$

So

$$\int_W \varphi(x)dx = \sum_{i=1}^{k} \mathrm{Vol}\,(\Phi(W_i))$$

$$= \mathrm{Vol}\,(\Phi(W)).$$

By continuously refining the subdivision, we obtain a sequence of step functions $(\varphi_n)_{n\in\mathbb{N}}$, which converge on W to Δ_Φ and which satisfy for all n

$$\int_W \varphi_n(x)dx = \mathrm{Vol}\,(\Phi(W)).$$

By lemma 18.5 we can apply the dominated convergence theorem and obtain thereby the lemma. □

Definition 18.7 Let $U \subset \mathbb{R}^d$ be open, $\Phi : U \to \mathbb{R}^d$ injective and continuous. We say that Φ has a density function d_Φ provided for every cube W with $\overline{W} \subset U$,

$$\mathrm{Vol}\,(\Phi(W)) = \int_W d_\Phi(x)dx.$$

Lemma 18.6 thus states that a map with bounded measure derivative has this as its density function.

Theorem 18.8 Let $U, V \subset \mathbb{R}^d$ be open, $\Phi : U \to V$ a homeomorphism which has a density function d_Φ. If $g : V \to \mathbb{R} \cup \{\pm\infty\}$ is integrable then $g(\Phi(x))d_\Phi(x)$ is integrable over U, and one has

$$\int_{\Phi(U)} g(y)dy = \int_U g(\Phi(x))d_\Phi(x)dx. \tag{10}$$

Proof. From the definition of a density function, the statement follows in case $g \circ \Phi$ is the characteristic function of a cube W with $\overline{W} \subset U$, and by linearity of the integral also if $g \circ \Phi$ is a step function with support in U. Now let g be continuous with compact support K in V. Now $\Phi^{-1}(K)$, being the image of a compact set under a continuous map, is again compact, and $g \circ \Phi$ has therefore compact support in U. The function $g \circ \Phi$ is then a uniform limit of step functions φ_n with compact support K' in U, and likewise g is the limit of $\varphi_n \circ \Phi^{-1}$. Furthermore, for a suitable constant c, the functions $\varphi_n \cdot d_\Phi$ are bounded in absolute value by the integrable function $c\chi_K \cdot d_\Phi$. The dominated convergence theorem then implies the statement for continuous functions with compact support. We now decompose $g = g^+ - g^-$ and can therefore limit ourselves to the case $g \geq 0$.

If $g \in H_I(\mathbb{R}^d)$, then g is the limit of a monotonically increasing sequence $(f_n)_{n \in \mathbb{N}} \subset C_c(\mathbb{R}^d)$. We set

$$f'_n(x) := \begin{cases} f_n(x) & \text{if } d(x, \partial V) \geq \frac{2}{n} \quad \text{and } x \in V \\ n(d(x, \partial V) - \frac{1}{n})f_n(x) & \text{if } \frac{1}{n} \leq d(x, \partial V) < \frac{2}{n} \text{ and } x \in V \\ 0 & \text{if } d(x, \partial V) < \frac{1}{n} \quad \text{or } x \notin V \end{cases}$$

Here, $d(x, \partial V)$ denotes the distance of x from the closed set ∂V : this is a continuous function. Now the functions f'_n even have compact support contained in V and converge monotonically increasing to $g \cdot \chi_V$. Similarly, $(f'_n \circ \Phi)d_\Phi$ converge monotonically increasing to $(g \circ \Phi)d_\Phi$, for, we can assume without restriction that $d_\Phi \geq 0$.

The monotone convergence theorem of B. Levi therefore yields the assertion for $g \in H_I$ and likewise for $g \in H_S$. Finally, for arbitrary integrable g one has

$$\int_{\Phi(U)} g(y)dy = \inf\{\int_{\Phi(U)} h(y)dy, h \in H_I, h \geq g\}$$

$$= \sup\{\int_{\Phi(U)} f(y)dy, f \in H_S, f \leq g\}.$$

From what has already been proved it follows that

$$\int_{\Phi(U)} g(y)dy = \inf\{\int_U h(\Phi(x))d_\Phi(x)dx, h \in H_I, h \geq g\}$$

$$= \sup\{\int_U f(\Phi(x))d_\Phi(x)dx, f \in H_S, f \leq g\}.$$

Now, as for $h \in H_I, f \in H_S$ with $f \leq g \leq h$

$$f(\Phi(x))d_\Phi(x) \leq g(\Phi(x))d_\Phi(x) \leq h(\Phi(x))d_\Phi(x)$$

(notice that $d_\Phi \geq 0$ and we have restricted ourselves to the case $g \geq 0$), it follows easily that $g(\Phi(x))d_\Phi(x)$ is integrable over U, and that (10) holds.   □

Proof of theorem 18.2 The proof is now easy. Lemmas 18.3 and 18.6 imply that a diffeomorphism Φ has a density function $d_\Phi = |\det D\Phi|$ defined everywhere. Theorem 18.8 therefore allows us to conclude the integrability of $(f \circ \Phi)|\det D\Phi|$ over U from that of $f : V \to \mathbb{R} \cup \{\pm\infty\}$, together with the corresponding integral formulae. For the other direction we use that Φ^{-1} is likewise a diffeomorphism with density function

$d_{\Phi^{-1}} = |\det D(\Phi^{-1})| = |\det D\Phi|^{-1} \circ \Phi^{-1}$. Therefore if $g = (f \circ \Phi)|\det D\Phi|$ is integrable, so is also $(g \circ \Phi^{-1})|\det D(\Phi^{-1})| = f$, by theorem 18.8, again with the asserted integral formula. □

Examples. We shall now consider transformations to polar coordinates. First the planar case:

We consider the mapping

$$\Phi : \{(r, \varphi) : r \geq 0, \varphi \in \mathbb{R}\} \to \mathbb{R}^2$$

$$(r, \varphi) \mapsto (r \cos \varphi, r \sin \varphi) \text{ (in complex coordinates } (r, \varphi) \mapsto re^{i\varphi}).$$

We compute for the differential

$$D\Phi = \begin{pmatrix} \cos \varphi & -r \sin \varphi \\ \sin \varphi & r \cos \varphi \end{pmatrix},$$

so in particular

$$\det D\Phi = r. \tag{11}$$

From this we see, with the help of the inverse function theorem, that Φ is locally invertible for $r > 0$; however Φ is not globally invertible as for $k \in \mathbb{Z}$ $\Phi(r, \varphi+2\pi k) = \Phi(r, \varphi)$. In order to ensure the injectivity of Φ, one can restrict the domain of definition to

$$\{(r, \varphi) : r > 0, 0 \leq \varphi < 2\pi\};$$

the image is then $\mathbb{R}^2 \backslash \{0\}$.

From theorem 18.2 we deduce

Corollary 18.9 $f : \mathbb{R}^2 \to \mathbb{R} \cup \{\pm\infty\}$ *is integrable precisely when the function*

$$(r, \varphi) \mapsto r f(\Phi(r, \varphi))$$

is integrable over $\{(r, \varphi) : r \geq 0, 0 \leq \varphi \leq 2\pi\}$. *In this case we have*

$$\int_{\mathbb{R}^2} f(x, y) dx dy = \int_0^{2\pi} \int_0^{\infty} f(r \cos \varphi, r \sin \varphi) r dr d\varphi. \tag{12}$$

Proof. We set in theorem 18.2

$$U := \{(r, \varphi) : r > 0, 0 < \varphi < 2\pi\}$$

and $V := \Phi(U) = \mathbb{R}^2 \backslash \{(x, 0) : x \geq 0\}$. Φ then establishes, as required in theorem 18.2, a diffeomorphism between the open sets U and V. The result follows from theorem 18.2 as the complements $\mathbb{R}^2 \backslash V$ and $\{r \geq 0, 0 \leq \varphi \leq 2\pi\} \backslash \{r > 0, 0 < \varphi < 2\pi\}$ are null sets. □

In the same way we define spatial polar coordinates by the map

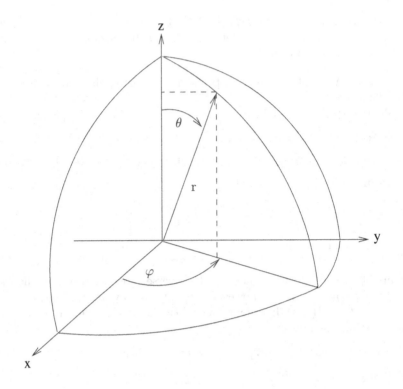

$$\Psi : \{(r, \theta, \varphi) : r \geq 0, \theta, \varphi \in \mathbb{R}\} \to \mathbb{R}^3$$
$$(r, \theta, \varphi) \mapsto (r \sin \theta \cos \varphi, r \sin \theta \sin \varphi, r \cos \theta).$$

Now

$$D\Psi = \begin{pmatrix} \sin \theta \cos \varphi & r \cos \theta \cos \varphi & -r \sin \theta \sin \varphi \\ \sin \theta \sin \varphi & r \cos \theta \sin \varphi & r \sin \theta \cos \varphi \\ \cos \theta & -r \sin \theta & 0 \end{pmatrix},$$

and

$$\det D\Psi = r^2 \sin \theta.$$

Again Ψ is not globally invertible; we have $\Psi(r, \theta, \varphi + 2\pi k) = \Psi(r, \theta, \varphi)$ for $k \in \mathbb{Z}$ and $\Psi(r, \theta + \pi, \varphi) = \Psi(r, \pi - \theta, \varphi + \pi)$. The map Ψ is injective on $\{(r, \theta, \varphi) : r > 0, 0 < \theta < \pi, 0 \leq \varphi < 2\pi\}$ and surjective on $\{(r, \theta, \varphi) : r \geq 0, 0 \leq \theta \leq \pi, 0 \leq \varphi < 2\pi\}$.

Exactly as for corollary 18.9 one proves

Corollary 18.10 $f : \mathbb{R}^3 \to \mathbb{R} \cup \{\pm\infty\}$ *is integrable precisely when the function*

$$(r, \theta, \varphi) \mapsto r^2 \sin \theta f(\Psi(r, \theta, \varphi))$$

is integrable on $\{r \geq 0, 0 \leq \theta \leq \pi, 0 \leq \varphi \leq 2\pi\}$ *and*

$$\int_{\mathbb{R}^3} f(x, y, z) dx dy dz = \int_0^{2\pi} \int_0^\pi \int_0^\infty f(\Psi(r, \theta, \varphi)) r^2 \sin \theta \, dr \, d\theta \, d\varphi. \qquad (13)$$

From the formulae (12) and (13) we recover for $d = 2$ and $d = 3$, respectively, the formulae obtained earlier for integrals of rotationally symmetric functions.

Exercises for § 18

1)   Try to simplify the proof of theorem 18.2 in case f is continuous with compact support.

2)   Derive and discuss the transformation formula for cylinder coordinates in $\mathbb{R}^3$, i.e. for

$$(r, \varphi, z) \mapsto (r \cos \varphi, r \sin \varphi, z).$$

Chapter V.

L^p and Sobolev Spaces

19. The L^p-Spaces

The L^p-spaces of functions the pth power of whose absolute value is integrable are introduced. Minkowski's and Hölder's inequality are shown, and the L^p-spaces are seen to be Banach spaces. Also, the approximation of L^p-functions by smooth ones is discussed. These mollifications are also used to show the existence of partitions of unity.

We recall first that we have called two functions equivalent if they differ only on a null set (see §15). For purposes of integration, equivalent functions are completely the same as all their integrals coincide.

Definition 19.1 Let $\Omega \subset \mathbb{R}^d$ be open, $p \geq 1$ ($p \in \mathbb{R}$). $L^p(\Omega)$ is defined to be the set of all equivalence classes of measurable functions $f : \Omega \to \mathbb{R} \cup \{\pm\infty\}$ for which $|f|^p$ is integrable over Ω. For $f \in L^p(\Omega)$ we set

$$\|f\|_{L^p(\Omega)} := \Big(\int_\Omega |f(x)|^p dx \Big)^{\frac{1}{p}}.$$

Remark on notation: Sometimes, we also need to consider vector valued functions $F = (f^1, \ldots, f^d) : \Omega \to (\mathbb{R} \cup \{\pm\infty\}^c$. We shall sometimes write $|\cdot|$ in place of $\|\cdot\|$ for the Euclidean norm in $\mathbb{R}^c$, and we put

$$\|F\|_{L^p(\Omega)} := \Big(\sum_{i=1}^c \|f^i\|_{L^p(\Omega)}^p \Big)^{\frac{1}{p}}.$$

Obviously $\|f\|_{L^p(\Omega)}$ depends only on the equivalence class of f and one has

$$\|f\|_{L^p(\Omega)} \geq 0 \text{ for all } f \tag{1}$$

and

$$\|f\|_{L^p(\Omega)} = 0 \iff f \text{ is a null function .} \tag{2}$$

Thus $\|f\|_{L^p(\Omega)}$ is positive definite on the space of equivalence classes. In what follows we shall often, for the sake of simplicity, talk of measurable functions when what is really meant is an equivalence class of measurable functions. If a pointwise property is attributed to an equivalence class of measurable functions, it means that there is an element of the equivalence class which has this property. For example, $f \equiv 0$ means, with this agreement, that f is a null function.

Definition 19.2 For a measurable $f : \Omega \to \mathbb{R} \cup \{\pm\infty\}$ we set

$$\operatorname*{ess\,sup}_{x \in \Omega} f(x) := \inf\{a \in \mathbb{R} \cup \{\infty\} : f(x) \le a \text{ for almost all } x \in \Omega\}$$

and analogously

$$\operatorname*{ess\,inf}_{x \in \Omega} f(x) := \sup\{a \in \mathbb{R} \cup \{-\infty\} : a \le f(x) \quad \text{for almost all } x \in \Omega\}$$

(essential supremum and infimum, respectively).

$L^\infty(\Omega)$ is the space of all (equivalence classes) of measurable functions with

$$\|f\|_{L^\infty(\Omega)} := \operatorname*{ess\,sup}_{x \in \Omega} |f(x)| < \infty.$$

We have, as before,

$$\|f\|_{L^\infty(\Omega)} \ge 0 \tag{3}$$

and

$$\|f\|_{L^\infty(\Omega)} = 0 \iff f \equiv 0 \text{ (i.e. } f \text{ is null function).} \tag{4}$$

Example. As an example we consider the function

$$f(x) = \|x\|^\alpha \quad (\alpha \in \mathbb{R}),$$

and the domains $\Omega_1 = \{x \in \mathbb{R}^d : \|x\| < 1\}$ and $\Omega_2 = \{x \in \mathbb{R}^d : \|x\| > 1\}$. Obviously,

$$f \in L^\infty(\Omega_1) \iff \alpha \ge 0$$

and

$$f \in L^\infty(\Omega_2) \iff \alpha \le 0.$$

To investigate to which L^p-spaces f belongs we must examine if the integral

$$\int_\Omega \|x\|^{\alpha p} dx$$

exists for $\Omega = \Omega_1$ and Ω_2 resp.

By theorem 13.21 we have, with $r = \|x\|$

$$\int_{\Omega_1} \|x\|^{\alpha p} dx = d\omega_d \int_0^1 r^{\alpha p} \cdot r^{d-1} dr = d\omega_d \int_0^1 r^{\alpha p + d - 1} dr,$$

and this integral exists precisely when $\alpha p + d > 0$.

Thus

$$f \in L^p(\Omega_1) \iff \alpha p + d > 0.$$

Similarly,

$$\int_{\Omega_2} \|x\|^{\alpha p} dx = d\omega_d \int_1^\infty r^{\alpha p + d - 1} dr,$$

so

$$f \in L^p(\Omega_2) \iff \alpha p + d < 0.$$

□

We shall now investigate the L^p-spaces.
For $1 \le p \le \infty, \lambda \in \mathbb{R}, f \in L^p(\Omega)$ we have

$$\|\lambda f\|_{L^p(\Omega)} = |\lambda| \|f\|_{L^p(\Omega)}.$$

Thus with f, λf also lies in $L^p(\Omega)$. We shall now show that $\|\cdot\|_{L^p(\Omega)}$ defines a norm: for this, there remains only to verify the triangle inequality. For $p = \infty$ this is clear:

$$\|f + g\|_{L^\infty(\Omega)} \le \|f\|_{L^\infty(\Omega)} + \|g\|_{L^\infty(\Omega)}, \tag{5}$$

and similarly for $p = 1$ (due to the monotonicity of the integral)

$$\|f + g\|_{L^1(\Omega)} \le \|f\|_{L^1(\Omega)} + \|g\|_{L^1(\Omega)}. \tag{6}$$

For $1 < p < \infty$, we need still some preparations. Here and later, when there is no danger of confusion, we shall often simply write $\|\cdot\|_p$ instead of $\|\cdot\|_{L^p(\Omega)}$.
We recall first the so-called *Young inequality*.

Lemma 19.3 *For $a, b \ge 0, p, q > 1, \frac{1}{p} + \frac{1}{q} = 1$,*

$$ab \le \frac{a^p}{p} + \frac{b^q}{q}. \tag{7}$$

We now prove the *Hölder inequality*:

Theorem 19.4 *Let $p, q \ge 1, \frac{1}{p} + \frac{1}{q} = 1$ (if $p = 1$ then q is to be set $= \infty$ and conversely), $f \in L^p(\Omega), g \in L^q(\Omega)$. Then $fg \in L^1(\Omega)$ and*

$$\|fg\|_1 \le \|f\|_p \cdot \|g\|_q. \tag{8}$$

Proof. First of all, by corollary 17.13, with f and g also $f \cdot g$ is measurable. In case $p = 1$ and $q = \infty$ we have

$$\int_\Omega |f(x)g(x)| dx \le \operatorname*{ess\,sup}_{x \in \Omega} |g(x)| \int_\Omega |f(x)| dx$$

$$= \|g\|_\infty \cdot \|f\|_1.$$

There remains the case $p, q > 1$ to be handled. First of all, the assertion holds clearly in case $f = 0$ or $g = 0$.

Moreover, the inequality (8) is homogeneous in the sense that if it holds for f and g then it holds also for $\lambda f, \mu g$ $(\lambda, \mu \in \mathbb{R})$. We can therefore assume that

$$\|f\|_p = 1 = \|g\|_q \text{ (so } \int_\Omega |f(x)|^p dx = 1 = \int_\Omega |g(x)|^q dx). \tag{9}$$

By Young's inequality (lemma 19.3) we have

$$|fg| \leq \frac{|f|^p}{p} + \frac{|g|^q}{q}$$

and therefore by (9)

$$\int_\Omega |f(x)g(x)| dx \leq \frac{1}{p} + \frac{1}{q} = 1.$$

$\square$

We can now prove the triangle inequality for $L^p(\Omega)$, which is also called the *Minkowski inequality.*

Corollary 19.5 *Let* $1 \leq p \leq \infty$, $f, g \in L^p(\Omega)$. *Then* $f + g \in L^p(\Omega)$ *and*

$$\|f + g\|_{L^p(\Omega)} \leq \|f\|_{L^p(\Omega)} + \|g\|_{L^p(\Omega)}.$$

Proof. The cases $p = 1, \infty$ have already been dealt with. So let $1 < p < \infty$ and $q = \frac{p}{p-1}$, so that $\frac{1}{p} + \frac{1}{q} = 1$, and $\eta := |f + g|^{p-1}$. Now $\eta^q = |f + g|^p$ and it is clear that if $f, g \in L^p(\Omega)$ then $f + g$ is also in $L^p(\Omega)$ and therefore $\eta \in L^q(\Omega)$ with

$$\|\eta\|_q = \|f + g\|_p^{\frac{p}{q}}.$$

Furthermore,

$$|f + g|^p = |f + g|\eta \leq |f\eta| + |g\eta| \tag{10}$$

and therefore from the Hölder inequality by integrating (10), we obtain

$$\|f + g\|_p^p \leq \|f\eta\|_1 + \|g\eta\|_1$$
$$\leq (\|f\|_p + \|g\|_p)\|\eta\|_q$$
$$= (\|f\|_p + \|g\|_p)\|f + g\|_p^{\frac{p}{q}},$$

so

$$\|f + g\|_p = \|f + g\|_p^{p - \frac{p}{q}} \leq \|f\|_p + \|g\|_p.$$

$\square$

We summarize our considerations above as

Corollary 19.6 $L^p(\Omega)$ *is a vector space, and* $\|\cdot\|_{L^p(\Omega)}$ *is a norm on* $L^p(\Omega)$ *for all* $1 \leq p \leq \infty$. $\square$

For later purposes we note also the following generalization of theorem 19.4, which can be proved easily by induction:

Let $p_1, \ldots, p_n \geq 1, \frac{1}{p_1} + \ldots + \frac{1}{p_n} = 1, f_i \in L^{p_i}(\Omega) \ (i = 1, \ldots, n)$. Then

$$\int_\Omega |f_1(x) \ldots f_n(x)| dx \leq \|f_1\|_{p_1} \ldots \|f_n\|_{p_n}. \tag{11}$$

Further consequences of the Hölder inequality are

Corollary 19.7 *Let* $\Omega \subset \mathbb{R}^d$ *be open,* $\text{Vol}\,(\Omega) < \infty, 1 \leq p \leq q \leq \infty$. *Then* $L^q(\Omega) \subset L^p(\Omega)$, *and for* $f \in L^q(\Omega)$ *we have*

$$\text{Vol}\,(\Omega)^{-\frac{1}{p}} \|f\|_{L^p(\Omega)} \leq \text{Vol}\,(\Omega)^{-\frac{1}{q}} \|f\|_{L^q(\Omega)}.$$

Proof.

$$\|f\|^p_{L^p(\Omega)} = \int_\Omega |f(x)|^p dx = \int_\Omega 1 \cdot |f(x)|^p dx$$

$$\leq \||f|^p\|_{L^{\frac{q}{p}}(\Omega)} \cdot \|1\|_{L^{\frac{q}{q-p}}(\Omega)}$$

$$\left(\text{as } \frac{p}{q} + \frac{q-p}{q} = 1\right) \text{ by theorem 19.4}$$

$$= \|f\|^p_{L^q(\Omega)} \text{Vol}\,(\Omega)^{1-\frac{p}{q}},$$

and this implies the result. $\square$

Corollary 19.8 *Let* $1 \leq p \leq q \leq r, \frac{1}{q} = \frac{\lambda}{p} + \frac{(1-\lambda)}{r}$. *If* $f \in L^p(\Omega) \cap L^r(\Omega)$ *then* $f \in L^q(\Omega)$ *and*

$$\|f\|_q \leq \|f\|_p^\lambda \cdot \|f\|_r^{1-\lambda}.$$

Proof. $\int_\Omega |f(x)|^q dx = \int_\Omega |f(x)|^{\lambda q} \cdot |f(x)|^{(1-\lambda)q} dx$. For $p_1 = \frac{p}{\lambda q}, p_2 = \frac{r}{(1-\lambda)q}$ we have $\frac{1}{p_1} + \frac{1}{p_2} = 1$, so by theorem 19.4,

$$\int_\Omega |f(x)|^q dx \leq \||f|^{\lambda q}\|_{p_1} \cdot \||f|^{(1-\lambda)q}\|_{p_2}$$

$$= \|f\|_p^{\lambda q} \cdot \|f\|_r^{(1-\lambda)q}$$

$\square$

Corollary 19.9 *Let* $\Omega \subset \mathbb{R}^d$ *be open,* $\mathrm{Vol}\,(\Omega) < \infty$. *For* $f \in L^p(\Omega)$ *we set* $(1 \le p < \infty)$ $\Phi_p(f) := \left(\frac{1}{\mathrm{Vol}\,(\Omega)} \int_\Omega |f(x)|^p dx\right)^{\frac{1}{p}}$ $(1 \le p < \infty)$. *If* $|f|^p$ *is measurable but not integrable, we set* $\Phi_p(f) = \infty$. *Then for every measurable* $f : \Omega \to \mathbb{R} \cup \{\pm\infty\}$

$$\lim_{p \to \infty} \Phi_p(f) = \operatorname*{ess\,sup}_{x \in \Omega} |f(x)|.$$

Proof. For $f \in L^p(\Omega)$,

$$\Phi_p(f) = \mathrm{Vol}\,(\Omega)^{-\frac{1}{p}} \|f\|_{L^p(\Omega)}.$$

By corollary 19.7, $\Phi_p(f)$ is monotonically increasing in p and for $1 \le p < \infty$ we have

$$\Phi_p(f) \le \operatorname*{ess\,sup}_{x \in \Omega} |f(x)|.$$

Thereby $\lim\limits_{p \to \infty} \Phi_p(f)$ exists in $\mathbb{R} \cup \{\infty\}$ and it now remains only to show that

$$\operatorname*{ess\,sup}_{x \in \Omega} |f(x)| \le \lim_{p \to \infty} \Phi_p(f).$$

For $K \in \mathbb{R}$ let

$$A_K := \{x \in \Omega : |f(x)| \ge K\}.$$

A_K is measurable since f is measurable (see theorem 17.11).

Moreover, for $K < \operatorname*{ess\,sup}_{x \in \Omega} |f(x)|$,

$$\mathrm{Vol}\,(A_K) > 0.$$

Also

$$\Phi_p(f) \ge \mathrm{Vol}\,(\Omega)^{-\frac{1}{p}} \left(\int_{A_K} |f(x)|^p dx\right)^{\frac{1}{p}} \ge \mathrm{Vol}\,(\Omega)^{-\frac{1}{p}} \mathrm{Vol}\,(A_K)^{\frac{1}{p}} \cdot K,$$

since $|f(x)| \ge K$ for $x \in A_K$. As $\mathrm{Vol}\,(A_K) > 0$, it follows from this that

$$\lim_{p \to \infty} \Phi_p(f) \ge K.$$

As this holds for all $K < \operatorname*{ess\,sup}_{x \in \Omega} |f(x)|$, we have

$$\lim \Phi_p(f) \ge \operatorname*{ess\,sup}_{x \in \Omega} |f(x)|.$$

This completes the proof of the corollary. $\square$

We shall now show that $L^p(\Omega)$ is even a Banach space. To understand this statement, we shall first compare the different convergence concepts which are available to us.

First, we have the pointwise convergence almost everywhere. Then there is the L^∞-convergence:

$$f_n \xrightarrow{L^\infty} f \iff \|f_n - f\|_\infty \to 0.$$

Obviously, the L^∞-convergence implies pointwise convergence almost everywhere. Finally, we have the L^p-convergence for $1 \le p < \infty$:

$$f_n \xrightarrow{L^p} f \iff \|f_n - f\|_p \to 0;$$

this convergence is also called strong convergence (we shall later introduce yet another convergence – the weak convergence). For $p = 2$, one also says mean quadratic convergence. We now give two examples which show that for $1 \le p < \infty$ neither pointwise convergence implies L^p-convergence nor does L^p-convergence imply pointwise convergence.

Examples.

1)   This example is already known to us from our study of the convergence theorems. Let $f_n : (0,1) \to \mathbb{R}$ be defined by

$$f_n(x) := \begin{cases} n & \text{for } 0 < x \le \frac{1}{n} \\ 0 & \text{for } \frac{1}{n} < x < 1. \end{cases}$$

Now f_n converges pointwise to 0, however

$$\|f_n\|_{L^p} = n^{\frac{p-1}{p}} (= n \text{ for } p = \infty)$$

and therefore

$$\|f_n\| \to \infty \text{ as } n \to \infty \quad (p > 1).$$

For $p = 1$ we indeed have $\|f_n\| = 1$ for all n but still there is no $f \in L^1(\Omega)$ with

$$\|f_n - f\|_1 \to 0, \tag{12}$$

for as $f_n(x) = 0$ for $x > \frac{1}{n}$, f must then equal 0, which again cannot lead to (12) as $\|f_n\| = 1$.

Besides, one can also set

$$g_n(x) := \begin{cases} n^2 & \text{for } 0 \le x \le \frac{1}{n} \\ 0 & \text{otherwise} \end{cases} ;$$

then also $\|g_n\|_1 \to \infty$ as $n \to \infty$.

2)   We consider the following sequence

$$\chi_{[0,1]}, \chi_{[0,\frac{1}{2}]}, \chi_{[\frac{1}{2},1]}, \chi_{[0,\frac{1}{4}]}, \chi_{[\frac{1}{4},\frac{1}{2}]}, \cdots$$

For $n \in \mathbb{N}$, if k denotes the natural number uniquely determined by

$$2^k \le n < 2^{k+1}$$

and

$$I_n := [n2^{-k} - 1, (n+1)2^{-k} - 1]$$

then we can represent the sequence by

$$f_n = \chi_{I_n}.$$

Then, for $1 \le p < \infty$,

$$\|f_n\|_p = 2^{-\frac{k}{p}},$$

so $\|f_n\|_p \to 0$ as $n \to \infty$, but $f_n(x)$ does not converge for any $x \in (0,1)$, because for every x there exist infinitely many $m \in \mathbb{N}$ with $f_m(x) = 1$ and infinitely many $\ell \in \mathbb{N}$ with $f_\ell(x) = 0$.

We now prove

Lemma 19.10 *Let* $(f_n)_{n \in \mathbb{N}} \subset L^p(\Omega), 1 \le p < \infty$ *and*

$$\sum_{n=1}^{\infty} \|f_n\|_p =: M < \infty.$$

Then Σf_n *converges in* L^p *as well as pointwise almost everywhere to a function* $h \in L^p(\Omega)$.

Proof. Let

$$g_n := \sum_{j=1}^{n} |f_j|$$

Then $(g_n)_{n \in \mathbb{N}} \subset L^p(\Omega)$ is monotonically increasing and nonnegative and, on account of $\|g_n\|_p \le \sum_{j=1}^{n} \|f_j\|_p$ (corollary 19.5), we have $\|g_n\|_p \le M$, so $\int_\Omega (g_n(x))^p dx \le M^p$. By the dominated convergence theorem, $(g_n^p)_{n \in \mathbb{N}}$ converges pointwise almost everywhere to an integrable k with

$$\int_\Omega k(x) dx \le M^p.$$

This means that $(g_n)_{n \in \mathbb{N}}$ converges to a non-negative measurable function g with $g^p = k$, so $g \in L^p(\Omega)$ and $\|g\|_p \le M$.

Since the series Σf_n converges absolutely (almost everywhere), Σf_n converges pointwise everywhere to a function h with $|h| \le g$. By theorem 17.3 (iii), h is measurable and it follows that $h \in L^p(\Omega)$,

$$\|h\|_p \le M.$$

Thus $\sum\limits_{j=1}^{n} f_j - h$ converges pointwise almost everywhere to 0 and, besides, as

$$|\sum_{j=1}^{n} f_j - h| \le 2|g|,$$

it follows from the dominated convergence theorem that

$$\|\sum_{j=1}^{n} f_j - h\|_p \to 0 \text{ for } n \to \infty.$$

This verifies the L^p-convergence.   □

Theorem 19.11 *For $1 \le p \le \infty$, $L^p(\Omega)$ is a Banach space.*

Proof. First, let $1 \le p < \infty$. Let $(f_n)_{n \in \mathbb{N}} \subset L^p(\Omega)$ be a Cauchy sequence. For every $k \in \mathbb{N}$ there exists an $N(k) \in \mathbb{N}$ such that

$$\|f_n - f_{N(k)}\|_p < 2^{-k} \text{ for } n \ge N(k); \tag{13}$$

we may assume, without loss of generality, that $N(k) < N(k+1)$. Then the series

$$\|f_{N(1)}\|_p + \|f_{N(2)} - f_{N(1)}\|_p + \|f_{N(3)} - f_{N(2)}\|_p + \dots$$

converges, as one sees by comparison with the geometric series. By lemma 19.10, the series

$$f_{N(1)} + (f_{N(2)} - f_{N(1)}) + (f_{N(3)} - f_{N(2)}) + \dots \tag{14}$$

converges to an $f \in L^p(\Omega)$, and indeed in the L^p-norm as well as pointwise almost everywhere. The L^p-convergence however means that

$$\|f_{N(k)} - f\|_p \to 0 \text{ for } k \to \infty.$$

For $\varepsilon > 0$ we now choose k so large that

$$\|f_{N(k)} - f\|_p < \frac{\varepsilon}{2} \tag{15}$$

and

$$2^{-k} < \frac{\varepsilon}{2}. \tag{16}$$

Then, for $n \ge N(k)$, we have

$$\|f_n - f\|_p \le \|f_n - f_{N(k)}\| + \|f_{N(k)} - f\| < \varepsilon$$

(by (13), (15), (16)) and thereby f_n converges in the L^p-norm to f.

This takes care of the case $p < \infty$. Before we come to the proof for $p = \infty$, we record the following statement, which follows from the fact that convergence in (14) occurs also pointwise almost everywhere.

Theorem 19.12 *If $(f_n)_{n\in\mathbb{N}}$ converges to f in L^p then a subsequence of (f_n) converges pointwise almost everywhere to f.*

We now prove theorem 19.11 for $p = \infty$. So let $(f_n)_{n\in\mathbb{N}}$ be a Cauchy sequence in $L^\infty(\Omega)$. For every $k \in \mathbb{N}$ there exists then an $N(k) \in \mathbb{N}$ with

$$\operatorname*{ess\,sup}_{x\in\Omega} |f_n(x) - f_m(x)| < \frac{1}{k} \text{ for } n, m \geq N(k).$$

For every pair (n, m) with $n, m \geq N(k)$, let $A_{nm,k}$ be the null set of those $x \in \Omega$ with

$$|f_n(x) - f_m(x)| \geq \frac{1}{k}.$$

Then $A := \bigcup_{\substack{k\geq 1 \\ n,m\geq N(k)}} A_{nm,k}$ is also a null set (lemma 15.3). On $\Omega\backslash A$ the se-

quence $(f_n)_{n\in\mathbb{N}}$ converges even uniformly to a function f, as for $n, m \geq N(k)$ we have

$$\sup_{x\in\Omega\backslash A} |f_n(x) - f_m(x)| < \frac{1}{k}, \tag{17}$$

and $(f_n)_{n\in\mathbb{N}}$ thereby form a pointwise Cauchy sequence on $\Omega\backslash A$. Since A is a null set, the sequence f_n converges pointwise almost everywhere on Ω to f, thereby f, by theorem 17.3 (iii), is measurable. Besides, for $n \geq N(k)$ we have

$$\operatorname*{ess\,sup}_{x\in\Omega} |f_n(x) - f(x)| = \sup_{x\in\Omega\backslash A} |f_n(x) - f(x)| \leq \frac{1}{k},$$

which one obtains by letting $m \to \infty$ in (17). Thus f_n converges to f in $L^\infty(\Omega)$. $\qquad\square$

We shall now study the convolution of L^p-functions with kernel functions. For this, let ρ be a non-negative, bounded, integrable function with support in the unit ball of $\mathbb{R}^d$, so

$$\rho(x) \geq 0 \text{ for all } x \in \mathbb{R}^d,$$
$$\rho(x) = 0 \text{ for } \|x\| \geq 1$$

and, moreover,

$$\int_{\mathbb{R}^d} \rho(x)dx = 1. \tag{18}$$

Now let $f \in L^1(\Omega)$. The convolution of f (with parameter h) is defined as

$$f_h(x) := \frac{1}{h^d} \int_\Omega \rho\left(\frac{x-y}{h}\right)f(y)dy \quad \text{for } x \in \Omega, h < \operatorname{dist}(x, \partial\Omega).$$

In the following we shall use the symbol $A \subset\subset \Omega$ to denote that A is bounded and its closure is contained in Ω.

Theorem 19.13 *If $f \in C^0(\Omega)$, then $f_h \to f$ as $h \to 0$, uniformly on every $\Omega' \subset\subset \Omega$.*

Proof. We have

$$f_h(x) = \frac{1}{h^d} \int\limits_{|x-y|\leq h} \rho(\frac{x-y}{h}) f(y) dy = \int\limits_{|z|\leq 1} \rho(z) f(x - hz) dz$$

$$\text{(substituting } z = \frac{x-y}{h}).$$

Now if $\Omega' \subset\subset \Omega$ and $h < \frac{1}{2} \operatorname{dist}(\Omega', \partial\Omega)$, it follows that

$$\sup_{x\in\Omega'} |f(x) - f_h(x)| = \sup_{x\in\Omega'} | \int\limits_{|z|\leq 1} \rho(z)(f(x) - f(x - hz)) dz|,$$

(for, by (18), $\int\limits_{|z|\leq 1} \rho(z) f(x) dz = f(x)$)

$$\leq \sup_{x\in\Omega'} \int\limits_{|z|\leq 1} \rho(z)|f(x) - f(x - hz)| dz$$

$$\leq \sup_{x\in\Omega'} \sup_{|z|\leq 1} |f(x) - f(x - hz)|, \text{ again by (18).}$$

As f is uniformly bounded on the compact set

$$\{x : \operatorname{dist}(x, \Omega') \leq h\} \subset \Omega \quad \text{(by choice of } h),$$

this tends to 0 as $h \to 0$. □

Lemma 19.14 (Approximation of L^p-functions). *For any $f \in L^p(\Omega)$, $1 \leq p < \infty$, and any $\varepsilon > 0$, there exists a $\varphi \in C_c(\mathbb{R}^d)$ with $\|f - \varphi\|_p < \varepsilon$.*

Proof. We extend f by zero on $\mathbb{R}^d \backslash \Omega$. It suffices to consider the case $f \geq 0$. For $n \in \mathbb{N}$ we set

$$f_n(x) := \begin{cases} \min(f(x), n) & \text{for } \|x\| \leq n \\ 0 & \text{for } \|x\| > n. \end{cases}$$

Then $f_n \in L^1(\mathbb{R}^d)$ for all $n \in \mathbb{N}$. As $|f_n - f|^p \leq |f|^p$, the dominated convergence theorem gives

$$f_n \xrightarrow{L^p} f \text{ for } n \to \infty.$$

Therefore, for $\varepsilon > 0$ there exists an $n \in \mathbb{N}$ with

$$\|f_n - f\|_{L^p(\mathbb{R}^d)} < \frac{\varepsilon}{2}. \tag{19}$$

Furthermore, by theorem 14.8 there exists a $\varphi \in C_c(\mathbb{R}^d)$ with

$$\|f_n - \varphi\|_{L^1(\mathbb{R}^d)} < \frac{\varepsilon^p}{2^p n^{p-1}}. \tag{20}$$

As $0 \le f_n \le n$, we can also assume here that $0 \le \varphi \le n$, in that, if necessary, we replace φ by med $(0, \varphi, n)$. Then $|f_n - \varphi| \le n$ so

$$|f_n - \varphi|^p \le n^{p-1} |f_n - \varphi|,$$

and therefore

$$\|f_n - \varphi\|_{L^p(\Omega)}^p \le \|f_n - \varphi\|_{L^p(\mathbb{R}^d)}^p \le n^{p-1} \|f_n - \varphi\|_{L^1(\mathbb{R}^d)} < (\frac{\varepsilon}{2})^p \text{ by (20)}. \tag{21}$$

From (19) and (20) it follows that

$$\|f - \varphi\|_{L^p(\Omega)} < \varepsilon.$$

$\square$

Lemma 19.15 *For any $f \in L^p(\Omega), 1 \le p < \infty$, we may find a sequence $(\varphi_n)_{n \in \mathbb{N}} \in C_c(\mathbb{R}^d)$ that converges to f almost everywhere in Ω.*

Proof. By lemma 19.14, there exists such a sequence that converges to f in $L^p(\Omega)$.

By theorem 19.12, a subsequence of this sequence then converges to f almost everywhere.

$\square$

We now have the fundamental theorem about the convergence of the regularizations or approximations to the original function f:

Theorem 19.16 *Let $f \in L^p(\Omega), 1 \le p < \infty$. Then f_h converges to f in $L^p(\Omega)$ as well as pointwise almost everywhere in Ω as $h \to 0$.*
(here we have extended f by zero on $\mathbb{R}^d \backslash \Omega$ in order to define f_h on all of Ω).
For the pointwise emergence almost everywhere it suffices, in fact, that $f \in L^1_{loc}(\Omega)$, i.e. that f is locally integrable on Ω.

Proof. We write again

$$f_h(x) = \int_{|z| \le 1} \rho(z) f(x - hz) dz$$

$$= \int_{|z| \le 1} (\rho(z)^{1 - \frac{1}{p}})(\rho(z)^{\frac{1}{p}} f(x - hz)) dz$$

$$\le (\int_{|z| \le 1} \rho(z) dz)^{1 - \frac{1}{p}} (\int_{|z| \le 1} \rho(z) |f(x - hz)|^p dz)^{\frac{1}{p}}$$

(by Hölder's inequality)

$$= (\int_{|z| \le 1} \rho(z) |f(x - hz)|^p dz)^{\frac{1}{p}} \text{ by (18)}.$$

Thus

$$\int_\Omega |f_h(x)|^p dx \le \int_\Omega \int_{|z|\le 1} \rho(z)|f(x-hz)|^p dz dx$$

$$= \int_{|z|\le 1} \rho(z)(\int_\Omega |f(x-hz)|^p dx) dz \text{ by Fubini}$$

$$\le \int_{|z|\le 1} \rho(z)\left(\int_{\mathbb{R}^d} |f(y)|^p dy\right) dz = \int_{\mathbb{R}^d} |f(y)|^p dy \text{ by (18).} \quad (22)$$

$$= \int_\Omega |f(y)|^p dy.$$

Now let $\varepsilon > 0$. We first choose n so large that, using corollary 16.8,

$$\left(\int_{\Omega \cap \{x\in\mathbb{R}^d:\|x\|\ge n\}} |f(x) - f_h(x)|^p dx\right)^{\frac{1}{p}} < \frac{\varepsilon}{4} \quad (23)$$

(here by e.g. (22), n is independent of h).

Then we choose, using lemma 19.14, a $\varphi \in C_c(\mathbb{R}^d)$ with

$$\|f - \varphi\|_{L^p(\mathbb{R}^d)} < \frac{\varepsilon}{4}. \quad (24)$$

By theorem 19.13, for h sufficiently small, we have

$$\|\varphi - \varphi_h\|_{L^p(\Omega\cap\{x\in\mathbb{R}^d:\|x\|\le n\})} \quad (25)$$

$$\le (\text{Vol}(\Omega \cap \{\|x\| \le n\}))^{\frac{1}{p}}\left(\sup_{x\in\Omega,\|x\|\le n} |\varphi(x) - \varphi_h(x)|\right) < \frac{\varepsilon}{4}.$$

We now apply (22) to $f - \varphi$ instead of f and obtain altogether

$$\|f - f_h\|_{L^p(\Omega)} \le \|f - f_h\|_{L^p(\Omega\cap\{x\in\mathbb{R}^d:\|x\|\le n\})}$$

$$+ \|f - f_h\|_{L^p(\Omega\cap\{x\in\mathbb{R}^d:\|x\|\ge n\})}$$

$$\le \|f - \varphi\|_{L^p(\mathbb{R}^d)} + \|\varphi - \varphi_h\|_{L^p(\Omega\cap\{x\in\mathbb{R}^d:\|x\|\le n\})}$$

$$+ \|\varphi_h - f_h\|_{L^p(\mathbb{R}^d)} + \frac{\varepsilon}{4} \text{ by (23)}$$

$$< \varepsilon, \text{ by (22), (24), (25).}$$

This shows the L^p-convergence. It remains to show the pointwise convergence almost everywhere.

For this, we need some more preparations. This will also provide us with the opportunity to introduce a very useful technical tool, namely covering theorems.

In order to derive such pointwise results about integrable functions, we need a covering theorem stating that we can cover an open set efficiently by balls. The setting is the following: We consider a collection $\mathbf{B}$ of closed balls $B = B(x,r) = \{y \in \mathbb{R}^d : |x - y| \leq r\}$ for points $x \in \mathbb{R}^d$ and radii $r > 0$. A set $A \subset \mathbb{R}^d$ is covered by $\mathbf{B}$ if

$$A \subset \bigcup_{B \in \mathbf{B}} B.$$

A is finely covered by $\mathbf{B}$ when also for each $y \in A$

$$\inf\{r \mid y \in B(x,r) \in \mathbf{B}\} = 0,$$

that is, each point in A is contained in arbitrarily small balls.

The **Vitali covering theorem** then says

Theorem 19.17 *Let $\mathbf{B}$ be a collection of closed ball $B = B(x,r)$ in $\mathbb{R}^d$ as above, with $\rho := \sup\{r \mid B = B(x,r) \in \mathbf{B}\} < \infty$. Then there exists a countable family of disjoint balls $B(x_n, r_n)$, $n \in \mathbb{N}$, with*

$$A := \bigcup_{B \in \mathbf{B}} B \subset \bigcup_{n \in \mathbb{N}} B(x_n, 5r_n).$$

In words: We can select a countable family of *disjoint* balls from the original collection so that A is covered by these balls when their radii are enlarged by a factor of 5.

Proof. We let $\mathbf{B}_1$ be a maximal disjoint collection of balls in $\mathbf{B}$ with radii between $\frac{\rho}{2}$ and ρ. $\mathbf{B}_1$ is countable (and in fact finite when the union of our balls is bounded). Having iteratively selected $\mathbf{B}_j$ for $j = 1, ..., m - 1$, we take $\mathbf{B}_m$ as a maximal family of disjoint balls from $\mathbf{B}$, disjoint from all the balls in the $\mathbf{B}_j$ for $j < m$, among those with radii between $\frac{\rho}{2^m}$ and $\frac{\rho}{2^{m-1}}$. $\mathbf{N} := \bigcup_{m \in \mathbb{N}} \mathbf{B}_m$ then is a countable collection of closed balls from $\mathbf{B}$, and we claim that it satisfies the conclusion of the theorem. In fact, let $B(x,r) \in \mathbf{B}$. We choose m with $\frac{\rho}{2^m} < r \leq \frac{\rho}{2^{m-1}}$. By the maximality of the $\mathbf{B}_j$, there exists some $j \leq m$ and some ball $B(x_n, r_n) \in \mathbf{B}$ with $B(x,r) \cap B(x_n, r_n) \neq$ set, and by construction, $r \leq 2r_n$. Therefore, $B(x,r) \subset B(x_n, 5r_n)$. This implies the conclusion of the theorem. $\qquad\square$

An important consequence of the Vitali covering theorem is

Theorem 19.18 *For any open set $\Omega \subset \mathbb{R}^d$ and any $\epsilon > 0$, we can find a countable collection $\mathbf{N}$ of disjoint closed balls of radii $\leq \epsilon$ contained in Ω with*

$$\mathrm{Vol}(\Omega - \bigcup_{B \in \mathbf{N}} B) = 0.$$

Proof. We first treat the case $\text{Vol}(\Omega) < \infty$ as we can later easily reduce the general case to this one. The decisive step is to show that for some $0 < \eta < 1$, we can find finitely many balls $B_1, ..., B_{n_1} \subset \Omega$ of radii $\leq \epsilon$ with

$$\text{Vol}(\Omega - \bigcup_{\nu=1}^{n_1} B_\nu) \leq \eta \text{Vol}(\Omega). \tag{26}$$

Having achieved this, we put $\Omega_1 := \Omega - \bigcup_{\nu=1}^{n_1} B_\nu$, and having iteratively found

$$\Omega_j = \Omega - \bigcup_{\nu=1}^{n_j} B_\nu = \Omega_{j-1} - \bigcup_{\nu=n_{j-1}+1}^{n_j} B_\nu, \text{ with balls } B_{n_{j-1}+1}, ..., B_{n_j} \subset \Omega_{j-1}$$

of radii at most ϵ, and $\text{Vol}(\Omega - \bigcup_{\nu=1}^{n_j} B_\nu) = \text{Vol}(\Omega_j) \leq \eta^j \text{Vol}(\Omega)$, we see that we achieve our aim by letting $j \to \infty$. In order to show (26), we use the Vitali covering theorem to find, in the family of balls of radius at most ϵ contained in Ω, a countable disjoint family of balls $B_n = B(x_n, r_n)$ with $\Omega \subset \bigcup_n B(x_n, 5r_n)$. Therefore

$$\text{Vol}(\Omega) \leq \sum_n \text{Vol}(B(x_n, 5r_n)) = 5^d \sum_n \text{Vol}(B(x_n, r_n)) = 5^d \text{Vol}(\bigcup_n B(x_n, r_n))$$

since the latter balls are disjoint. From this,

$$\text{Vol}(\Omega - \bigcup_n B(x_n, r_n)) \leq (1 - 5^{-d}) \text{Vol}(\Omega).$$

When we now choose $1 - 5^{-d} < \eta < 1$, then, since the family of balls just chosen is countable, we can find finitely many of them so that (26) holds.

It remains to treat the case $\text{Vol}(\Omega) = \infty$, but in that case we simply apply the previous reasoning to the sets $\Omega_N := \{x \in \Omega : N < |x| < N+1\}$ which, as $N \to \infty$, cover Ω up to a nullset. $\qquad\square$

While this is an important result, for our subsequent purposes, we need a slightly different version:

Corollary 19.19 *Let $A \subset \mathbb{R}^d$ (not necessarily open), and let $\mathbf{B}$ be a collection of closed balls covering A finely. We can then find a countable collection $\mathbf{N}$ of disjoint closed balls from $\mathbf{B}$ with*

$$\text{Vol}(A - \bigcup_{B \in \mathbf{N}} B) = 0.$$

In fact, if A is open, since the covering is fine, all these balls can be chosen to be contained in A.

Proof. The proof is the same as the one of the preceding theorem, taking $\mathbf{B}$ in place of the collection of balls in Ω of radius $\leq \epsilon$. The fineness of the covering **B** is needed to guarantee that in the iterative step, the balls needed to cover Ω_j are contained in Ω_j and therefore disjoint from the ones chosen in previous steps.   □

Besides the Vitali covering theorem, there is another, more powerful, covering result, the **Besicovitch covering theorem** which we state here without proof as it will not be employed in this book:

Theorem. *There exists a positive integer N_d depending only on the dimension d with the property that for any family $\mathbf{B} = \{B(x,r) : x \in A\}$ of balls with centers in some set $A \subset \mathbb{R}^d$, with uniformly bounded radii, we can find at most N_d disjointed subfamilies $\mathbf{B}_j$, $j = 1, ..., N_d$, (that is, any two balls from the same $\mathbf{B}_j$ are disjoint), such that the set A of the centers of our balls can be covered by their union, that is,*

$$A \subset \bigcup_{j=1}^{N_d} \bigcup_{B \in \mathbf{B}_j} B.$$

The Besicovitch covering theorem simply says that from the original ball covering, we can find a subcovering of A, the set of the centers of the balls, for which each point is contained in at most N_d balls.

We now return to the convergence of the approximations of our L^p function f. We first consider a particular kernel function, namely

$$\rho_0(x) = \begin{cases} \frac{1}{\omega_d} & \text{for } x \in B(0,1), \\ 0 & \text{otherwise,} \end{cases}$$

where ω_d denotes the volume of the ball $B(0,1) \subset \mathbb{R}^d$, so that (18) is obviously satisfied. The convolution of $f \in L^1(\Omega)$ then is simply the averaged function

$$f_h(x) = \frac{1}{\omega_d h^d} \int_{B(x,h)} f(y)dy,$$

assuming $B(x,h) \subset \Omega$. It is also customary to use the notation

$$\fint_{B(x,h)} f(y)dy := \frac{1}{\omega_d h^d} \int_{B(x,h)} f(y)dy$$

for our present f_h. We shall now first verify theorem 19.16 for this particular choice of kernel function; before stating the result, we formulate a definition:

Definition 19.20 $x \in \Omega$ is called a Lebesgue point for $f \in L^1(\Omega)$ if

$$f(x) = \lim_{h \to 0} \frac{1}{\omega_d h^d} \int\limits_{B(x,h)} f(y)dy, \quad (h < \text{dist } (x, \partial\Omega))$$

i.e., if the value of f at x is the limit of the averages of f over the balls $B(x,h)$ for $h \to 0$.

Theorem 19.21 *Let $f \in L^1_{loc}(\Omega)$. Then almost every point $x \in \Omega$ is a Lebesgue point for f.*

Proof.

$$|\fint\limits_{B(x,h)} f(y)dy - f(x)| \leq \fint\limits_{B(x,h)} |f(y) - \varphi(y)|dy + \fint\limits_{B(x,h)} |\varphi(y) - f(x)|dy$$

where we take φ as a continuous function so that the last term converges to $|\varphi(x) - f(x)|$ for $h \to 0$. Therefore

$$\{x \in \Omega : \limsup_{h \to 0} |\fint\limits_{B(x,h)} f(y)dy - f(x)| > \epsilon\}$$

$$\subset \{x \in \Omega : \limsup_{h \to 0} \fint\limits_{B(x,h)} |f(y) - \varphi(y)|dy > \frac{\epsilon}{2}\} \cup \{x \in \Omega : |\varphi(x) - f(x)| > \frac{\epsilon}{2}\}$$

$$=: U_\epsilon \cup V_\epsilon. \tag{27}$$

We now claim that

$$\int\limits_{U_\epsilon} |f(y) - \varphi(y)|dy \geq \frac{\epsilon}{2}\text{Vol}(U_\epsilon). \tag{28}$$

As it suffices to show this for bounded subsets of U_ϵ, we may assume that $\text{Vol}(U_\epsilon) < \infty$. We let Ω be any open set containing U_ϵ. Let $\delta > 0$. By definition of U_ϵ, for any $x \in U_\epsilon$, we can find a sequence of closed balls $B(x, r_n) \subset \Omega$, with $r_n \to 0$ – and, consequently, they cover U_ϵ finely – and with

$$\int\limits_{B(x,r_n)} |f(y) - \varphi(y)|dy \geq (\frac{\epsilon}{2} - \delta)\text{Vol}(B(x, r_n)).$$

By corollary 19.19, we then find a countable disjoint family **N** of these balls that cover U_ϵ up to a nullset. Thus,

$$(\frac{\epsilon}{2}-\delta)\text{Vol}(U_\epsilon) \leq (\frac{\epsilon}{2}-\delta) \sum_{B \in \mathbf{N}} \text{Vol}(B) \leq \sum_{B \in \mathbf{N}} \int\limits_B |f(y)-\varphi(y)|dy \leq \int\limits_\Omega |f(y)-\varphi(y)|dy.$$

Since Ω and δ are arbitrary, we obtain (28).

Obviously, also

$$\int_{V_\epsilon} |f(y) - \varphi(y)| dy \geq \frac{\epsilon}{2} \mathrm{Vol}(V_\epsilon).$$

Therefore, by lemma 19.14, by an appropriate choice of φ, we can make $\mathrm{Vol}(U_\epsilon)$ and $\mathrm{Vol}(V_\epsilon)$ arbitrarily small.

Therefore, by (27), $\mathrm{Vol}(\{x \in \Omega : \limsup_{h \to 0} | \fint_{B(x,h)} f(y)dy - f(x)| > \epsilon\}) = 0$ for every $\epsilon > 0$, and the theorem follows. $\square$

We can now also easily complete the *proof of theorem 19.16* by showing that $f_h(x)$ converges to $f(x)$ for $h \to 0$ at every Lebesgue point x of f. Namely,

$$|f_h(x) - f(x)| \leq \frac{1}{h^d} \int_{|x-y| \leq h} \rho(\frac{x-y}{h})|f(y) - f(x)| dy \leq \sup \rho \frac{1}{h^d} \int_{|x-y| \leq h} |f(y) - f(x)| dy$$

which goes to 0 for $h \to 0$ for all Lebesgue points x of f. Since almost every point is a Lebesgue point by the preceding theorem, the proof of theorem 19.16 is complete. $\square$

We now consider the kernel

$$\rho_1(x) := \begin{cases} c \, \exp(\frac{1}{\|x\|^2 - 1}) & \text{for } \|x\| < 1 \\ 0 & \text{for } \|x\| \geq 1 \end{cases},$$

where the constant $c > 0$ is so chosen that (18) holds.

In contrast to the kernel ρ_0, the kernel ρ_1 is of class $C^\infty(\mathbb{R}^d)$.

We now let f_h denote the convolution with this smooth kernel ρ_1, i.e.,

$$f_h(x) = \frac{1}{h^d} \int_\Omega \rho_1(\frac{x-y}{h}) f(y) dy \quad \text{for } x \in \Omega, \ h < \mathrm{dist}\,(x, \partial\Omega),$$

and obtain

Lemma 19.22 *If $\Omega' \subset\subset \Omega$ and $h < \mathrm{dist}\,(\Omega', \partial\Omega)$ then*

$$f_h \in C^\infty(\Omega').$$

Proof. As $\rho_1 \in C_0^\infty(\mathbb{R}^d)$, this follows directly from theorem 16.10, as for every k there exists a c_k with $\sup_{x \in \mathbb{R}^d} |D^k \rho_1(x)| \leq c_k$, where D^k stands abbreviatively for all the k-th partial derivatives. $\square$

Definition 19.23 For $k = 0, 1, 2, 3, \ldots, \infty$ we define

$$C_0^k(\Omega) := \{f \in C^k(\Omega) : \operatorname{supp} f \subset\subset \Omega\}.$$

Corollary 19.24 $C_0^\infty(\Omega)$ *is dense in* $L^p(\Omega)$ *for* $1 \leq p < \infty$.

Proof. The assertion means that for any $f \in L^p(\Omega)$ and $\varepsilon > 0$ one can find a $\varphi \in C_0^\infty(\Omega)$ with

$$\|f - \varphi\|_{L^p(\Omega)} < \varepsilon.$$

First we choose a $\Omega' \subset\subset \Omega$ with $\|f\|_{L^p(\Omega\setminus\overline{\Omega}')} < \frac{\varepsilon}{3}$ (see corollary 16.8).

We set

$$\tilde{f}(x) := \begin{cases} f(x) & \text{for } x \in \overline{\Omega}' \\ 0 & \text{for } x \in \mathbb{R}^d\setminus\overline{\Omega}' \end{cases}.$$

By theorem 19.16 there exists an $h < \operatorname{dist}(\Omega', \partial\Omega)$ with

$$\|\tilde{f} - \tilde{f}_h\|_{L^p(\Omega)} < \frac{\varepsilon}{3}.$$

As $\tilde{f}(x) = 0$ for $x \in \Omega\setminus\overline{\Omega}'$, it follows in particular that

$$\|\tilde{f}_h\|_{L^p(\Omega\setminus\overline{\Omega}')} < \frac{\varepsilon}{3}.$$

It follows that

$$\|f - \tilde{f}_h\|_{L^p(\Omega)} \leq \|f\|_{L^p(\Omega\setminus\overline{\Omega}')} + \|\tilde{f}_h\|_{L^p(\Omega\setminus\overline{\Omega}')}$$
$$+ \|\tilde{f} - \tilde{f}_h\|_{L^p(\Omega')} < \varepsilon.$$

By lemma 19.22, $\tilde{f}_h \in C^\infty(\mathbb{R}^d)$, and by choice of h, $\tilde{f}_h$ has compact support in Ω. $\qquad\square$

Corollary 19.25 *Let* $f \in L^2(\Omega)$. *Assume that for all* $\varphi \in C_0^\infty(\Omega)$ *we have*

$$\int_\Omega f(x)\varphi(x)dx = 0.$$

Then $f \equiv 0$.

Proof. By corollary 19.24 there exists a sequence $(\varphi_n)_{n\in\mathbb{N}} \subset C_0^\infty(\Omega)$ with $\|f - \varphi_n\|_2 \to 0$, i.e.

$$\int_\Omega (f(x) - \varphi_n(x))^2 dx \to 0 \text{ for } n \to \infty. \tag{29}$$

The Hölder inequality implies

$$\int_\Omega f^2(x)dx - \int_\Omega f(x)\varphi_n(x)dx \leq (\int_\Omega f^2(x)dx)^{\frac{1}{2}} (\int_\Omega (f(x) - \varphi_n(x))^2 dx)^{\frac{1}{2}}.$$

It follows from (29) that

$$\int_\Omega f^2(x)dx = \lim_{n\to\infty} \int_\Omega f(x)\varphi_n(x)dx = 0,$$

by assumption.

Thus $\|f\|_2 = 0$ and therefore $f \equiv 0$.   □

Corollary 19.26 Let $f \in C^0(\Omega)$ and for all $\varphi \in C_0^\infty(\Omega)$ assume that

$$\int_\Omega f(x)\varphi(x)dx = 0.$$

Then $f \equiv 0$.

Proof. For every $\Omega' \subset\subset \Omega$, $f_{|\Omega'} \in L^2(\Omega')$, and by the previous corollary we have $f_{|\Omega'} \equiv 0$. Thus $f \equiv 0$.   □

We finally use the regularization method introduced in this § to show the existence of a so-called partition of unity.

Theorem 19.27 Let Ω be open in $\mathbb{R}^d$ and be covered by the family $\{U_i\}_{i\in I}$ of open sets, i.e. $\Omega = \bigcap_{i\in I} U_i$. Then there exists a subordinate partition of unity, i.e. a system $\{\phi_j\}_{j\in J}$ of functions in $C_0^\infty(\mathbb{R}^d)$ satisfying:
(i)   For each $j \in J$, supp $\phi_j \subset U_i$ for some $i \in I$.
(ii) $0 \leq \phi_j(x) \leq 1$ for all $j \in J$, $x \in \mathbb{R}^d$.
(iii) $\sum_{j\in J} \phi_j(x) = 1$ for all $x \in \Omega$.
(iv)   For each compact $K \in \Omega$, there are only finitely many ϕ_j that are not identically zero on K.

Proof. For each open ball $U(y, r)$, $y \in \mathbb{R}^d$, $r > 0$, we may find a function $\rho_{y,r}$ with

$$\rho_{y,r} \begin{cases} > 0 & \text{if } x \in U(y, r) \\ = 0 & \text{otherwise,} \end{cases}$$

by translating and scaling the above function ρ_1. For $n \in \mathbb{N}$, let

$$K_n := \{x \in \Omega : \|x\| \leq n, \ \text{dist}(x, \partial\Omega) \geq \frac{1}{n}\} \ (K_0 := K_{-1} := \emptyset).$$

The K_n are compact sets with $K_n \subset \overset{\circ}{K}_{n+1}$ for all n and

$$\Omega = \bigcup_{n\in\mathbb{N}} K_n.$$

Let
$$\Omega_n := \overset{\circ}{K}_{n+1} \setminus K_{n-2}, \ A_n := K_n \setminus \overset{\circ}{K}_{n-1}.$$
Then A_n is compact, Ω_n open, $A_n \subset \Omega_n$ for all n, and
$$\Omega = \bigcup_{n \in \mathbb{N}} A_n.$$
Any compact subset of Ω is contained in some K_n, and therefore intersects only finitely many of the sets Ω_n. For any $x \in A_n$, we find some open ball $U(x, r) \subset \Omega_n$, and also $U(x, r) \subset U_i$ for some $i \in I$. Since A_n is compact, it can be covered by finitely many such balls
$$U(x_\nu^n, r_\nu^n), \nu = 1, \ldots, k_n.$$
Then any compact $K \subset \Omega$ intersects only finitely many of these balls $U(x_\nu^n, r_\nu^n)$ ($n \in \mathbb{N}, \nu = 1, \ldots, k_n$), while these balls in turn cover Ω. We simplify the notation and denote the family of these balls by
$$\{U(x_j, r_j)\}_{j \in \mathbb{N}}.$$
For every j, we then take the function
$$\rho_j := \rho_{x_j, r_j}.$$
Since ρ_j vanishes outside $U(x_j, r_j)$, on any compact $K \subset \Omega$, only finitely many of the ρ_j are nonzero. Therefore,
$$\phi(x) := \sum_{j \in \mathbb{N}} \rho_j(x)$$
is convergent for every $x \in \Omega$, and positive as the balls $U(x_j, r_j)$ cover Ω and ρ_j is positive on $U(x_j, r_j)$. We put
$$\phi_j(x) := \frac{\rho_j(x)}{\phi(x)}$$
to get all the conclusions of the theorem (remember that we have chosen the balls $U(x_j, r_j)$ so that each of them was contained in some U_i). $\qquad\square$

Exercises for § 19

1) Let $\Omega \in \mathbb{R}^d$ be open, Vol $(\Omega) < \infty$, $f : \Omega \to \mathbb{R} \cup \{\pm\infty\}$ measurable with $|f|^p \in L^1(\Omega)$, and put

$$\Phi_p(f) := \Big(\frac{1}{\text{Vol}(\Omega)} \int\limits_{\Omega} |f(x)|^p dx\Big)^{\frac{1}{p}}.$$

Show
a) $\lim\limits_{p \to -\infty} \Phi_p(f) = \text{ess inf } |f|$.
b) $\lim\limits_{p \to 0} \Phi_p(f) = \exp\big(\frac{1}{\text{Vol}(\Omega)} \int\limits_{\Omega} \log |f(x)| dx\big)$
(provided these limits exist).

2) Let $\Omega \subset \mathbb{R}^d$ be open, $1 \le p < q \le \infty$. Construct a sequence $(f_n)_{n \in \mathbb{N}} \subset L^p(\Omega) \cap L^q(\Omega)$ that converges in $L^p(\Omega)$, but not in $L^q(\Omega)$. Do there also exist open sets Ω for which one may find a sequence that converges in $L^q(\Omega)$, but not in $L^p(\Omega)$? And what happens if Ω is bounded?

3) For $n \in \mathbb{N}$, define $f_n : (0,1) \to \mathbb{R}$ as follows

$$f_n(x) := \sin^n(\pi n x).$$

For $1 \le p < \infty$, show

$$\lim\limits_{n \to \infty} \|f_n\|_{L^p((0,1))} = 0.$$

4) Let $0 < s \le 1$, and let $L^s(\mathbb{R}^d)$ be the space of equivalence classes of measurable functions f with $|f|^s \in L^1(\mathbb{R}^d)$. Show that $L^s(\mathbb{R}^d)$ is a vector space. Put

$$\|f\|_s := \Big(\int\limits_{\mathbb{R}^d} |f(x)|^s dx\Big)^{\frac{1}{s}}.$$

Show that for *nonnegative* $f, g \in L^s(\mathbb{R}^d)$, we have

$$\|f\|_s + \|g\|_s \le \|f + g\|_s.$$

5) Let $1 \le p < \infty$, and suppose that $(f_n)_{n \in \mathbb{N}} \subset L^p(\mathbb{R}^d)$ converges to some function f pointwise almost everywhere.
a)   Show that if $\|f_n\|_p \le M < \infty$ for all n, then $f \in L^p(\mathbb{R}^d)$, and $\|f\|_p \le M$.
For $p = 1$, construct an example where $\|f_n\|_p$ does not converge to $\|f\|_p$.
b)   Suppose now that $\|f_n\|_p$ converges to $\|f\|_p$. Show that f_n then converges to f in L^p, i.e. $\|f_n - f\|_p$ converges to 0.

(Hint: Show that for every $\varepsilon > 0$, one may find a compact set $K \subset \mathbb{R}^d$ and $n_0 \in \mathbb{N}$ such that for $n \geq n_0$

$$\int_{\mathbb{R}^d \setminus K} |f_n(x)|^p dx < \varepsilon.)$$

(Hint: Use the theorem of Egorov.)

c)   What happens for $p = \infty$?

6)   Show that $L^\infty(\mathbb{R})$ is not separable (a metric space (X, d) is called separable if it contains a countable subset $(x_n)_{n \in \mathbb{N}}$ that is dense in X, i.e. for every open ball $U(p, r) \subset X$, we may find some $n_0 \in \mathbb{N}$ with $x_{n_0} \in U(p, r)$).
(Hint: Consider the subset of $L^\infty(\mathbb{R})$ consisting of chracteristic functions of intervals.)

7)   Let f be a nonnegative measurable function on Ω, and for $c > 0$, $A_c := \{x : f(x) \geq c\}$. Show that, if $f \in L^p$,

$$c^p \, \mathrm{Vol}\,(A_c) \leq \int f^p.$$

Conversely, if $\mathrm{Vol}\,(\Omega) < \infty$ and if there exist constants M and $\eta > 0$ with

$$\mathrm{Vol}\,(A_c) \leq Mc^{-p-\eta}$$

for all $c > 0$, then $f \in L^p(\Omega)$.

8)   For $0 < x < 1$, put $f(x) := (1 + x^{\frac{1}{p}})^p + (1 - x^{\frac{1}{p}})^p$.
a)   For $p \geq 2$, show the following inequalities $(0 < x < 1)$

(i) $f''(x) \leq 0$

(ii) $f(x) \leq f(y) + (x - y)f'(y) \quad (0 < y < 1)$

(iii) $f(x) \leq 2^{p-1}(x + 1)$

(iv) $f(x) \leq 2(1 + x^{\frac{q}{p}})^{\frac{p}{q}}, \quad$ where $\dfrac{1}{p} + \dfrac{1}{q} = 1.$

b)   For $1 \leq p < 2$, show the above inequalities with " $\geq$ " in place of " $\leq$ ".

9)

a)   Use the preceding exercise to show the following inequalities for $a, b \in \mathbb{R}, p \geq 2, \frac{1}{p} + \frac{1}{q} = 1$:

(i) $2(|a|^p + |b|^p) \leq |a + b|^p + |a - b|^p \leq 2^{p-1}(|a|^p + |b|^p)$

(ii) $|a + b|^p + |a - b|^p \leq 2(|a|^q + |b|^q)^{\frac{p}{q}}$

(iii) $2(|a|^p + |b|^p)^{\frac{q}{p}} \leq |a + b|^q + |a - b|^q$

Use these inequalities to show Clarkson's inequalities for $f, g \in L^p$ $(p \geq 2)$

$$2(\|f\|_p^p + \|g\|_p^p) \leq \|f + g\|_p^p + \|f - g\|_p^p$$
$$\leq 2^{p-1}(\|f\|_p^p + \|g\|_p^p)$$
$$\|f + g\|_p^p + \|f - g\|_p^p \leq 2(\|f\|_q^q + \|g\|_q^q)^{\frac{p}{q}}$$
$$2(\|f\|_p^p + \|g\|_p^p)^{\frac{q}{p}} \leq \|f + g\|_q^q + \|f - g\|_q^q.$$

b)   For $1 \leq p < 2$ show the above inequalities with "$\geq$" in place of "$\leq$".

(Hint: Use the Minkowski inequality in L^s for $0 < s \leq 1$ from Exercise 4.)

10)   Deduce corollary 19.19 from the Besicovitch covering theorem.

20. Integration by Parts. Weak Derivatives. Sobolev Spaces

Weak derivatives are introduced by taking the rule for integration by parts as a definition. Spaces of functions that are in L^p together with certain weak derivatives are called Sobolev spaces. Sobolev's embedding theorem says that such functions are continuous if their weak derivatives satisfy strong enough integrability properties. Rellich's compactness theorem says that integral bounds on weak derivatives implies convergence of subsequences of the functions itself in L^p.

Lemma 20.1 *Let $\Omega \subset \mathbb{R}^d$ be open, $1 \leq i \leq d$. For all $\varphi \in C_0^1(\Omega)$ we have*

$$\int_\Omega \frac{\partial \varphi(x)}{\partial x^i} dx = 0. \tag{1}$$

Proof. By setting $\varphi(x) = 0$ for $x \in \mathbb{R}^d \backslash \Omega$, we can work with $\varphi \in C_0^1(\mathbb{R}^d)$. Let
$\mathrm{supp}\varphi \subset [-M, M]^d$ for $M \in \mathbb{R}$. Without loss of generality, assume that $i = d$. Then we have for fixed $(x^1, \ldots, x^{d-1}) \in \mathbb{R}^{d-1}$

$$\int_\mathbb{R} \frac{\partial \varphi}{\partial x^d}(x^1, \ldots, x^d) dx^d = \varphi(x^1, \ldots, x^{d-1}, M) - \varphi(x^1, \ldots, x^{d-1}, -M) = 0,$$

and therefore

$$\int_{\mathbb{R}^d} \frac{\partial \varphi(x)}{\partial x^d} dx = 0.$$

$\square$

From (1) we deduce that for $f \in C^1(\Omega), \varphi \in C_0^1(\Omega)$, (so $f\varphi \in C_0^1(\Omega)$),

$$\int_\Omega \frac{\partial f}{\partial x^i}(x)\varphi(x) dx = - \int_\Omega f(x) \frac{\partial \varphi}{\partial x^i}(x) dx, \tag{2}$$

using the product rule for differentiation.
By iteration we obtain for $f \in C^2(\Omega), \varphi \in C_0^2(\Omega)$,

$$\int_\Omega \frac{\partial^2 f(x)}{(\partial x^i)^2}\varphi(x) dx = - \int_\Omega \frac{\partial f}{\partial x^i}(x) \frac{\partial \varphi}{\partial x^i}(x) = \int_\Omega f(x) \frac{\partial^2 \varphi(x)}{(\partial x^i)^2} dx, \tag{3}$$

and by summation over i

$$\int_\Omega \Delta f(x)\varphi(x)dx = -\int_\Omega \operatorname{grad} f(x) \cdot \operatorname{grad} \varphi(x)dx = \int f(x)\Delta\varphi(x)dx \quad (4)$$

where the dot in the middle integral denotes the scalar product in $\mathbb{R}^d$.

Definition 20.2 In the following we set

$$L^1_{\text{loc}}(\Omega) := \{f : \Omega \to \mathbb{R} \cup \{\pm\infty\} : f \in L^1(\Omega') \text{ for every } \Omega' \subset\subset \Omega\}.$$

We will use the above formulae as a motivation for introducing a concept of differentiation for functions which are not necessarily differentiable in the classical sense.

Definition 20.3 Let $f \in L^1_{\text{loc}}(\Omega)$. A function $v \in L^1_{\text{loc}}(\Omega)$ is called the weak derivative of f in the direction $x^i (x = (x^1, \ldots, x^d) \in \mathbb{R}^d)$ if

$$\int_\Omega v(x)\varphi(x)dx = -\int_\Omega f(x)\frac{\partial\varphi(x)}{\partial x^i}dx \quad (5)$$

holds for all $\varphi \in C^1_0(\Omega)$ (such a function is also called a test-function). We write $v = D_i f$. In case f has weak derivatives $D_i f$ for $i = 1, \ldots, d$, we write $Df = (D_1 f, \ldots, D_d f)$.

Obviously every $f \in C^1(\Omega)$ has weak derivatives, namely

$$D_i f = \frac{\partial f}{\partial x^i},$$

as a comparison of (2) and (5) shows. However, there are functions which have weak derivatives, but are not in $C^1(\Omega)$. On the other hand, not every function in L^1_{loc} has weak derivatives.

Examples. Let $\Omega = (-1, 1) \subset \mathbb{R}$
1) $f(x) := |x|$. Then f has the weak derivative

$$Df(x) = \begin{cases} 1 & \text{for } 0 \leq x < 1 \\ -1 & \text{for } -1 < x < 0, \end{cases}$$

because for every $\varphi \in C^1_0((-1, 1))$ we have

$$\int_{-1}^0 (-\varphi(x))dx + \int_0^1 \varphi(x)dx = -\int_{-1}^1 \varphi'(x) \cdot |x|dx.$$

2)

$$f(x) := \begin{cases} 1 & \text{for } 0 \le x < 1 \\ 0 & \text{for } -1 < x < 0 \end{cases}$$

has no weak derivative, for $Df(x)$ must then be zero for $x \ne 0$, thus as an L^1_{loc}-function, $Df \equiv 0$, but it does not hold for every $\varphi \in C^1_0(-1, 1)$ that

$$0 = \int_{-1}^{1} \varphi(x) \cdot 0 \, dx = -\int_{-1}^{1} \varphi'(x) f(x) dx = -\int_{0}^{1} \varphi'(x) dx = \varphi(0).$$

The weak derivatives of higher order are defined analogously.

Definition 20.4 Let $f \in L^1_{\text{loc}}(\Omega)$, $\alpha := (\alpha_1, \dots \alpha_d)$, $\alpha_i \ge 0$ $(i = 1, \dots d)$, $|\alpha| := \sum_{i=1}^{d} \alpha_i > 0$,

$$D_\alpha \varphi := \left(\frac{\partial}{\partial x^1}\right)^{\alpha_1} \cdots \left(\frac{\partial}{\partial x^d}\right)^{\alpha_d} \varphi \text{ for } \varphi \in C^{|\alpha|}(\Omega).$$

A function $v \in L^1_{\text{loc}}(\Omega)$ is called the α-th weak derivative of f, in symbols $v = D_\alpha f$ if

$$\int_\Omega \varphi v dx = (-1)^{|\alpha|} \int_\Omega f \cdot D_\alpha \varphi dx \; \forall \; \varphi \in C^{|\alpha|}_0(\Omega).$$

Definition 20.5 For $k \in \mathbb{N}$, $1 \le p \le \infty$, we define the Sobolev space $W^{k,p}(\Omega)$ by $W^{k,p}(\Omega) := \{f \in L^p(\Omega) : D_\alpha f \text{ exists and is in } L^p(\Omega) \text{ for all } |\alpha| \le k\}$.

We set

$$\|f\|_{W^{k,p}(\Omega)} := \left(\sum_{|\alpha| \le k} \int_\Omega |D_\alpha f|^p\right)^{\frac{1}{p}} \text{ for } 1 \le p < \infty$$

and

$$\|f\|_{W^{k,\infty}(\Omega)} := \sum_{|\alpha| \le k} \operatorname*{ess\,sup}_{x \in \Omega} |D_\alpha f(x)|.$$

As for L^p spaces, one sees that the Sobolev spaces $W^{k,p}(\Omega)$ are normed spaces. We want to show that the spaces $W^{k,p}(\Omega)$ are Banach spaces.

In the rest of this chapter, we shall only consider the case $1 \le p < \infty$.

Example. As an example, we want to see when, for $\Omega = U(0,1) = \{x \in \mathbb{R}^d : \|x\| < 1\}$ the function

$$f(x) = \|x\|^\alpha \; (\alpha \in \mathbb{R})$$

lies in $W^{1,p}(\Omega)$.

First, we calculate that for $x \neq 0$, $i = 1, \ldots, d$

$$\frac{\partial}{\partial x^i} \|x\|^\alpha = \frac{\partial}{\partial x^i} \Big(\sum_{j=1}^{d} |x^j|^2\Big)^{\frac{\alpha}{2}} = \alpha x^i \Big(\sum_{j=1}^{d} |x^j|^2\Big)^{\frac{\alpha}{2}-1}.$$

We have already seen in § 19 that

$$\|x\|^\alpha \in L^p(\Omega) \iff \alpha p + d > 0.$$

Correspondingly, because $|\frac{\partial}{\partial x^i}\|x\|^\alpha| \leq |\alpha|\|x\|^{\alpha-1}$, $\frac{\partial}{\partial x^i}\|x\|^\alpha \in L^p(\Omega)$ for $(\alpha - 1)p + d > 0$.

We will now show that for $d + \alpha > 1$

$$v(x) := \begin{cases} \frac{\partial}{\partial x^i}\|x\|^\alpha & \text{for } x \neq 0 \\ 0 & \text{for } x = 0 \end{cases}$$

is the ith weak derivative of $\|x\|^\alpha$. Then we would have shown that

$$\|x\|^\alpha \in W^{1,p}(\Omega) \text{ for } (\alpha - 1)p + d > 0.$$

For the verification that v is the weak derivative of $\|x\|^\alpha$ we set

$$\eta_n(r) := \begin{cases} 1 & \text{for } r \geq \frac{2}{n} \\ n(r - \frac{1}{n}) & \text{for } \frac{1}{n} \leq r < \frac{2}{n} \\ 0 & \text{for } 0 \leq r < \frac{1}{n} \end{cases} \qquad (n \in \mathbb{N})$$

For $\varphi \in C_0^1(\Omega)$ we have

$$\int_\Omega \|x\|^\alpha \frac{\partial}{\partial x^i}(\eta_n(\|x\|)\varphi(x))dx = -\int_\Omega \frac{\partial\|x\|^\alpha}{\partial x^i}\eta_n(\|x\|)\varphi(x)dx,$$

as η_n, and thus the singularity of $\frac{\partial}{\partial x^i}\|x\|^\alpha$, vanishes in a neighborhood of 0.

We must certainly still justify our use of integration by parts without $\varphi \cdot \eta_n$ being a C^1-function. This can be done by means of an approximation argument or by lemma 19.14.

By the dominated convergence theorem, for $(\alpha - 1) + d > 0$

$$\int_\Omega \frac{\partial\|x\|^\alpha}{\partial x^i}\eta_n(\|x\|)\varphi(x)dx \to \int_\Omega v(x)\varphi(x)dx$$

as $n \to \infty$ and

$$\int_\Omega \|x\|^\alpha \frac{\partial}{\partial x^i}(\eta_n(\|x\|)\varphi(x))dx \to \int_\Omega \|x\|^\alpha \frac{\partial}{\partial x^i}\varphi(x)dx,$$

namely, as with $r = \|x\|$,

$$\left| \int_{\Omega} \|x\|^{\alpha} \varphi(x) D_i \eta_n(\|x\|) dx \right| c_0 \le c_0 \sup |\varphi| \cdot \int_{\frac{1}{n}}^{\frac{2}{n}} r^{\alpha} \cdot n r^{d-1} dr \quad \text{(theorem 13.21)}$$

with a certain constant c_0

$$= c_0 \frac{\sup |\varphi|}{d+\alpha} \cdot n \left((\frac{2}{n})^{d+\alpha} - (\frac{1}{n})^{d+\alpha} \right)$$

$$\to 0 \text{ as } n \to \infty \text{ and } d + \alpha > 1.$$

We have shown thereby that v is indeed the ith weak derivative of $\|x\|^{\alpha}$ for $d + \alpha > 1$.

Lemma 20.6 *Let* $f \in L^1_{\text{loc}}(\Omega)$ *and assume that* $v = D_i f$ *exists. If* dist $(x, \partial \Omega) > h$ *then*

$$D_i(f_h(x)) = (D_i f)_h(x),$$

where the convolution f_h *with the smooth kernel* ρ_1 *is defined as in § 19.*

Proof. By differentiation under the integral we obtain

$$D_i(f_h(x)) = \frac{1}{h^d} \int \frac{\partial}{\partial x^i} \rho(\frac{x-y}{h}) f(y) dy$$

$$= \frac{-1}{h^d} \int \frac{\partial}{\partial y^i} \rho(\frac{x-y}{h}) f(y) dy$$

$$= \frac{1}{h^d} \int \rho(\frac{x-y}{h}) D_i f(y) dy \quad \text{by definition of } D_i f$$

$$= (D_i f)_h(x).$$

$\square$

Theorem 20.7 *Let* $f, v \in L^p(\Omega)$, $1 \le p < \infty$. *Then* $v = D_i f$
$\Leftrightarrow \exists (f_n) \subset C^{\infty}(\Omega) : f_n \to f$ *in* $L^p(\Omega)$, $\frac{\partial}{\partial x^i} f_n \to v$ *in* $L^p(\Omega')$ *for any* $\Omega' \subset\subset \Omega$.

Proof. " $\Longrightarrow$ " We consider the convolution f_h as in the previous lemma, by which
$v_h = D_i(f_h)$ and by theorem 19.16 as $h \to 0$,

$$f_n \to f \text{ in } L^p(\Omega), v_n \to v \text{ in } L^p(\Omega') \text{ for any } \Omega' \subset\subset \Omega.$$

" $\Longleftarrow$ " Let $\varphi \in C^1_0(\Omega)$. Then $\varphi, D_i \varphi$ $(i = 1, \ldots, d)$ are bounded and from the dominated convergence theorem it follows that

$$\int_{\Omega} f(x) \frac{\partial}{\partial x^i} \varphi(x) dx = \lim_{n \to \infty} \int_{\Omega} f_n(x) \frac{\partial}{\partial x^i} \varphi(x) dx = - \lim_{n \to \infty} \int D_i f_n(x) \varphi(x) dx$$

$$= - \int_{\Omega} v(x) \varphi(x) dx.$$

Thus $v = D_i f$.   □

Definition 20.8 $H^{k,p}(\Omega)$ is the closure of $C^\infty(\Omega) \cap W^{k,p}(\Omega)$ relative to the $W^{k,p}$-norm, and $H_0^{k,p}(\Omega)$ is the closure of $C_0^\infty(\Omega)$ relative to the $W^{k,p}$-norm.

The preceding involves a slight abuse of notation, as $C^\infty(\Omega)$ is a space of functions, while $W^{k,p}(\Omega)$ is defined as a space of equivalence classes of functions, but the reader should have no difficulties sorting this out.

Thus $f \in H_0^{k,p}(\Omega)$ precisely when there exists a sequence $(f_n)_{n\in\mathbb{N}} \subset C_0^\infty(\Omega)$ with $\lim_{n\to\infty} \|f_n - f\|_{W^{k,p}(\Omega)} = 0$.

For $k \geq 1$, the functions in $H_0^{k,p}(\Omega)$ vanish on $\partial\Omega$ in a sense which will be made precise later.

Corollary 20.9 *Let $1 \leq p < \infty, k \in \mathbb{N}$. The space $W^{k,p}(\Omega)$ is complete with respect to $\|\cdot\|_{W^{k,p}}$, thus it is a Banach space.*

Proof. We consider the case $k = 1$: the general case then follows by induction. Let f_n be a Cauchy sequence in $W^{1,p}(\Omega)$. Then $f_n, D_i f_n$ $(i = 1, \ldots, d)$ are Cauchy sequences in $L^p(\Omega)$. As $L^p(\Omega)$ is complete, there exist $f, v^i \in L^p(\Omega)$ such that $f_n \to f, D_i f_n \to v^i$ $(i = 1, \ldots, d)$ in $L^p(\Omega)$.

Now for $\phi \in C_0^1(\Omega)$ we have

$$\int_\Omega D_i f_n \cdot \phi = - \int_\Omega f_n D_i \phi,$$

the left side converges to $\int_\Omega v^i \cdot \phi$ and the right to $- \int_\Omega f \cdot D_i \phi$. Thus $D_i f = v^i$ and therefore $f \in W^{1,p}(\Omega)$. This proves the completeness of $W^{k,p}(\Omega)$.   □

Corollary 20.10 *Let $1 \leq p < \infty, k \in \mathbb{N}$. Then*

$$W^{k,p}(\Omega) = H^{k,p}(\Omega).$$

Proof. We need to show that the space $C^\infty(\Omega) \cap W^{k,p}(\Omega)$ is dense in $W^{k,p}(\Omega)$. We put

$$\Omega_n := \{x \in \Omega : \|x\| < n, \text{dist}(x, \partial\Omega) > \frac{1}{n}\} \text{ for } n \in \mathbb{N} \ (\Omega_0 := \Omega_{-1} := \emptyset).$$

Then

$$\Omega_n \subset\subset \Omega_{n+1}, \bigcup_{n\in\mathbb{N}} \Omega_n = \Omega.$$

Using Theorem 19.27, we let $\{\phi_j\}_{j\in\mathbb{N}}$ be a partition of unity subordinate to the cover $\{\Omega_{n+2} \setminus \overline{\Omega_{n-1}}\}$ of Ω. Let $f \in W^{k,p}(\Omega)$. For $\epsilon > 0$, using Theorem 20.7 we may find positive numbers $h_n, n \in \mathbb{N}$, with

$$h_n \leq \text{dist}\,(\Omega_n, \partial\Omega_{n+1})$$

$$\|(\phi_n f)_{h_n} - \phi_n f\|_{W^{k,p}(\Omega)} < \frac{\epsilon}{2^n}.$$

By the properties of the ϕ_n, at most finitely many of the smooth functions $(\phi_n f)_{h_n}$ are nonzero on any $\Omega' \subset\subset \Omega$. Therefore

$$\widetilde{f} := \sum_n (\phi_n f)_{h_n} \in C^\infty(\Omega).$$

Since

$$\|f - \widetilde{f}\|_{W^{k,p}(\Omega)} \leq \sum_n \|(\phi_n f)_{h_n} - \phi_n f\| < \epsilon,$$

we conclude that every $W^{k,p}(\Omega)$ function can be approximated by smooth functions in $W^{k,p}(\Omega)$, as was to be shown.   □

We have seen in § 19 (corollary 19.24) that $C_0^\infty(\Omega)$ is dense in $L^p(\Omega)$ relative to the L^p-norm. The previous corollary means that $C^\infty(\Omega)$ is dense in $W^{k,p}(\Omega)$ relative to the $W^{k,p}(\Omega)$ norm. However, for $k \geq 1$, $C_0^\infty(\Omega)$ is no longer dense in $W^{k,p}(\Omega)$. We shall see later as an example that a non zero constant does not lie in $H_0^{k,p}(\Omega)$ and thus also not in the closure of $C_0^\infty(\Omega)$. On the other hand for every Ω with Vol $(\Omega) < \infty$, naturally any constant lies in $W^{k,p}(\Omega)$. To get a feeling for the spaces $H_0^{1,p}(\Omega)$, we consider the example $\Omega = U(0,1) = \{x \in \mathbb{R}^d : \|x\| < 1\}$. Let $f \in C^1(\Omega) \cap C^0(\overline{\Omega})$ and let $f = 0$ on $\partial\Omega$. We claim that f then lies in $H_0^{1,p}(\Omega)$. For a proof, we set, for $\lambda > 1$,

$$f_\lambda(x) := f(\lambda x).$$

Then $f_\lambda \in C_0^1(\Omega)$. Since f_λ has compact support, the regularizations $f_{\lambda,h}$ also have compact support for h sufficiently small (independent of λ), and therefore $f_\lambda \in H_0^{1,p}(\Omega)$. Now as $\lambda \to 1$, f_λ tends pointwise to f, and $D_i f_\lambda(x) = \lambda \frac{\partial}{\partial x^i} f(\lambda x)$ converges pointwise to $\frac{\partial}{\partial x^i} f(x)$. By the dominated convergence theorem, f_λ also tends to f in the $W^{1,p}$-norm and as $H_0^{1,p}(\Omega)$ is, by definition, closed, it follows that $f \in H_0^{1,p}(\Omega)$.

We now want to give still a few more rules for dealing with Sobolev functions.

Lemma 20.11 *Let $1 \leq p < \infty$. Let $f \in W^{1,p}(\Omega), \psi \in C^1(\mathbb{R})$ with*

$$\sup_{y \in \mathbb{R}} |\psi'(y)| < \infty.$$

Then

$$\psi \circ f \in W^{1,p}(\Omega), D(\psi \circ f) = \psi'(f)Df.$$

Proof. Let $(f_n)_{n \in \mathbb{N}} \subset C^\infty(\Omega), f_n \to f$ in $W^{1,p}(\Omega)$ as $n \to \infty$. Then

$$\int_\Omega |\psi(f_n) - \psi(f)|^p dx \le \sup |\psi'|^p \int_\Omega |f_n - f|^p dx \to 0$$

and

$$\int_\Omega |\psi'(f_n)Df_n - \psi'(f)Df|^p dx \le 2^p (\sup |\psi'|^p \int_\Omega |Df_n - Df|^p dx$$

$$+ \int_\Omega |\psi'(f_n) - \psi'(f)|^p |Df|^p dx).$$

By theorem 19.12, a suitable subsequence of f_n converges pointwise almost everywhere to f in Ω. Since ψ' is continuous, $\psi'(f_n)$ also converges pointwise almost everywhere to $\psi'(f)$ and the last integral therefore approaches zero, by the Lebesgue dominated convergence theorem.

Thus
$$\psi(f_n) \to \psi(f) \text{ in } L^p(\Omega)$$
$$D(\psi(f_n)) = \psi'(f_n)Df_n \to \psi'(f)Df \text{ in } L^p(\Omega)$$

and therefore, by theorem 20.7,

$$\psi \circ f \in W^{1,p}(\Omega) \text{ und } D(\psi \circ f) = \psi'(f)Df.$$

$\square$

Corollary 20.12 *If $f \in W^{1,p}(\Omega)(1 \le p < \infty)$ then so is $|f|$ and*

$$D_i|f|(x) = \begin{cases} D_i f(x) & \text{if } f(x) > 0 \\ 0 & \text{if } f(x) = 0 \quad (i = 1, \dots, d) \\ -D_i f(x) & \text{if } f(x) < 0 \end{cases}.$$

Proof. For $\varepsilon > 0$ we set

$$\psi_\varepsilon(f) := (f^2 + \varepsilon^2)^{\frac{1}{2}} - \varepsilon$$

Now ψ_ε fulfils the assumptions of lemma 20.11 and so for $\varphi \in C_0^1(\Omega)$

$$\int_\Omega \psi_\varepsilon(f(x))D_i\varphi(x)dx = -\int \frac{f(x)D_i f(x)}{(f^2(x) + \varepsilon^2)^{\frac{1}{2}}}\varphi(x)dx.$$

We let ε approach zero and apply as usual the dominated convergence theorem; it follows that

$$\int_\Omega |f(x)|D_i\varphi(x)dx = -\int v(x)\varphi(x)dx$$

with

$$v(x) = \begin{cases} D_i f(x) & \text{if } f(x) > 0 \\ 0 & \text{if } f(x) = 0 \\ -D_i f(x) & \text{if } f(x) < 0 \ . \end{cases}$$

This is the assertion that $|f|$ is weakly differentiable and its derivative has the required form. That f, as well as $|f|$, lies in $W^{1,p}$ then follows directly. $\qquad\square$

Corollary 20.13 *Let* $f \in W^{1,p}(\Omega)$, $A \subset \Omega$, $f \equiv$ *constant on* A. *Then* $Df = 0$ *almost everywhere on* A.

Proof. We may assume that $f \equiv 0$ on A. Then $Df = D|f| = 0$ on A by corollary 20.12. $\qquad\square$

Lemma 20.14 *Let* $1 \le p < \infty$. *Let* $\Omega_0 \subset \Omega$, $g \in W^{1,p}(\Omega)$, $h \in W^{1,p}(\Omega_0)$ *and* $h - g \in H_0^{1,p}(\Omega_0)$. *Then*

$$f(x) := \begin{cases} h(x) & x \in \Omega_0 \\ g(x) & x \in \Omega \backslash \Omega_0 \end{cases}$$

is in $W^{1,p}(\Omega)$ *and*

$$D_i f(x) = \begin{cases} D_i h(x), & x \in \Omega_0 \\ D_i g(x) & x \in \Omega \backslash \Omega_0, \end{cases} \quad i = 1, \dots, d.$$

Thus one can replace a Sobolev function in $W^{1,p}$ on an interior subset by another one with the same boundary values, without leaving the Sobolev space. Thus corners play no role for membership of $W^{1,p}$.

Proof. By considering $h - g$ and $f - g$ instead of h and f, we can assume that $g = 0$. We thus have to show the following:
  Let $h \in H_0^{1,p}(\Omega_0)$, then

$$f(x) = \begin{cases} h(x) & \text{for } x \in \Omega_0 \\ 0 & \text{for } x \in \Omega \backslash \Omega_0 \end{cases}$$

is also in $W^{1,p}(\Omega)$ and

$$D_i f(x) = \begin{cases} D_i h(x) & \text{for } x \in \Omega_0 \\ 0 & \text{for } x \in \Omega \backslash \Omega_0. \end{cases}$$

Let $(h_n)_{n \in \mathbb{N}} \subset C_0^\infty(\Omega_0)$ be a sequence that converges in the $W^{1,p}$-norm to h. Then we have

$$f_n(x) := \begin{cases} h_n(x) & \text{for } x \in \Omega_0 \\ 0 & \text{for } x \in \Omega \backslash \Omega_0 \end{cases}$$

in $C_0^\infty(\Omega)$, with derivative

$$\frac{\partial}{\partial x^i} f_n(x) = \begin{cases} \frac{\partial}{\partial x^i} h_n(x) & \text{for } x \in \Omega_0 \\ 0 & \text{for } x \in \Omega \backslash \Omega_0. \end{cases}$$

The assertion now follows directly from theorem 20.7.   □

We now consider, for a moment, the case $d = 1$, and so, we let our domain Ω be an open interval $I \subset \mathbb{R}$. We consider

$$f \in H^{1,1}(I).$$

By theorem 20.7, we may find functions $f_n \in C^\infty(I)$ that converge to f, and whose derivatives converge to the weak derivative Df, in $L^p(I')$ for any $I' \subset I$. Thus

$$\lim_{n \to \infty} \int_{I'} |f_n(x) - f(x)| dx + \int_{I'} |\frac{df_n}{dx}(x) - Df(x)| dx = 0.$$

Since f_n is smooth, for $a, b \in I$, and $\varepsilon > 0$, we obtain

$$|f_n(b) - f_n(a)| = |\int_a^b \frac{df_n}{dx}(x) dx| \le \int_a^b |\frac{df_n(x)}{dx}| dx$$

$$\le \int_a^b |Df(x)| dx + \frac{\varepsilon}{2}$$

for sufficiently large n, because of the above convergence.

Moreover, since $Df \in L^1(I)$, using corollary 16.9, we may find $\delta > 0$ such that whenever $a, b \in I$ satisfy $|a - b| < \delta$, then

$$\int_a^b |Df(x)| dx < \frac{\varepsilon}{2}.$$

In conclusion

$$|f_n(b) - f_n(a)| < \varepsilon$$

whenever n is sufficiently large and $|b - a| < \delta$. Therefore, the f_n are equicontinuous, and by the theorem 5.20 of Arzela-Ascoli, they converge uniformly towards their continuous limit f.

Moreover, by passing to the limit $n \to \infty$ in the above inequality, we then also get

$$|f(b) - f(a)| \le \int_a^b |Df(x)| dx$$

for all $a, b \in I$.

Thus, in particular, for $d = 1$, every $H^{1,1}$-function is uniformly continuous.

We now come to the important Sobolev embedding theorem, which constitutes a certain extension of the preceding result to the case $d > 1$.

Theorem 20.15 *Let $\Omega \subset \mathbb{R}^d$ be bounded and open and $f \in H_0^{1,p}(\Omega)$. Then for $p < d$*

$$f \in L^{\frac{dp}{d-p}}(\Omega)$$

and for $p > d$

$$f \in C^0(\overline{\Omega}).$$

Moreover, there exist constants $c = c(p,d)$ with the property that for all $f \in H_0^{1,p}(\Omega)$

$$\|f\|_{L^{\frac{dp}{d-p}}(\Omega)} \leq c\|Df\|_{L^p(\Omega)} \quad \text{for } p < d \tag{6}$$

and

$$\sup_{x \in \Omega} |f(x)| \leq c \operatorname{Vol}(\Omega)^{\frac{1}{d}-\frac{1}{p}} \|Df\|_{L^p(\Omega)} \quad \text{for } p > d. \tag{7}$$

Proof. Since the case $d = 1$ has already been analyzed, we now consider $d > 1$. We first assume that $f \in C_0^1(\Omega)$ and treat the case $p = 1$. We write, as usual, $x = (x^1, \ldots, x^d)$. For $i = 1, \ldots, d$ we have

$$|f(x)| \leq \int_{-\infty}^{x^i} |D_i f(x)| dx^i.$$

Here we have used the compactness of the support of f and therefore

$$|f(x)|^d \leq \prod_{i=1}^{d} \int_{-\infty}^{\infty} |D_i f| dx^i$$

where this expression stands as an abbreviation for

$$\prod_{i=1}^{d} \int_{-\infty}^{\infty} |D_i f(x^1, \ldots, x^{i-1}, y, x^{i+1}, \ldots, x^d)| dy,$$

and

$$|f(x)|^{\frac{d}{d-1}} \leq \left(\prod_{i=1}^{d} \int_{-\infty}^{\infty} |D_i f| dx^i \right)^{\frac{1}{d-1}}.$$

It follows that

$$\int_{-\infty}^{\infty} |f(\xi)|^{\frac{d}{d-1}} d\xi^1 \leq \left(\int_{-\infty}^{\infty} |D_1 f| dx^1 \right)^{\frac{1}{d-1}}$$

$$\cdot \int_{-\infty}^{\infty} \left(\prod_{i=2}^{d} \int_{-\infty}^{\infty} |D_i f| dx^i \right)^{\frac{1}{d-1}} d\xi^1,$$

thus

$$\int\limits_{-\infty}^{\infty} |f(x)|^{\frac{d}{d-1}} dx^1 \leq \left(\int\limits_{-\infty}^{\infty} |D_1 f| dx^1 \right)^{\frac{1}{d-1}} \cdot \left(\prod_{i\neq 1} \int\limits_{-\infty}^{\infty} \int\limits_{-\infty}^{\infty} \right) |D_i f| dx^i dx^1 \Big)^{\frac{1}{d-1}},$$

where we have used (11) from §19 (the generalized Hölder inequality) for $p_1 = \ldots = p_{d-1} = d - 1$.

Further, with the same argument we have

$$\int\limits_{-\infty}^{\infty} \int\limits_{-\infty}^{\infty} |f(x)|^{\frac{d}{d-1}} dx^1 dx^2 \leq \left(\int\limits_{-\infty}^{\infty} \int\limits_{-\infty}^{\infty} |D_1 f| dx^1 dx^2 \right)^{\frac{1}{d-1}}$$

$$\cdot \left(\int\limits_{-\infty}^{\infty} \int\limits_{-\infty}^{\infty} |D_2 f| dx^1 dx^2 \right)^{\frac{1}{d-1}}$$

$$\cdot \left(\prod_{i=3}^{d} \int\limits_{-\infty}^{\infty} \int\limits_{-\infty}^{\infty} \int\limits_{-\infty}^{\infty} |D_i f| dx^i dx^1 dx^2 \right)^{\frac{1}{d-1}}.$$

Iteratively, it follows that

$$\int\limits_{\Omega} |f(x)|^{\frac{d}{d-1}} dx \leq \left(\prod_{i=1}^{d} \int\limits_{\Omega} |D_i f| dx \right)^{\frac{1}{d-1}},$$

thus

$$\|f\|_{\frac{d}{d-1}} \leq \left(\prod_{i=1}^{d} \int\limits_{\Omega} |D_i f| dx \right)^{\frac{1}{d}}$$

$$\leq \frac{1}{d} \int\limits_{\Omega} \sum_{i=1}^{d} |D_i f| dx,$$

as the geometric mean is not bigger than the arithmetic mean, hence

$$\|f\|_{\frac{d}{d-1}} \leq \frac{1}{d} \|Df\|_1, \tag{8}$$

which is (6) for $p = 1$.

If one applies (8) to $|f|^{\gamma} (\gamma > 1)$, then one obtains

$$\||f|^{\gamma}\|_{\frac{d}{d-1}} \leq \frac{\gamma}{d} \int\limits_{\Omega} |f|^{\gamma-1} |Df| dx, \tag{9}$$

(using the chain rule of lemma 20.11)

$$\leq \frac{\gamma}{d} \||f|^{\gamma-1}\|_q \cdot \|Df\|_p \quad \text{for} \quad \frac{1}{p} + \frac{1}{q} = 1$$

(by the Hölder inequality).

If $p < d$, then $\gamma = \frac{(d-1)p}{d-p}$ satisfies $\frac{dp}{d-p} = \frac{\gamma d}{d-1} = \frac{(\gamma-1)p}{p-1}$ and (9) yields, taking into account $q = \frac{p}{p-1}$, that

$$\|f\|_{\frac{\gamma d}{d-1}}^{\gamma} \leq \frac{\gamma}{d}\|f\|_{\frac{\gamma d}{d-1}}^{\gamma-1} \cdot \|Df\|_p,$$

so

$$\|f\|_{\frac{\gamma d}{d-1}} \leq \frac{\gamma}{d}\|Df\|_p, \tag{10}$$

which is (6).

We now treat the case $p > d$. We assume that

$$\mathrm{Vol}\,(\Omega) = 1 \tag{11}$$

and

$$\|Df\|_{L^p(\Omega)} = 1. \tag{12}$$

Then (9) becomes

$$\||f|^{\gamma}\|_{\frac{d}{d-1}} \leq \frac{\gamma}{d}\||f|^{\gamma-1}\|_{\frac{p}{p-1}},$$

and therefore

$$\|f\|_{\gamma\frac{d}{d-1}} \leq \left(\frac{\gamma}{d}\right)^{\frac{1}{\gamma}}\|f\|_{(\gamma-1)\frac{p}{p-1}}^{\frac{\gamma-1}{\gamma}}$$

and then

$$\|f\|_{\gamma\frac{d}{d-1}} \leq \left(\frac{\gamma}{d}\right)^{\frac{1}{\gamma}}\|f\|_{\gamma\frac{p}{p-1}}^{\frac{\gamma-1}{\gamma}} \tag{13}$$

by corollary 19.7, as $\mathrm{Vol}\,(\Omega) = 1$.

We now set

$$\gamma = \frac{d}{d-1} \cdot \frac{p-1}{p} > 1, \text{ as } p > d.$$

By substituting γ^n for γ in (13) we get

$$\|f\|_{\gamma^n\frac{d}{d-1}} \leq \left(\frac{\gamma^n}{d}\right)^{\frac{1}{\gamma^n}}\|f\|_{\gamma^{n-1}\frac{d}{d-1}}^{\frac{\gamma^n-1}{\gamma^n}} \tag{14}$$

because

$$\gamma^n\frac{p}{p-1} = \gamma^{n-1}\frac{d}{d-1}.$$

Now if for infinitely many $n \in \mathbb{N}$ we have

$$\|f\|_{\gamma^n\frac{d}{d-1}} \leq 1,$$

then by corollary 19.9

$$\operatorname*{ess\,sup}_{x\in\Omega} |f(x)| \leq 1, \tag{15}$$

because of the normalization (11).

If, on the contrary, for all $n \geq n_0$ (n_0 being chosen minimal)

$$\|f\|_{\gamma^n \frac{d}{d-1}} \geq 1,$$

it follows from (14) that for $n \geq n_0 + 1$

$$\|f\|_{\gamma^n \frac{d}{d-1}} \leq (\frac{\gamma^n}{d})^{\frac{1}{\gamma^n}} \|f\|_{\gamma^{n-1} \frac{d}{d-1}} \tag{16}$$

and then from (16) by iteration

$$\|f\|_{\gamma^n \frac{d}{d-1}} \leq \frac{\gamma^{\sum\limits_{\nu=n_0+1}^{n} \nu \frac{1}{\gamma^\nu}}}{d^{\sum\limits_{\nu=n_0+1}^{n} \frac{1}{\gamma^\nu}}} \|f\|_{\gamma^{n_0} \frac{d}{d-1}}. \tag{17}$$

We then apply (14) for $n = n_0$ and obtain on the right side $\|f\|_{\gamma^{n_0-1} \frac{d}{d-1}}$. We may assume that $\|f\|_{\gamma^{n_0-1} \frac{d}{d-1}} \leq 1$ or $n_0 = 1$.

For the last case, taking into consideration (8) as well as the normalization $\|Df\|_p = 1$, it follows in every case from corollary 19.9, by taking the limit as $n \to \infty$, that

$$\operatorname{ess\,sup}_{x \in \Omega} |f(x)| \leq c(p,d). \tag{18}$$

We shall now get rid of the normalizations (11) and (12). In case $\|Df\|_p \neq 0$, we can consider $g := \frac{f}{\|Df\|_p}$. Then g fulfils (12) and from (18) applied to g it follows that

$$\operatorname{ess\,sup}_{x \in \Omega} |f(x)| \leq c(p,d)\|Df\|_p. \tag{19}$$

A C_0^1-function f with $\|Df\|_p = 0$ is already identically zero and so in that case the assertion follows trivially.

Finally, to remove (11), we consider the coordinate transformation

$$y = y(x) = \operatorname{Vol}(\Omega)^{\frac{1}{d}} x \tag{20}$$

and

$$\tilde{\Omega} := \{y : x \in \Omega\}$$
$$\tilde{f}(y) := f(x).$$

Then

$$\operatorname{Vol} \tilde{\Omega} = 1$$

and

$$D_i f(x) = D_i \tilde{f}(y) \operatorname{Vol}(\Omega)^{\frac{1}{d}}$$

and finally

$$\left(\int_\Omega |D_i f(x)|^p dx\right)^{\frac{1}{p}} = \left(\int_{\tilde{\Omega}} |D_i \tilde{f}(y)|^p \frac{dy}{\operatorname{Vol}\Omega}\right)^{\frac{1}{p}} \cdot \operatorname{Vol}(\Omega)^{\frac{1}{d}},$$

so applying (19) to $\tilde{f}$ shows that

$$\underset{x \in \Omega}{\text{ess sup}} |f(x)| \le c(p,d) \operatorname{Vol}(\Omega)^{\frac{1}{d}-\frac{1}{p}} \|Df\|_p, \tag{21}$$

as claimed.

Up till now we have only handled the case $f \in C_0^1(\Omega)$. If now $f \in H_0^{1,p}(\Omega)$, then we approximate f in the $W^{1,p}$-norm by C_0^∞-functions f_n and apply (6) and (7) to the differences $f_n - f_m$. Since $(Df_n)_{n \in \mathbb{N}}$ is a Cauchy sequence in L^p it follows that it is a Cauchy sequence in $L^{\frac{dp}{d-p}}(\Omega)$ (for $p < d$) and in $C^0(\overline{\Omega})$ (for $p > d$). Therefore f also lies in the corresponding space and fulfils (6) and (7), respectively. □

As a corollary, we obtain the so-called Poincaré inequality

Corollary 20.16 *Let $\Omega \subset \mathbb{R}^d$ be open and bounded, $f \in H_0^{1,2}(\Omega)$. Then*

$$\|f\|_{L^2(\Omega)} \le \text{const.} \operatorname{Vol}(\Omega)^{\frac{1}{d}} \|Df\|_{L^2(\Omega)}.$$

Proof. By theorem 20.15 we have

$$\|f\|_2 \le \text{const.} \|Df\|_{\frac{2d}{d+2}} \quad \left(\text{taking } p = \frac{2d}{d+2}\right)$$

$$\le \text{const.} \operatorname{Vol}(\Omega)^{\frac{d+2}{2d}-\frac{1}{2}} \|Df\|_2 \quad \text{by corollary 19.7}$$

$$= \text{const.} \operatorname{Vol}(\Omega)^{\frac{1}{d}} \|Df\|_2.$$

□

A consequence of the last corollary is that on $H_0^{1,2}(\Omega)$, the norms $\|f\|_{W^{1,2}}$ and $\|Df\|_{L^2}$ are equivalent.

Moreover, it follows from the Sobolev embedding theorem that a non-zero constant does not lie in $H_0^{1,p}(\Omega)$. Thus $C_0^\infty(\Omega)$ is dense in $L^p(\Omega)$ but not in $W^{1,p}(\Omega)$.

Example. We shall now show by an example that for $d > 1$, a function in $H^{1,d}(\Omega)$ need not necessarily be bounded and thus, the analysis carried out before theorem 20.15 for $d = 1$ does not extend to higher dimensions; indeed, let

$$f(x) := \log(\log \frac{1}{r}) \text{ with } r = \|x\|$$

$$U(0,R) := \{x \in \mathbb{R}^d : \|x\| < R\} \text{ with } R < 1$$

Now for $x \neq 0$

$$\frac{\partial}{\partial r} f(x) = \frac{+1}{r \log r}$$

so

$$\int\limits_{U(0,R)} |\frac{\partial}{\partial r} f(x)|^d dx = d\omega_d \int\limits_0^R \frac{r^{d-1}}{r^d (\log r)^d} dr = \frac{-d\omega_d}{(d-1)(\log r)^{d-1}} \Big|_0^R$$

$$= \frac{-d\omega_d}{(d-1)(\log R)^{d-1}} < \infty.$$

It remains to show that $\frac{\partial}{\partial r} f(x)$ is also the weak derivative of f, for there is again a singularity at zero. Once this is shown, it would follow that $f \in H^{1,d}(U(0,R))$ but f is not in $L^\infty(U(0,r))$.

For this we consider, as before,

$$\eta_n(r) := \begin{cases} 1 & \text{for } r \geq \frac{2}{n} \\ n(r - \frac{1}{n}) & \text{for } \frac{1}{n} \leq r < \frac{2}{n} \\ 0 & \text{for } 0 \leq r < \frac{1}{n}. \end{cases}$$

For $\varphi \in C_0^1(U(0,R))$ we have

$$\int\limits_{U(0,R)} \log \log \frac{1}{r} \cdot \frac{\partial}{\partial x^i} (\eta_n(r)\varphi(x)) dx$$

$$= - \int\limits_{U(0,R)} \frac{\partial}{\partial x^i} (\log \log \frac{1}{r}) \cdot \eta_n(r)\varphi(x) dx$$

and by going to the limit as $n \to \infty$ we obtain as in the example of $\|x\|^\alpha$

$$\int\limits_{U(0,R)} \log \log \frac{1}{r} \frac{\partial}{\partial x^i} \varphi(x) dx$$

$$= - \int\limits_{U(0,R)} \frac{\partial}{\partial x^i} (\log \log \frac{1}{r}) \varphi(x) dx.$$

The critical term is now

$$|\int\limits_{U(0,R)} \log \log \frac{1}{r} \varphi(x) \frac{\partial}{\partial x^i} \eta_n(r) dx|$$

$$\leq \text{const. sup} |\varphi(x)| \int\limits_{\frac{1}{n}}^{\frac{2}{n}} \log \log \frac{1}{r} n r^{d-1} dr \to 0 \quad \text{for } n \to \infty.$$

This completes the verification that $\log(\log \frac{1}{r}) \in H^{1,d}(U(0,R))$ for $R < 1$.

Finally, a function in $H^{1,d}(\Omega) \cap L^\infty(\Omega)$ $(\Omega \subset \mathbb{R}^d, d \geq 2)$ need not be continuous, as the example

$$f(x) = \sin \log \log \frac{1}{r} \quad (r = \|x\|) \quad \text{on } U(0,R) \text{ for } R < 1$$

shows.

However, we do have from theorem 20.15 and corollary 19.7

Corollary 20.17 *Let $\Omega \subset \mathbb{R}^d$ be open and bounded, $f \in H_0^{1,d}(\Omega)$. Then*

$$f \in L^p(\Omega) \quad for\ 1 \le p < \infty.$$

$\square$

Corollary 20.18 *Let $\Omega \subset \mathbb{R}^d$ be open and bounded. Then*

$$H_0^{k,p}(\Omega) \subset \begin{cases} L^{\frac{dp}{(d-kp)}}(\Omega) & for\ kp < d \\ C^m(\overline{\Omega}) & for\ 0 \le m < k - \frac{d}{p} \end{cases}$$

$(C^m(\overline{\Omega}) := \{f \in C^m(\Omega) : f$ *and all its derivatives of order up to and including m are continuous in $\overline{\Omega}$*$\})$.

Proof. The first embedding follows by using theorem 20.15 iteratively, the second then from the first and the case $p > d$ in theorem 20.15. A detailed proof can be given for $k = 2$ according to the following scheme.

For the case $k = 2$, it follows first of all that

$$H_0^{2,p} \subset H^{1,\frac{dp}{d-p}} \quad for\ p < d.$$

To see this, let $(f_n)_{n \in \mathbb{N}} \subset C_0^\infty$ be a sequence converging to f in the $W^{2,p}$-norm. By the Sobolev embedding theorem

$$\|f_n - f\|_{W^{1,\frac{dp}{d-p}}} \le \text{const.} \|f_n - f\|_{W^{2,p}},$$

thus $f \in H_0^{1,\frac{dp}{d-p}}$. Hence $H_0^{2,p} \subset H_0^{1,\frac{dp}{d-p}}$ for $p < d$.

If now even $p < d$, so $\frac{dp}{d-p} < d$, we use theorem 20.15 with $\frac{dp}{d-p}$ in place of d and obtain the required inequality

$$H_0^{2,p}(\Omega) \subset L^{\frac{dp}{d-2p}}(\Omega).$$

Similarly as $L^r \subset L^q$ for $1 \le q \le r \le \infty$ (corollary 19.7; notice that here Ω is assumed to be bounded),

$$H_0^{2,d} \subset H_0^{1,q} \quad for\ every\ q < \infty.$$

Finally, it follows for $f \in H_0^{2,p}$ with $p > d$ that f as well as Df is continuous on $\overline{\Omega}$. Then Df is also the classical derivative of f. This follows, for example, from theorem 19.15 and lemma 20.6, for one only has to observe that for a sequence $(f_n)_{n \in \mathbb{N}} \subset C^1$, for which f_n as well as Df_n converge uniformly, $\lim Df_n$ is the derivative of $\lim f_n$. This simply means that C^1 is a Banach space and it was shown in theorem 5.12.

Thus if $p > d$ then $H_0^{2,p}(\Omega) \subset C^1(\overline{\Omega})$; if $p < d$ but $\frac{dp}{d-p} > d$ (i.e. $2p > d$) then $H_0^{2,p}(\Omega) \subset C^0(\overline{\Omega})$; if $p = d$ then, likewise, $H_0^{2,d}(\Omega) \subset C^0(\overline{\Omega})$, by the preceding considerations and theorem 20.15. In this way, one shows the assertion easily for arbitrary k.   □

Corollary 20.19 *Let $\Omega \subset \mathbb{R}^d$ be open and bounded. If $u \in H_0^{k,p}(\Omega)$ for a fixed p and all $k \in \mathbb{N}$, then $u \in C^\infty(\overline{\Omega})$.*   □

Theorem 20.15 shows, amongst other things, that for $p < d$ the space $H_0^{1,p}(\Omega)$ is embedded continuously in the space $L^{\frac{dp}{d-p}}(\Omega)$, i.e. there exists an injective bounded (= continuous) linear map

$$i : H_0^{1,p}(\Omega) \to L^{\frac{dp}{d-p}}(\Omega).$$

We recall that a continuous linear map is said to be compact if the image of every bounded sequence contains a convergent subsequence.

We recall also that for a bounded Ω, there does exist a continuous embedding

$$j : L^r(\Omega) \to L^q(\Omega) \quad \text{for } 1 \leq q \leq r \leq \infty$$

(corollary 19.7). In particular, there thus exists a continuous embedding

$$j \circ i : H_0^{1,p}(\Omega) \to L^q(\Omega) \quad \text{for } 1 \leq q \leq \frac{dp}{d-p}.$$

The *compactness theorem of Rellich-Kondrachov* which we now state says that this embedding is compact for $1 \leq q < \frac{dp}{d-p}$.

Theorem 20.20 *Let $\Omega \subset \mathbb{R}^d$ be open and bounded. Then for $p < d$ and $1 \leq q < \frac{dp}{d-p}$, and $p \geq d$ and $1 \leq q < \infty$, respectively, the space $H_0^{1,p}(\Omega)$ is compactly embedded in $L^q(\Omega)$.*

We shall apply this theorem later mostly in the following form: If $(f_n)_{n \in \mathbb{N}} \subset H_0^{1,2}(\Omega)$ is bounded, so

$$\|f_n\|_{W^{1,2}(\Omega)} \leq K,$$

then (f_n) contains a convergent subsequence in the L^2-norm (and then by theorem 19.12, also a subsequence which converges pointwise almost everywhere). (This in fact is the statement originally proved by Rellich.)

Proof. Again by corollary 19.7, $H_0^{1,p}(\Omega) \subset H_0^{1,r}(\Omega)$ for $1 \leq r \leq p < \infty$, so we can limit ourselves to the case $p < d$. (Namely, that case implies that for every $r < d$ and any $1 \leq q < \frac{dr}{d-r}$, $H_0^{1,r}(\Omega)$ is compactly embedded in $L^q(\Omega)$. Since for every $1 \leq q < \infty$, we may find $r < d$ with $q < \frac{dr}{d-r}$, we thus obtain the compactness of the embedding of $H_0^{1,p}(\Omega) \subset H_0^{1,r}(\Omega)$ in $L^q(\Omega)$ for every q in the other case $p \geq d$ as well.)

We first consider the case $q = 1$. So let $(f_n)_{n \in \mathbb{N}} \subset H_0^{1,p}(\Omega)$,

$$\|f_n\|_{W^{1,p}(\Omega)} \leq K. \tag{22}$$

We shall show that $(f_n)_{n \in \mathbb{N}}$ in $L^1(\Omega)$ is totally bounded, that is, for every $\varepsilon > 0$ there exist finitely many $u_1, \ldots, u_k \in L^1(\Omega)$ with the following property: for every $n \in \mathbb{N}$ there exists an $i \in \{1, \ldots, k\}$ with

$$\|f_n - u_i\|_{L^1(\Omega)} < \varepsilon. \tag{23}$$

Thus $(f_n)_{n \in \mathbb{N}}$ would be covered by finitely many open balls

$$U_i := \{h \in L^1(\Omega) : \|h - u_i\|_{L^1} < \varepsilon\}.$$

Then by theorem 7.40 the closure of $(f_n)_{n \in \mathbb{N}}$ must be compact in $L^1(\Omega)$ and thus $(f_n)_{n \in \mathbb{N}}$ would contain a convergent subsequence relative to the L^1-norm.

First of all for every $n \in \mathbb{N}$ and $\varepsilon > 0$, by the definition of $H_0^{1,p}(\Omega)$ there exists a $g_n \in C_0^1(\Omega)$ with

$$\|f_n - g_n\|_{L^1(\Omega)} \leq \text{ const. } \|f_n - g_n\|_{W^{1,p}(\Omega)} \tag{24}$$

$$< \frac{\varepsilon}{3}.$$

For $g \in C_0^1(\Omega)$ we consider the convolution

$$g_h(x) = \frac{1}{h^d} \int_\Omega \rho(\frac{x-y}{h}) g(y) dy.$$

Now for every $g \in C_0^1(\Omega)$ we have

$$|g(x) - g_h(x)| = \frac{1}{h^d} |\int_\Omega \rho(\frac{x-y}{h})(g(x) - g(y)) dy|$$

$$= |\int_{|z| \leq 1} \rho(z)(g(x) - g(x - hz)) dz|$$

(see the calculation in §19)

$$\leq \int_{|z| \leq 1} \rho(z) \int_0^{h|z|} |\frac{\partial}{\partial r} g(x - r\omega)| dr dz \quad \text{mit } \omega = \frac{z}{|z|}.$$

Integrating with respect to x gives

$$\int_{\Omega} |g(x) - g_h(x)| dx \leq \int_{|z| \leq 1} \rho(z) h |z| \int_{\Omega} |Dg(x)| dx dz$$

$$(Dg = (D_1 g, \ldots D_d g)) \tag{25}$$

$$\leq h \int_{\Omega} |Dg(x)| dx.$$

On account of (22) and (24) there exists a K' with

$$\|g_n\|_{W^{1,1}(\Omega)} \leq \text{const.} \|g_n\|_{W^{1,p}(\Omega)} \leq K' \tag{26}$$

for all n.

From (25) it then follows that we can choose $h > 0$ so small that for all $n \in \mathbb{N}$

$$\|g_n - g_{n,h}\|_{L^1(\Omega)} < \frac{\varepsilon}{3}. \tag{27}$$

Furthermore,

$$|g_h(x)| = \frac{1}{h^d} \left| \int_{\Omega} \rho(\frac{x-y}{h}) g(y)\right| \leq \frac{1}{h^d} \sup_z |\rho(z)| \|g\|_{L^1(\Omega)} \tag{28}$$

and

$$|D_i g_h(x)| = \frac{1}{h^d} \left| \int_{\Omega} D_i \rho(\frac{x-y}{h}) g(y) dy \right| \leq \frac{1}{h^{d+1}} \sup |\frac{\partial}{\partial z^i} \rho(z)| \|g\|_{L^1(\Omega)}. \tag{29}$$

Applying (28) and (29) to g_n it follows, by (26), that

$$\|g_{n,h}\|_{C^1(\Omega)} \leq \text{const.} \tag{30}$$

The constant here depends on h, but we have chosen h to be fixed and in fact so that (27) holds. The sequence $(g_{n,h})_{n \in \mathbb{N}}$ is therefore uniformly bounded and equicontinuous and therefore contains, by the Arzela-Ascoli theorem, a uniformly convergent subsequence. This subsequence then converges also in $L^1(\Omega)$. The closure of $\{g_{n,h} : n \in \mathbb{N}\}$ is thus compact in $L^1(\Omega)$ and hence totally bounded. Consequently there exist finitely many $u_1, \ldots, u_k \in L^1(\Omega)$ with the property that for any $n \in \mathbb{N}$ there is an $i \in \{1, \ldots, k\}$ with

$$\|g_{n,h} - u_i\|_{L^1(\Omega)} < \frac{\varepsilon}{3}. \tag{31}$$

By (24), (27) and (31), for any $n \in \mathbb{N}$ there is an $i \in \{1, \ldots k\}$ with

$$\|f_n - u_i\|_{L^1(\Omega)} < \varepsilon.$$

this is (23) and with it the case $q = 1$ is settled.

Now for arbitrary $q < \frac{dp}{d-p}$ we have for $f \in H_0^{1,p}(\Omega)$

$$\|f\|_{L^q(\Omega)} \leq \|f\|_{L^1(\Omega)}^{\lambda} \|f\|_{L^{\frac{dp}{d-p}}(\Omega)}^{1-\lambda} \tag{32}$$

with $\dfrac{1}{q} = \lambda + (1-\lambda)(\dfrac{1}{p} - \dfrac{1}{d})$

(by corollary 19.8)

$$\leq \text{ const. } \|f\|_{L^1(\Omega)}^{\lambda} \|Df\|_{L^p(\Omega)}^{1-\lambda}$$

by theorem 20.15.

We have already shown that $(f_n)_{n\in\mathbb{N}}$, after a choice of a subsequence, converges in $L^1(\Omega)$ and thus it is a Cauchy sequence, i.e. $\forall \varepsilon > 0 \exists N \in \mathbb{N} \forall n, m \geq N$

$$\|f_n - f_m\|_{L^1(\Omega)} < \varepsilon.$$

Taking into account (22), (32) then gives that

$$\|f_n - f_m\|_{L^q(\Omega)} \leq \text{ const. } \varepsilon^{\lambda}.$$

Thus $(f_n)_{n\in\mathbb{N}}$ is also a Cauchy sequence in $L^q(\Omega)$. Thereby, on account of the completeness of $L^q(\Omega)$ (theorem 19.11), the proof of the theorem is complete.

□

Exercises for § 20

1) Which of the following functions are weakly differentiable? (Note that the answer may depend on d!)

a) $f : \mathbb{R}^d \to \mathbb{R}, f(x) := \|x\|$.

b) The components of $f : \mathbb{R}^d \to \mathbb{R}^d, f(x) := \begin{cases} \frac{x}{\|x\|} & \text{for } x \neq 0 \\ 0 & \text{for } x = 0. \end{cases}$

c) $f : \mathbb{R}^d \to \mathbb{R}, f(x) := \begin{cases} \log \|x\| & \text{for } x \neq 0 \\ 0 & \text{for } x = 0 \end{cases}$

d) $f : \mathbb{R}^d \to \mathbb{R}, f(x) := \begin{cases} 1 & \text{for } x^1 \geq 0 \\ -1 & \text{for } x^1 < 0 \end{cases}$ $(x = (x^1, \ldots, x^d))$

e) $f : \mathbb{R} \to \mathbb{R}, f(x) = e^{\sin^5 x} + 7x^9$.

2) For which values of α, k, p, d is

$$\|x\|^{\alpha} \in W^{k,p}(U(0,1))?$$

$(U(0,1) = \{x \in \mathbb{R}^d : \|x\| < 1\}).$

3) Let $f, g \in H^{1,2}_{\text{loc}}(\Omega)$ where this space is defined analogously to L^2_{loc}. Show that $f \cdot g$ is weakly differentiable, with

$$D(fg) = fDg + gDf.$$

(Hint: Consider first the case where one of the functions is in $C^1(\Omega)$.)

4)   Let $\psi : \mathbb{R} \to \mathbb{R}$ be continuous and piecewise continuously differentiable, with $\sup |\psi'| < \infty$. Let $f \in W^{1,p}(\Omega)$. Show that $\psi \circ f \in W^{1,p}(\Omega)$, with

$$D(\psi \circ f)(x) = \begin{cases} \psi'(f(x))Df(x) & \text{if } \psi \text{ is differentiable at } f(x) \\ 0 & \text{otherwise} \end{cases}.$$

Chapter VI.

Introduction to the Calculus of Variations and Elliptic Partial Differential Equations

21. Hilbert Spaces. Weak Convergence

Hilbert spaces are Banach spaces with norm derived from a scalar product. A sequence in a Hilbert space is said to converge weakly if its scalar product with any fixed element of the Hilbert space converges. Weak convergence satisfies important compactness properties that do not hold for ordinary convergence in an infinite dimensional Hilbert space. In particular, any bounded sequence contains a weakly convergent subsequence.

Definition 21.1 A real (complex) Hilbert space H is a vector space over $\mathbb{R}(\mathbb{C})$ which is equipped with a scalar product

$$(\cdot,\cdot) : H \times H \to \mathbb{R}(\mathbb{C})$$

with the following properties

i) $(x,y) = (y,x)$ $((x,y) = \overline{(y,x)})$ respectively) for all $x,y \in H$

ii) $(\lambda_1 x_1 + \lambda_2 x_2, y) = \lambda_1(x_1,y) + \lambda_2(x_2,y)$ $\forall$ $\lambda_1, \lambda_2 \in \mathbb{R}(\mathbb{C})$, $x_1, x_2, y \in H$

iii) $(x,x) > 0$ for all $x \neq 0$, $x \in H$

iv) H is complete relative to the norm $\|x\| := (x,x)^{1/2}$ (i.e. every Cauchy sequence relative to $\| \cdot \|$ has a limit point in H).

Thus, every Hilbert space is a Banach space (one notes the triangle inequality in lemma 21.2 infra). Naturally, all the finite dimensional Euclidean vector spaces, and thus $\mathbb{R}^d$ with the usual scalar product, are Hilbert spaces. The significance and use of the concept of Hilbert space however lies in making possible an infinite dimensional generalization of the Euclidean space and its geometry. For us, the most important Hilbert spaces are the spaces $L^2(\Omega)$ with

$$(f,g)_{L^2(\Omega)} := \int_\Omega f(x)g(x)dx$$

(this expression is finite for $f,g \in L^2(\Omega)$ because of the Hölder inequality), and more generally the Sobolev spaces $W^{k,2}(\Omega)$ with

$$(f,g)_{W^{k,2}(\Omega)} := \sum_{|\alpha| \leq k} \int_\Omega D_\alpha f(x) \cdot D_\alpha g(x)dx.$$

That $L^2(\Omega)$ and $W^{k,2}(\Omega)$, with these scalar products, are in fact Hilbert spaces follows from the theorems in §19 and §20, where e.g. the completeness was proved. We shall consider here mainly real vector spaces; however in physics complex vector spaces are also important, e.g. $L^2(\Omega, \mathbb{C}) := \{f : \Omega \to \mathbb{C}, \operatorname{Re} f, \operatorname{Im} f \in L^2(\Omega)\}$ with

$$(f, g)_{L^2(\Omega, \mathbb{C})} := \int_\Omega f(x)\overline{g(x)}dx.$$

The theory of real and complex Hilbert spaces is, to a large extent, similar.

Lemma 21.2 *The following inequalities hold for a Hilbert space H :*

—   *the Schwarz inequality*

$$|(x, y)| \leq \|x\| \cdot \|y\| \tag{1}$$

   with equality only for linearly dependent x and y

—   *the triangle inequality*

$$\|x + y\| \leq \|x\| + \|y\| \tag{2}$$

—   *the parallelogram law*

$$\|x + y\|^2 + \|x - y\|^2 = 2(\|x\|^2 + \|y\|^2) \tag{3}$$

Proof. (1) follows from $\|x + \lambda y\|^2 \geq 0$ with $\lambda = \frac{-(x,y)}{\|y\|^2}$, and (2) then follows from (1). (3) is obtained by a direct calculation.   □

Definition 21.3 $x, y \in H$ (H is a Hilbert space) are called orthogonal if $(x, y) = 0$. If F is a subspace of H then so is

$$F^\perp := \{x \in H : (x, y) = 0 \text{ for all } y \in F\}.$$

Theorem 21.4 *Let F be a closed subspace of a Hilbert space H. Then every $x \in H$ has a unique decomposition*

$$x = y + z \text{ with } y \in F, z \in F^\perp. \tag{4}$$

Proof. We give a proof for a real Hilbert space; the complex case is analogous.
   Let

$$d := \inf_{y \in F} \|x - y\|$$

and $(y_n)_{n \in \mathbb{N}} \subset F$ be a minimal sequence, thus

$$\|x - y_n\| \to d. \tag{5}$$

From (3) it follows that

$$4\|x - \frac{1}{2}(y_m + y_n)\|^2 + \|y_m - y_n\|^2 \tag{6}$$
$$= 2(\|x - y_m\|^2 + \|x - y_n\|^2).$$

As with y_n, y_m also $\frac{1}{2}(y_n + y_m) \in F$, it follows that (y_n) is a Cauchy sequence; as H is complete, it has a limit, say y. Since F is closed, $y \in F$ and $d = \|x - y\|$.

We set $z = x - y$ and have to show that $z \in F^\perp$. If $y' \in F, \alpha \in \mathbb{R}$ then $y + \alpha y' \in F$. Therefore

$$d^2 \le \|x - y - \alpha y'\|^2 = (z - \alpha y', z - \alpha y')$$
$$= \|z\|^2 - 2\alpha(y', z) + \alpha^2\|y'\|^2.$$

As $\|z\| = d$, it follows for all $\alpha > 0$

$$|(y', z)| \le \frac{\alpha}{2}\|y'\|^2$$

and therefore

$$(y', z) = 0 \text{ for all } y' \in F,$$

so $z \in F^\perp$.

The uniqueness of y follows by (6) from

$$\|x - \frac{1}{2}(y_1 + y_2)\| < d$$

if $\|x - y_1\| = d = \|x - y_2\|$ and $y_1 \neq y_2$, which would contradict the minimality of d. $\square$

Corollary 21.5 *For every closed subspace F of a Hilbert space H there exists a unique linear map*

$$\pi : H \to F$$

with

$$\|\pi\| = \sup_{x \neq 0} \frac{\|\pi x\|}{\|x\|} = 1 \tag{7}$$

$$\pi^2 = \pi \quad (\pi \text{ is a projection}) \tag{8}$$

$$\ker \pi = F^\perp. \tag{9}$$

Proof. For $x = y + z$ as in (4) we set

$$\pi x = y.$$

All the assertions then follow immediately. $\square$

We now prove the *Riesz representation theorem*

Theorem 21.6 *Let L be a bounded linear functional on a Hilbert space H (so $L : H \to \mathbb{R}$ (resp. $\mathbb{C}$) is linear with*

$$\|L\| := \sup_{x \neq 0} \frac{|Lx|}{\|x\|} < \infty).$$

Then there exists a uniquely determined $y \in H$ with

$$L(x) = (x, y) \text{ for all } x \in H. \tag{10}$$

Moreover

$$\|L\| = \|y\|. \tag{11}$$

Proof. Let $N := \ker L := \{x : Lx = 0\}$. If $N = H$ then we set $y = 0$. So let $N \neq H$. N is closed, and so, by theorem 21.4, there exists a $z \neq 0, z \in H$ with

$$(x, z) = 0 \text{ for all } x \in N.$$

Therefore $Lz \neq 0$, and for $x \in H$ we have

$$L(x - \frac{Lx}{Lz}z) = Lx - \frac{Lx}{Lz}Lz = 0,$$

thus $x - \frac{Lx}{Lz}z \in N$, so

$$(x - \frac{Lx}{Lz}z, z) = 0,$$

and

$$(x, z) = \frac{Lx}{Lz}\|z\|^2,$$

and with

$$y := \frac{Lz}{\|z\|^2} \cdot z$$

we have

$$Lx = (x, y).$$

If there were y_1, y_2 with this property, then we would have

$$(x, y_1 - y_2) = 0 \quad \forall x,$$

so $\|y_1 - y_2\|^2$ would be zero and therefore $y_1 = y_2$.
    Furthermore, we have, by the Schwarz inequality,

$$\|L\| = \sup_{x \neq 0} \frac{|(x, y)|}{\|x\|} \leq \|y\|$$

and on the other hand

$$\|y\|^2 = (y, y) = Ly \le \|L\| \cdot \|y\|,$$

so altogether

$$\|y\| = \|L\|.$$

$\square$

Definition 21.7 Let H be a Hilbert space. We say that $(x_n)_{n \in \mathbb{N}} \subset H$ is weakly convergent to $x \in H$ if

$$(x_n, y) \to (x, y) \text{ for all } y \in H.$$

In symbols, $x_n \rightharpoonup x$.

As

$$|(x_n, y) - (x, y)| \le \|x_n - x\| \cdot \|y\| \text{ by (1)}$$

a sequence $(x_n)_{n \in \mathbb{N}}$ which converges in the usual sense is also weakly convergent.

However, we shall see in the following that a weakly convergent sequence need not be convergent in the usual sense. To distinguish the usual Hilbert space convergence from weak convergence, we shall usually call the former **strong convergence**.

In a finite dimensional Euclidean space, weak convergence simply means convergence componentwise and thus here weak and strong convergence are equivalent.

Theorem 21.8 *Every bounded sequence $(x_n)_{n \in \mathbb{N}}$ in a Hilbert space H has a weakly convergent subsequence.*

Proof. Let $\|x_n\| \le M$. It suffices to show that for a suitable subsequence (x_{n_k}) and some $x \in H$, $(x_n, y) \to (x, y)$ for all y which lie in the closure $\bar{S}$ of the linear subspace S of H spanned by x_n. For, by theorem 21.4, $y \in H$ can be split as

$$y = y_0 + y_1, y_0 \in \bar{S}, y_1 \in \bar{S}^\perp$$

and $(x_n, y_1) = 0$ for all n.

For a fixed m, (x_n, x_m) has a bound independent of n as $\|x_n\| \le M$ and therefore $((x_n, x_m))_{n \in \mathbb{N}}$ has a convergent subsequence. By the Cantor diagonal process, one obtains a subsequence (x_{n_k}) of (x_n) for which (x_{n_k}, x_m) converges for every $m \in \mathbb{N}$ as $k \to \infty$. Then (x_{n_k}, y) also converges when $y \in S$. If $y \in \bar{S}$ then for $y' \in S$

$$|(x_{n_j} - x_{n_k}, y)| \le |(x_{n_j}, y - y')| + |(x_{n_j} - x_{n_k}, y')|$$
$$+ |(x_{n_k}, y' - y)| \le 2M\|y - y'\|$$
$$+ |(x_{n_j} - x_{n_k}, y')|.$$

For a preassigned $\varepsilon > 0$ one chooses $y' \in S$ with $\|y' - y\| < \frac{\varepsilon}{4M}$ and then j and k so large that

$$|(x_{n_j} - x_{n_k}, y')| < \frac{\varepsilon}{2}.$$

It follows that $|(x_{n_j} - x_{n_k}, y)| \to 0$ for all $y \in \overline{S}$ as $j, k \to \infty$.

Thus (x_{n_k}, y) is a Cauchy sequence and consequently has a limit; we set

$$Ly := \lim_{k \to \infty} (x_{n_k}, y).$$

L is linear and bounded because of (1) and $\|x_n\| \le M$.

By theorem 21.6 there is an $x \in \overline{S}$ with

$$(x, y) = Ly \text{ for all } y \in \overline{S}$$

and then also $Ly = (x, y) = 0$ for $y \in \overline{S}^\perp$. Thus x_{n_k} converges weakly to x. $\qquad\square$

Corollary 21.9 *If $(x_n)_{n \in \mathbb{N}}$ converges weakly to x, then*

$$\|x\| \le \liminf_{n \to \infty} \|x_n\|.$$

(So the norm is lower semicontinuous with respect to weak convergence.)
$(x_n)_{n \in \mathbb{N}}$ converges to x precisely when it converges weakly to x and

$$\|x\| = \lim_{n \to \infty} \|x_n\|.$$

Proof. We have

$$0 \le (x_n - x, x_n - x) = (x_n, x_n) - 2(x_n, x) + (x, x).$$

As $(x_n, x) \to (x, x)$ as $n \to \infty$, it follows that

$$0 \le \liminf \|x_n\|^2 - \|x\|^2.$$

If $\|x\| = \lim_{n \to \infty} \|x_n\|$, it follows from the above considerations that $\lim_{n \to \infty} (x_n - x, x_n - x) = 0$, thus strong convergence. It is clear that, conversely, strong convergence implies convergence of norms and we have seen above that strong convergence implies weak convergence. $\qquad\square$

Example. We consider an orthonormal family $(e_n)_{n \in \mathbb{N}} \subset H$, so

$$(e_n, e_m) = \delta_{nm} \left(= \begin{cases} 1 & \text{for } n = m \\ 0 & \text{for } n \ne m \end{cases} \right)$$

Such an orthonormal family is called complete if every $x \in H$ satisfies

$$x = \sum_n (x, e_n) e_n.$$

As an example we consider $L^2((-\pi, \pi))$ and the functions $\frac{1}{\sqrt{2\pi}}$, $\frac{1}{\sqrt{\pi}} \cos nx$, $\frac{1}{\sqrt{\pi}} \sin nx$
$(n \in \mathbb{N})$. We have

$$(\frac{1}{\sqrt{\pi}} \cos nx, \frac{1}{\sqrt{\pi}} \cos mx) = \frac{1}{\pi} \int_{-\pi}^{\pi} \cos nx \cdot \cos mx dx = \delta_{mn}$$

$$(\frac{1}{\sqrt{\pi}} \sin nx, \frac{1}{\sqrt{\pi}} \sin mx) = \frac{1}{\pi} \int_{-\pi}^{\pi} \sin nx \cdot \sin mx dx = \delta_{mn}$$

$$(\frac{1}{\sqrt{\pi}} \sin nx, \frac{1}{\sqrt{\pi}} \cos mx) = \frac{1}{\pi} \int_{-\pi}^{\pi} \sin nx \cos mx dx = 0$$

$$(\frac{1}{\sqrt{2\pi}}, \frac{1}{\sqrt{2\pi}}) = \frac{1}{2\pi} \int_{-\pi}^{\pi} dx = 1$$

$$(\frac{1}{\sqrt{2\pi}}, \frac{1}{\sqrt{\pi}} \cos nx) = \frac{1}{\pi\sqrt{2}} \int_{-\pi}^{\pi} \cos nx = 0 \ (n \geq 1)$$

$$(\frac{1}{\sqrt{2\pi}}, \frac{1}{\sqrt{\pi}} \sin nx) = \frac{1}{\pi\sqrt{2}} \int_{-\pi}^{\pi} \sin nx = 0 \ (n \geq 1).$$

Thus we obtain an orthonormal family in $L^2((-\pi, \pi))$

$$e_1 := \frac{1}{\sqrt{2\pi}}, e_{2n} := \frac{1}{\sqrt{\pi}} \sin nx, e_{2n+1} := \frac{1}{\sqrt{\pi}} \cos nx \ (n \geq 1).$$

This orthonormal family is basic for the expansion of an $L^2((-\pi, \pi))$-function in a Fourier series. We shall come accross another L^2-orthogonal series when we treat the eigenvalue problem for the Laplace operator.

Now let H be an infinite dimensional Hilbert space and $(e_n)_{n \in \mathbb{N}}$ an orthonormal sequence in H. We claim that (e_n) converges weakly to zero. Otherwise, by choosing a subsequence from the e_n, there would be an $x \in H$ with

$$|(x, e_n)| \geq \varepsilon \text{ for all } n \in \mathbb{N}, \text{ and some } \varepsilon > 0. \tag{12}$$

Now $(x, e_m)e_m$ is the projection of x onto the subspace spanned by e_m as by theorem 21.4 $x = \alpha e_m + f$, with $(f, e_m) = 0$ and since $(e_m, e_m) = 1$, $\alpha = (x, e_m)$. Similarly

$$\sum_{n=1}^{N} (x, e_n)e_n$$

is the projection of x onto the subspace spanned by $e_1, \dots, e_N$. Therefore, if (12) holds,

$$\|x\|^2 = \|x - \sum_{n=1}^{N}(x,e_n)e_n\|^2 + \|\sum_{n=1}^{N}(x,e_n)e_n\|^2 \geq \sum_{n=1}^{N}(x,e_n)^2 \geq N\varepsilon^2.$$

Thus, (12) is not possible, and we conclude that $e_n \rightharpoonup 0$.

As $\|e_n\| = 1$ for all n, one sees that in Corollary 21.9 one cannot expect equality. Furthermore, (e_n) does not converge strongly (i.e. in the norm) to 0. Thus with respect to compactness arguments, weak convergence is the suitable analogue of convergence in finite dimensional spaces.

Corollary 21.10 (Banach-Saks) *Let $(x_n)_{n\in\mathbb{N}} \subset H$ with $\|x_n\| \leq K$ (independent of n). then there exists a subsequence $(x_{n_j}) \subset (x_n)$ and an $x \in H$ such that*

$$\frac{1}{k}\sum_{j=1}^{k} x_{n_j} \to x \ (convergence \ in \ norm) \quad as \ k \to \infty.$$

Proof. Let x be the weak limit of a subsequence $(x_{n_i}) \subset (x_n)$ (theorem 21.8) and let

$$y_i := x_{n_i} - x.$$

Then $y_i \rightharpoonup 0$ and $\|y_i\| \leq K'$ for a fixed K'.

One chooses successively $y_{i_j} \in (y_i)$ with

$$|(y_{i_\ell}, y_{i_j})| \leq \frac{1}{j} \ \text{for} \ \ell < j.$$

That this is possible follows from $y_i \rightharpoonup 0$ on account of which $(z, y_i) \to 0$ for any $z \in H$, and therefore also for the finitely many

$$y_{i_\ell}, \ell = 1, \ldots, j-1, \ \text{in} \ H, (y_{i_\ell}, y_i) \to 0.$$

Now

$$\|\frac{1}{k}\sum_{j=1}^{k} y_{i_j}\|^2 = \frac{1}{k^2}\left(\sum_{\ell,j=1}^{k}(y_{i_\ell}, y_{i_j})\right)$$

$$= \frac{1}{k^2}\left(\sum_{j=1}^{k}((y_{i_j}, y_{i_j}) + 2\sum_{\ell=1}^{j-1}(y_{i_\ell}, y_{i_j}))\right)$$

$$\leq \frac{1}{k^2}(kK'^2 + 2\sum_{j=1}^{k} j \cdot \frac{1}{j})$$

$$\leq \frac{K'^2 + 2}{k} \to 0 \ \text{as} \ k \to \infty;$$

hence the assertion.   □

We now prove further

Lemma 21.11 *Every weakly convergent sequence $(x_n) \subset H$ is bounded.*

Proof. It suffices to show that the bounded linear functions

$$L_n y := (x_n, y)$$

are uniformly bounded on $\{y : \|y\| \leq 1\}$. We then have that $\|x_n\| = (x_n, \frac{x_n}{\|x_n\|})$ is bounded independently of n.

For this it suffices, by the linearity of L_n, to verify their uniform boundedness on any ball. The existence of such a ball is now proved by contradiction. Otherwise, if $(L_n)_{n \in \mathbb{N}}$ is not bounded on any ball, then there exists a sequence (K_i) of closed balls

$$K_i := \{y : |y - y_i| \leq r_i\},$$
$$K_{i+1} \subset K_i$$
$$r_i \to 0$$

as well as a subsequence $(x_{n_i}) \subset (x_n)$ with

$$|L_{n_i} y| > i \; \forall \; y \in K_i. \tag{13}$$

Now the (y_i) form a Cauchy sequence and therefore have a limit $y_0 \in H$; we have

$$y_0 \in \bigcap_{i=1}^{\infty} K_i$$

and therefore by (13)

$$|L_{n_i} y_0| > i \; \forall \; i \in \mathbb{N},$$

in contradiction to the weak convergence of (x_{n_i}), which implies the convergence of $L_{n_i} y_0$. □

Corollary 21.12 *If A is a closed convex (i.e. $x, y \in A \implies tx + (1-t)y \in A$ for $0 \leq t \leq 1$) subset of a Hilbert space, e.g. a closed subspace of A, then A is also closed with respect to weak convergence.*

Proof. Let $(x_n)_{n \in \mathbb{N}} \subset A$ and assume (x_n) converges weakly to $x \in H$. By lemma 21.11, the sequence (x_n) is bounded and by corollary 21.10, for a subsequence $(x_{n_j})_{j \in \mathbb{N}}$, $\frac{1}{k} \sum_{j=1}^{k} x_{n_j}$ converges to x. Since A is convex, $\frac{1}{k} \sum_{j=1}^{k} x_{n_j} \in A$ for all j and then as A is closed, $x \in A$. □

Exercises for § 21

1

    a)   Let S be a closed convex subset of some Hilbert space H, $x \in H$. Show that there is a unique $y \in S$ with smallest distance to x.

    b)   Let $F : H \to \mathbb{R}$ be linear, continuous, $\|F\| = 1$. Show that there exists a unique $y \in \{x \in H : \|x\| = 1\}$ with $F(y) = 1$.

2)   In a *finite dimensional* Euclidean space, strong and weak convergence are equivalent.

3)   Let $(f_n)_{n \in \mathbb{N}} \subset C^0([a,b])$ $(a, b \in \mathbb{R})$ be a bounded sequence. Show that there exist a subsequence $(f_{n_k})_{k \in \mathbb{N}}$ of (f_n) and some $f \in L^2((a,b))$ with

$$\int_a^b f_{n_k}(x)\varphi(x)dx \to \int_a^b f(x)\varphi(x)dx$$

for every $\varphi \in C^0([a,b])$.

4)   Let $L : H \to K$ be a continuous linear map between Hilbert spaces. Suppose that $(x_n)_{n \in \mathbb{N}} \subset H$ converges weakly to $x \in H$. Then $(Lx_n)_{n \in \mathbb{N}}$ converges weakly to Lx. If L is compact (cf. Def. 7.48), then Lx_n converges strongly to Lx.

5)   Let H be an infinite dimensional Hilbert space with an orthonormal family $(e_n)_{n \in \mathbb{N}}$. Does there exist a linear map

$$L : H \to H$$

with

$$L(e_n) = ne_n \quad \text{for all } n \in \mathbb{N}?$$

22. Variational Principles and Partial Differential Equations

Dirichlet's principle consists in constructing harmonic functions by minimizing the Dirichlet integral in an appropriate class of functions. This idea is generalized, and minimizers of variational integrals are weak solutions of the associated differential equations of Euler and Lagrange. Several examples are discussed.

We shall first consider a special example, in order to make prominent the basic idea of the following considerations. The generalization of these reflections will then later present no great difficulty.

The equation to be treated in this example is perhaps the most important partial differential equation for mathematics and physics, namely the Laplace equation.

In the following, Ω will be an open, bounded subset of $\mathbb{R}^d$. A function $f : \Omega \to \mathbb{R}$ is said to be harmonic if it satisfies in Ω the Laplace equation

$$\Delta f(x) = \frac{\partial^2 f(x)}{(\partial x^1)^2} + \ldots + \frac{\partial^2 f(x)}{(\partial x^d)^2} = 0.$$

Harmonic functions occur, for example, in complex analysis. If $\Omega \subset \mathbb{C}$ and $z = x + iy \in \Omega$, and if $f(z) = u(z) + iv(z)$ is holomorphic on Ω, then the so-called Cauchy-Riemann differential equations

$$\frac{\partial u}{\partial x} = \frac{\partial v}{\partial y}, \ \frac{\partial u}{\partial y} = -\frac{\partial v}{\partial x} \tag{1}$$

hold, and as a holomorphic function is in the class C^∞, we can differentiate (1) and obtain

$$\frac{\partial^2 u}{\partial x^2} + \frac{\partial^2 u}{\partial y^2} = \frac{\partial^2 v}{\partial x \partial y} - \frac{\partial^2 v}{\partial y \partial x} = 0$$

and similarly

$$\frac{\partial^2 v}{\partial x^2} + \frac{\partial^2 v}{\partial y^2} = 0.$$

Thus the real and imaginary parts of a holomorphic function are harmonic.

Conversely, two harmonic functions which satisfy (1) are called conjugate and a pair of conjugate harmonic functions gives precisely the holomorphic function $f = u + iv$.

In case (1) holds, one can interpret $(u(x, y), -v(x, y))$ as the velocity field of a two dimensional rotation-free incompressible fluid. For $d = 3$ the harmonic functions describe likewise the velocity field of a rotation-free incompressible fluid, as well as electrostatic and gravitational fields (outside attracting or repelling charges or attracting masses), temperature distribution in thermal equilibrium, equilibrium states of elastic membranes, etc.

The most important problem in harmonic functions is the Dirichlet problem: Here, a function $g : \partial\Omega \to \mathbb{R}$ is given and one seeks $f : \overline{\Omega} \to \mathbb{R}$ with

$$\Delta f(x) = 0 \quad \text{for } x \in \Omega \tag{2}$$
$$f(x) = g(x) \quad \text{for } x \in \partial\Omega.$$

For example, this models the state of equilibrium of a membrane which is fixed at the boundary of Ω.

There exist various methods to solve the Dirichlet problem for harmonic functions. Perhaps the most important and general is the so-called Dirichlet principle, which we want to introduce now.

In order to pose (2) sensibly, one must make certain assumptions on Ω and g. For the moment we only assume that $g \in W^{1,2}(\Omega)$. As already said, Ω is an open and bounded subset of $\mathbb{R}^d$. Further restrictions will follow in due course in our study of the boundary condition $f = g$ on $\partial\Omega$.

The Dirichlet principle consists in finding a solution of

$$\Delta f = 0 \quad \text{in } \Omega$$

$$f = g \quad \text{on } \partial\Omega \text{ (in the sense that } f - g \in H_0^{1,2}(\Omega))$$

by minimizing the Dirichlet integral

$$\frac{1}{2} \int_\Omega |Dv|^2 \quad (\text{here } Dv = (D_1 v, \ldots D_d v))$$

over all $v \in H^{1,2}(\Omega)$ for which $v - g \in H_0^{1,2}(\Omega)$.

We shall now verify that this method really works.

Let

$$m := \inf\{\frac{1}{2} \int_\Omega |Dv|^2 : v \in H^{1,2}(\Omega), v - g \in H_0^{1,2}(\Omega)\}.$$

We must show that m is assumed and that the function for which it is assumed is harmonic. (Notation: By corollary 20.10, $W^{1,2} = H^{1,2}$ and in the sequel we shall mostly write $H^{1,2}$ for this space.)

Let $(f_n)_{n \in \mathbb{N}}$ be a minimizing sequence, so $f_n - g \in H_0^{1,2}(\Omega)$ and

$$\int_\Omega |Df_n|^2 \to 2m.$$

By corollary 20.16 we have

$$\|f_n\|_{L^2(\Omega)} \leq \|g\|_{L^2(\Omega)} + \|f_n - g\|_{L^2(\Omega)}$$
$$\leq \|g\|_{L^2(\Omega)} + \text{const.} \|Df_n - Dg\|_{L^2(\Omega)}$$
$$\leq \|g\|_{L^2(\Omega)} + c_1\|Dg\|_{L^2(\Omega)} + c_2\|Df_n\|_{L^2(\Omega)}$$
$$\leq \text{const.} + c_2\|Df_n\|_{L^2(\Omega)},$$

as g has been chosen to be fixed.

Without loss of generality let

$$\|Df_n\|^2_{L^2(\Omega)} \leq m + 1.$$

It follows that

$$\|f_n\|_{H^{1,2}(\Omega)} \leq \text{const. (independent of } n).$$

By theorem 21.8 f_n converges weakly, after a choice of a subsequence, to an $f \in H^{1,2}(\Omega)$ with $f - g \in H_0^{1,2}(\Omega)$ (this follows from corollary 21.12) and corollary 21.9 gives

$$\int_\Omega |Df|^2 \leq \liminf_{n\to\infty} \int_\Omega |Df_n|^2 = 2m.$$

By the theorem of Rellich (theorem 20.20) the remaining term of $\|f_n\|^2_{H^{1,2}}$, namely $\int |f_n|^2$ is even continuous, so $\int_\Omega |f|^2 = \lim_{n\to\infty} \int_\Omega |f_n|^2$, after choosing a subsequence of (f_n).

Because of $f - g \in H_0^{1,2}(\Omega)$, it follows from the definition of m that

$$\int_\Omega |Df|^2 = 2m.$$

Furthermore, for every $v \in H_0^{1,2}, t \in \mathbb{R}$ we have

$$m \leq \int_\Omega |D(f + tv)|^2 = \int_\Omega |Df|^2 + 2t \int_\Omega Df \cdot Dv + t^2 \int_\Omega |Dv|^2$$

(where $Df \cdot Dv := \sum_{i=1}^{d} D_i f \cdot D_i v$) and differentiation by t at $t = 0$ gives

$$0 = \frac{d}{dt} \int_\Omega |D(f + tv)|^2|_{t=0} = 2 \int_\Omega Df \cdot Dv$$

for all $v \in H_0^{1,2}(\Omega)$.

By the way, this calculation also shows that the map

$$E : H^{1,2}(\Omega) \to \mathbb{R}$$

$$f \mapsto \int_\Omega |Df|^2$$

is differentiable, with

$$DE(f)(v) = 2 \int_\Omega Df \cdot Dv.$$

Definition 22.1 A function $f \in H^{1,2}(\Omega)$ is called weakly harmonic or a weak solution of the Laplace equation if

$$\int_\Omega Df \cdot Dv = 0 \quad \text{for all } v \in H_0^{1,2}(\Omega). \tag{3}$$

Obviously, every harmonic function satisfies (3). In order to obtain a harmonic function by applying the Dirichlet principle, one has now to show conversely that a solution of (3) is twice continuously differentiable and therefore, in particular, harmonic. This will be achieved in §23.

However, we shall presently treat a more general situation:

Definition 22.2 Let $\varphi \in L^2(\Omega)$. A function $f \in H^{1,2}(\Omega)$ is called a weak solution of the Poisson equation $(\Delta f = \varphi)$ if for all $v \in H_0^{1,2}(\Omega)$

$$\int_\Omega Df \cdot Dv + \int_\Omega \varphi \cdot v = 0 \tag{4}$$

holds.

Remark. For a preassigned boundary value g (in the sense of $f - g \in H_0^{1,2}(\Omega)$) a solution of (4) can be obtained by minimizing

$$\frac{1}{2} \int_\Omega |Dw|^2 + \int_\Omega \varphi \cdot w$$

in the class of all $w \in H^{1,2}(\Omega)$ for which $w - g \in H_0^{1,2}(\Omega)$. One notices that this expression is bounded from below by the Poincaré inequality (corollary 20.16), as we have fixed the boundary value g.

Another possibility of finding a solution of (4) for a preassigned $f - g \in H_0^{1,2}$ is the following:

If one sets $w := f - g \in H_0^{1,2}$, then w has to solve

$$\int_\Omega Dw \cdot Dv = -\int_\Omega \varphi \cdot v - \int_\Omega Dg \cdot Dv \tag{5}$$

for all $v \in H_0^{1,2}$.

The Poincaré inequality (corollary 20.16) implies that a scalar product on $H_0^{1,2}(\Omega)$ is already given by

$$((f, v)) := (Df, Dv)_{L^2(\Omega)} = \int_\Omega Df \cdot Dv.$$

With this scalar product, $H_0^{1,2}(\Omega)$ becomes a Hilbert space. Furthermore,

$$\int_\Omega \varphi \cdot v \leq \|\varphi\|_{L^2} \cdot \|v\|_{L^2} \leq \text{const.} \|\varphi\|_{L^2} \cdot \|Dv\|_{L^2},$$

again by corollary 20.16. It follows that

$$Lv := -\int_\Omega \varphi \cdot v - \int_\Omega Dg \cdot Dv$$

defines a bounded linear functional on $H_0^{1,2}(\Omega)$. By theorem 21.6 there exists a uniquely determined $w \in H_0^{1,2}(\Omega)$ with

$$((w, v)) = Lv \quad \text{for all } v \in H_0^{1,2},$$

and w then solves (5).

This argument also shows that a solution of (4) is unique. This also follows from the following general result.

Lemma 22.3 Let $f_i, i = 1, 2$, be weak solutions of $\Delta f_i = \varphi_i$ with $f_1 - f_2 \in H_0^{1,2}(\Omega)$. Then

$$\|f_1 - f_2\|_{W^{1,2}(\Omega)} \leq \text{const.} \|\varphi_1 - \varphi_2\|_{L^2(\Omega)}.$$

In particular, a weak solution of $\Delta f = \varphi, f - g \in H_0^{1,2}(\Omega)$ is uniquely determined by g and φ.

Proof. We have

$$\int_\Omega D(f_1 - f_2)Dv = -\int_\Omega (\varphi_1 - \varphi_2)v$$

for all $v \in H_0^{1,2}(\Omega)$ and therefore in particular

$$\int_\Omega D(f_1 - f_2)D(f_1 - f_2) = -\int_\Omega (\varphi_1 - \varphi_2)(f_1 - f_2)$$

$$\leq \|\varphi_1 - \varphi_2\|_{L^2(\Omega)}\|f_1 - f_2\|_{L^2(\Omega)}$$

$$\leq \text{const.} \|\varphi_1 - \varphi_2\|_{L^2(\Omega)}\|Df_1 - Df_2\|_{L^2(\Omega)}$$

by corollary 20.16, and consequently

$$\|Df_1 - Df_2\|_{L^2(\Omega)} \leq \text{const.} \|\varphi_1 - \varphi_2\|_{L^2(\Omega)}.$$

The assertion follows by another application of corollary 20.16.   □

We have thus obtained the existence and uniqueness of weak solutions of the Poisson equation in a very simple manner.

The aim of the regularity theory consists in showing that (for sufficiently well behaved φ) a weak solution is already of class C^2, and thus also a classical solution of $\Delta f = \varphi$. In particular we shall show that a solution of $\Delta f = 0$ is even of class $C^\infty(\Omega)$.

Besides, we must investigate in which sense, if for example $\partial\Omega$ is of class C^∞ – in a sense yet to be made precise – and $g \in C^\infty(\overline{\Omega})$, the boundary condition $f - g \in H_0^{1,2}(\Omega)$ is realized. It turns out that in this case, a solution of $\Delta f = 0$ is of class C^∞ and for all $x \in \partial\Omega$ $f(x) = g(x)$ holds.

We shall now endeavour to make a generalization of the above ideas. For this we shall first summarize the central idea of these considerations:

In order to minimize the Dirichlet integral, we had first observed that there exists a bounded minimizing sequence in $H^{1,2}$. From this we could then choose a weakly convergent subsequence. As the Dirichlet integral is lower semicontinuous with respect to weak convergence the limit of this sequence then yields a minimum. Thus, with this initial step, the existence of a minimum is established. The second important observation then was that a minimum must satisfy, at least in a weak form, a partial differential equation.

We shall now consider a variational problem of the form

$$I(f) := \int_\Omega H(x, f(x), D(f(x)))dx \to \min.$$

under yet to be specified conditions on the real valued function H; here Ω is always an open, bounded subset of $\mathbb{R}^d$ and f is allowed to vary in the space $H^{1,2}(\Omega)$.

Similar considerations could be made in the spaces $H^{1,p}(\Omega)$, but we have introduced the concept of weak convergence only in Hilbert and not in general Banach spaces.

Theorem 22.4 *Let $H : \Omega \times \mathbb{R}^d \to \mathbb{R}$ be non-negative, measurable in the first and convex in the second argument, so $H(x, tp + (1-t)q) \leq tH(x,p) + (1-t)H(x,q)$ holds for all $x \in \Omega$, $p, q \in \mathbb{R}^d$ and $0 \leq t \leq 1$.*

For $f \in H^{1,2}(\Omega)$ we define

$$I(f) := \int_\Omega H(x, Df(x))dx \leq \infty.$$

Then

$$I : H^{1,2}(\Omega) \to \mathbb{R} \cup \{\infty\}$$

is convex and lower semicontinuous (relative to strong convergence, i.e. if $(f_n)_{n\in\mathbb{N}}$ converges in $H^{1,2}$ to f then

$$I(f) \leq \liminf_{n \to \infty} I(f_n)).$$

(As H is continuous in the second argument (see below) and Df is measurable, $H(x, Df(x))$ is again measurable (by corollary 17.12), so $I(f)$ is well-defined).

Proof. The convexity of I follows from that of H, as the integral is a linear function: Let $f, g \in H^{1,2}(\Omega), 0 \leq t \leq 1$. Then

$$I(tf + (1-t)g) = \int_\Omega H(x, tDf(x) + (1-t)Dg(x))dx$$

$$\leq \int_\Omega \{tH(x, Df(x)) + (1-t)H(x, Dg(x))\}dx$$

$$= tI(f) + (1-t)I(g).$$

It remains to show the lower semicontinuity. Let $(f_n)_{n \in \mathbb{N}}$ converge to f in $H^{1,2}$. By choosing a subsequence, we may assume that $\liminf_{n \to \infty} I(f_n) = \lim_{n \to \infty} I(f_n)$. By a further choice of a subsequence, Df_n then converges pointwise to Df almost everywhere. By theorem 19.12 this follows from the fact that Df_n converges in L^2 to Df. As H is continuous in the second variable (see lemma 22.5 infra), $H(x, Df_n(x))$ converges pointwise to $H(x, Df(x))$ almost everywhere on Ω. By the assumption $H \geq 0$ we can apply Fatou's lemma and obtain

$$I(f) = \int_\Omega H(x, Df(x))dx = \int_\Omega \lim_{n \to \infty} H(x, Df_n(x))dx$$

$$\leq \liminf_{n \to \infty} \int_\Omega H(x, Df_n(x))dx$$

$$= \liminf_{n \to \infty} I(f_n).$$

(As $\lim I(f_n) = \liminf I(f_n)$, by choice of the first subsequence, $\liminf I(f_n)$ does not change anymore in choosing the second subsequence). Thereby, the lower semicontinuity has been shown. $\square$

We append further the following result:

Lemma 22.5 *Let $\varphi : \mathbb{R}^d \to \mathbb{R}$ be convex. Then φ is continuous.*

Proof. We must control the difference $|\varphi(y + h) - \varphi(y)|$ for $h \to 0$. We set $\ell := \frac{h}{|h|}$ (we may assume $h \neq 0$) and choose $t \in [0, 1]$ with

$$h = (1 - t)\ell.$$

By convexity, we have

$$\varphi(ty + (1 - t)(y + \ell)) \leq t\varphi(y) + (1 - t)\varphi(y + \ell)$$

so

$$\varphi(y + h) \leq t\varphi(y) + (1 - t)\varphi(y + \ell),$$

and therefore

$$\varphi(y + h) - \varphi(y) \leq \frac{1 - t}{t}(-\varphi(y + h) + \varphi(y + \ell)). \tag{6}$$

The convexity of φ also gives

$$\varphi(y) \leq t\varphi(y + h) + (1 - t)\varphi(y - t\ell),$$

so

$$\varphi(y + h) - \varphi(y) \geq \frac{1 - t}{t}(\varphi(y) - \varphi(y - t\ell)). \tag{7}$$

We now let h approach 0, so $t \to 1$, and obtain the continuity of φ at y from (6) and (7). $\qquad\square$

We now prove

Lemma 22.6 *Let A be a convex subset of a Hilbert space, $I : A \to \mathbb{R} \cup \{\pm\infty\}$ be convex and lower semicontinuous. Then I is also lower semicontinuous relative to weak convergence.*

Proof. Let $(f_n)_{n \in \mathbb{N}} \subset A$ be weakly convergent to $f \in A$. We then have to show that

$$I(f) \leq \liminf_{n \to \infty} I(f_n). \tag{8}$$

By choosing a subsequence, we may assume that $I(f_n)$ is convergent, say

$$\liminf_{n \to \infty} I(f_n) = \lim_{n \to \infty} I(f_n) =: \omega. \tag{9}$$

By choosing a further subsequence and using the Banach-Saks lemma (corollary 21.10) the convex combination

$$g_k := \frac{1}{k} \sum_{\nu=1}^{k} f_{N+\nu}$$

converges strongly to f as $k \to \infty$, and indeed for every $N \in \mathbb{N}$.
The convexity of I gives

$$I(g_k) \leq \frac{1}{k} \sum_{\nu=1}^{k} I(f_{N+\nu}). \tag{10}$$

Now we choose, for $\varepsilon > 0$, N so large that for all $\nu \in \mathbb{N}$

$$I(f_{N+\nu}) < \omega + \varepsilon$$

holds (compare (9)). By (10) it then follows that

$$\limsup_{k \to \infty} I(g_k) \leq \omega.$$

The lower semicontinuity of I relative to strong convergence now gives

$$I(f) \leq \liminf_{k \to \infty} I(g_k) \leq \limsup_{k \to \infty} I(g_k) \leq \omega = \liminf_{n \to \infty} I(f_n).$$

Thereby (8) has been verified. □

We obtain now the important

Corollary 22.7 *Let* $H : \Omega \times \mathbb{R}^d \to \mathbb{R}$ *be non-negative, measurable in the first and convex in the second argument. For* $f \in H^{1,2}(\Omega)$*, let*

$$I(f) := \int_\Omega H(x, Df(x)) dx.$$

Then I is lower semicontinuous relative to weak convergence in $H^{1,2}$.
Let A be a closed convex subset of $H^{1,2}(\Omega)$.
If there exists a bounded minimizing sequence $(f_n)_{n \in \mathbb{N}} \subset A$, that is,

$$I(f_n) \to \inf_{g \in A} I(g) \text{ with } \|f\|_{H^{1,2}} \leq K,$$

then I assumes its minimum on A, i.e. there is an $f \in A$ with

$$I(f) = \inf_{g \in A} I(g).$$

Proof. The lower semicontinuity follows from theorem 22.4 and lemma 22.6. Now let $(f_n)_{n \in \mathbb{N}}$ be a bounded minimizing sequence. By theorem 21.8, after choosing a subsequence, the sequence f_n converges weakly to an f, which by corollary 21.12 is in A. Due to weak lower semicontinuity it follows that

$$I(f) \leq \liminf_{n \to \infty} I(f_n) = \inf_{g \in A} I(g),$$

and, as trivially $\inf_{g \in A} I(g) \leq I(f)$ holds, the assertion follows. □

Remarks.
1)   In corollary 22.7, H depends only on x and $Df(x)$, but not on $f(x)$.
      In fact, in the general case

$$I(f) = \int_\Omega H(x, f(x), Df(x)) dx$$

there are lower semicontinuity results under suitable assumptions on H, but these are considerably more difficult to prove. The only exception is the following statement:

Let $H : \Omega \times \mathbb{R} \times \mathbb{R}^d \to \mathbb{R}$ be measurable in the first and jointly convex in the second and third argument, i.e. for $x \in \Omega, f, g \in \mathbb{R}, p, q \in \mathbb{R}^d, 0 \le t \le 1$ one has

$$H(x, tf + (1-t)g, tp + (1-t)q) \le tH(x, f, p) + (1-t)H(x, g, q).$$

Then the results of corollary 22.7 also hold for

$$I(f) := \int_{\Omega} H(x, f(x), Df(x))dx.$$

The proof of this result is the same as that of corollary 22.7.

2)   Weak convergence was a suitable concept for the above considerations due to the following reasons. One needs a convergence concept which, on the one hand, should allow lower semicontinuity statements and so should be as strong as possible, and on the other hand, it should admit a selection principle, so that every bounded sequence contains a convergent subsequence and therefore should be as weak as possible. The concept of weak convergence unites these two requirements.

Example. We now want to consider an important example:

For $i, j = 1, \ldots, d$, let $a_{ij} : \Omega \to \mathbb{R}$ be measurable functions with

$$\sum_{i,j=1}^{d} a_{ij}(x)\xi^i\xi^j \ge \lambda|\xi|^2 \tag{11}$$

for all $x \in \Omega$, $\xi = (\xi^1, \ldots, \xi^d) \in \mathbb{R}^d$, with a $\lambda > 0$.

The condition (11) is called an ellipticity condition.

We consider

$$I(f) := \int_{\Omega} \sum_{i,j=1}^{d} a_{ij}(x)D_i f(x)D_j f(x)dx$$

for $f \in H^{1,2}(\Omega)$.

We shall also assume that

$$\operatorname*{ess\,sup}_{\substack{x \in \Omega \\ i,j=1,\ldots d}} |a_{ij}(x)| \le m. \tag{12}$$

Then $I(f) < \infty$ for all $f \in H^{1,2}(\Omega)$.

By (11) and (12),

$$\lambda \int_\Omega |Df(x)|^2 dx \leq I(f) \leq md \int_\Omega |Df(x)|^2 dx \tag{13}$$

holds.

We now observe that

$$\langle f,g \rangle := \frac{1}{2} \int_\Omega \sum_{i,j=1}^{d} a_{ij}(x)(D_i f(x)D_j g(x) + D_j f(x)D_i g(x))dx$$

is bilinear, symmetric and positive semi-definite (so $\langle f,f \rangle \geq 0$ for all f) on $H^{1,2}(\Omega)$. Therefore the Schwarz inequality holds:

$$\langle f,g \rangle \leq I(f)^{\frac{1}{2}} \cdot I(g)^{\frac{1}{2}}. \tag{14}$$

It now follows easily that I is convex:

$$I(tf + (1-t)g) = \int_\Omega \sum_{i,j=1}^{d} a_{ij}(x)(t^2 D_i f(x)D_j f(x)$$
$$+ t(1-t)(D_i f(x)D_j g(x) + D_j f(x)D_i g(x))$$
$$+ (1-t)^2 D_i g(x)D_j g(x))dx,$$

thus

$$I(tf + (1-t)g) = t^2 I(f) + 2t(1-t)\langle f,g \rangle + (1-t)^2 I(g)$$
$$\leq t^2 I(f) + 2t(1-t)I(f)^{\frac{1}{2}}I(g)^{\frac{1}{2}} + (1-t)^2 I(g) \text{ by (14)}$$
$$\leq t^2 I(f) + t(1-t)(I(f) + I(g)) + (1-t)^2 I(g)$$
$$= tI(f) + (1-t)I(g).$$

Finally, we also observe that if we restrict ourselves to the space $H_0^{1,2}(\Omega)$, then every minimizing sequence for I is bounded. Namely, for $f \in H_0^{1,2}(\Omega)$, the Poincaré inequality (corollary 20.16) holds:

$$\|f\|_{H^{1,2}(\Omega)}^2 \leq c \int_\Omega |Df(x)|^2 dx \text{ where } c \text{ is a constant} \tag{15}$$
$$\leq \frac{c}{\lambda} I(f) \text{ by (13)},$$

and thereby a minimizing sequence is bounded in $H^{1,2}(\Omega)$.

In general, for a fixed $g \in H^{1,2}(\Omega)$, we can also consider the space

$$A_g := \{f \in H^{1,2}(\Omega) : f - g \in H_0^{1,2}(\Omega)\}$$

The space A_g is closed and convex and for $f \in A_g$ we have

$$\|f\|_{H^{1,2}(\Omega)} \leq \|f - g\|_{H^{1,2}(\Omega)} + \|g\|_{H^{1,2}(\Omega)}$$
$$\leq (\frac{c}{\lambda} I(f - g))^{\frac{1}{2}} + \|g\|_{H^{1,2}(\Omega)} \text{ since } f - g \in H_0^{1,2}(\Omega)$$
$$\leq (\frac{c}{\lambda}(I(f) + I(g))^2)^{\frac{1}{2}} + \|g\|_{H^{1,2}(\Omega)}$$

(using the triangle inequality implied by the Schwarz inequality for $I(f)^{\frac{1}{2}} = \langle f, f \rangle^{\frac{1}{2}})$

$$= (\frac{c}{\lambda})^{\frac{1}{2}} I(f) + (\frac{c}{\lambda})^{\frac{1}{2}} I(g) + \|g\|_{H^{1,2}(\Omega)}.$$

As g is fixed, the $H^{1,2}$-norm for a minimizing sequence for I in A_g is again bounded.

We deduce from corollary 22.7 that I assumes its minimum on A_g, i.e. for any $g \in H^{1,2}(\Omega)$ there exists an $f \in H^{1,2}(\Omega)$ with $f - g \in H_0^{1,2}(\Omega)$ and

$$I(f) = \inf\{I(h) : h \in H^{1,2}(\Omega), h - g \in H_0^{1,2}(\Omega)\}.$$

This generalizes the corresponding statements for the Dirichlet integral. In the same manner we can treat, for a given $\varphi \in L^2(\Omega)$

$$J(f) = \int_\Omega \Big(\sum_{i,j=1}^d a_{ij}(x) D_i f(x) D_j f(x) + \varphi(x) f(x) \Big) dx$$

and verify the existence of a minimum with given boundary conditions.

However, not every variational problem admits a minimum:

Examples.

1)   We consider, for $f : [-1, 1] \to \mathbb{R}$

$$I(f) := \int_{-1}^{1} (f'(x))^2 x^4 dx$$

with boundary conditions $f(-1) = -1, f(1) = 1$. Consider

$$f_n(x) = \begin{cases} -1 & \text{for } -1 \leq x < -\frac{1}{n} \\ nx & \text{for } -\frac{1}{n} \leq x \leq \frac{1}{n} \\ 1 & \text{for } \frac{1}{n} < x \leq 1 \end{cases}$$

Then $\lim_{n\to\infty} I(f_n) = 0$, but for every f we have $I(f) > 0$. Thus the infimum of $I(f)$, with the given boundary conditions, is not assumed.

2)   We shall now consider an example related to the question of realization of boundary values:
Let $\Omega := U(0, 1)\setminus\{0\} = \{x \in \mathbb{R}^d : 0 < \|x\| < 1\}, d \geq 2$.
We choose $g \in C^1(\overline{\Omega})$ with

$$g(x) = 0 \quad \text{for } \|x\| = 1$$
$$g(0) = 1$$

We want to minimize the Dirichlet integral over
$A_g = \{ f \in H^{1,2}(\Omega) : f - g \in H_0^{1,2}(\Omega) \}$. Consider, for $0 < \varepsilon < 1$,
$(r = \|x\|)$,

$$f_\varepsilon(r) := \begin{cases} 1 & \text{for } 0 \le r \le \varepsilon \\ \frac{\log(r)}{\log(\varepsilon)} & \text{for } \varepsilon < r \le 1. \end{cases}$$

By the computation rules given in §20, $f_\varepsilon(r)$ is in $H_0^{1,2}(\Omega)$ and

$$\int_\Omega |Df_\varepsilon(r)|^2 dx = \frac{1}{(\log \varepsilon)^2} \int_{\varepsilon \le r \le 1} \frac{1}{r^2} dx$$

$$= \frac{d\omega_d}{(\log \varepsilon)^2} \int_\varepsilon^1 \frac{r^{d-1}}{r^2} dr \quad \text{(theorem 13.21)}$$

$$= \begin{cases} \frac{2\pi}{\log \frac{1}{\varepsilon}} & \text{for } d = 2 \\ \frac{1}{(\log \varepsilon)^2} \frac{d\omega_d}{d-2} (1 - \varepsilon^{d-2}) & \text{for } d > 2 \end{cases}$$

It follows that

$$\lim_{\varepsilon \to 0} \int_\Omega |Df_\varepsilon|^2 = 0,$$

and thereby

$$\inf \{ \int_\Omega |Df|^2, f \in A_g \} = 0.$$

Now, it follows from the Poincaré inequality (corollary 20.16) as usual
that for a minimizing sequence $(f_n)_{n \in \mathbb{N}} \subset A_g$

$$\|f_n\|_{H^{1,2}} \to 0 \quad \text{for } n \to \infty.$$

Thus f_n converges in $H^{1,2}$ to zero. So the limit $f \equiv 0$ does not fulfil the
prescribed boundary condition $f(0) = 1$. The reason for this is that
an isolated point is really too small to play a role in the minimizing
of Dirichlet integrals. We shall later even see that there exists no
function h at all such that

$$h : B(0,1) \to \mathbb{R}, \ \Delta h(x) = 0 \text{ for } 0 < \|x\| < 1,$$
$$h(x) = 0 \text{ for } \|x\| = 1 \text{ and } h(0) = 1$$

(see example after theorem 24.4).
The phenomenon which has just appeared can be easily formulated
abstractly.

Definition 22.8 Let Ω be open in $\mathbb{R}^d$, $K \subset \Omega$ compact. We define the
capacity of K with respect to Ω by

$$\mathrm{cap}_\Omega(K) := \inf\{\int_\Omega |Df|^2 : f \in H_0^{1,2}(\Omega), f \geq 1 \text{ on } K\}.$$

So the capacity of an isolated point in $\mathbb{R}^d$ vanishes for $d \geq 2$.

In general we have

Theorem 22.9 *Let $\Omega \subset \mathbb{R}^d$ be open, $K \subset \Omega$ compact with $\mathrm{cap}_\Omega(K) = 0$. Then the Dirichlet principle cannot give a solution of the problem*

$$f : \overline{\Omega \backslash K} \to \mathbb{R}$$
$$\Delta f(x) = 0 \text{ for } x \in \Omega \backslash K$$
$$f(x) = 0 \text{ for } x \in \partial\Omega$$
$$f(x) = 1 \text{ for } x \in \partial K.$$

□

For an arbitrary $A \subset \Omega$ one can also define

$$\mathrm{cap}_\Omega(A) := \sup_{\substack{K \subset A \\ K \text{ compact}}} \mathrm{cap}(K)$$

(as for an A with e.g. $\mathrm{vol}(A) = \infty$ there is no $f \in H_0^{1,2}(\Omega)$ with $f \geq 1$ on A, we cannot define the capacity directly as in definition 22.8).

We shall now derive the so-called Euler-Lagrange differential equations as necessary conditions for the existence of a minimum of a variational problem.

Theorem 22.10 *Consider $H : \Omega \times \mathbb{R} \times \mathbb{R}^d \to \mathbb{R}$, with H measurable in the first and differentiable in the other two arguments. We set*

$$I(f) := \int_\Omega H(x, f(x), Df(x))dx$$

for $f \in H^{1,2}(\Omega)$. Assume that

$$|H(x, f, p)| \leq c_1|p|^2 + c_2|f|^2 + c_3 \tag{16}$$

with constants c_1, c_2, c_3 for almost all $x \in \Omega$ and all $f \in \mathbb{R}, p \in \mathbb{R}^d$. ($I(f)$ is therefore finite for all $f \in H^{1,2}(\Omega)$).

(i)   Let $A \subset H^{1,2}(\Omega)$ and let $f \in A$ satisfy

$$I(f) = \inf\{I(g) : g \in A\}.$$

Let A be such that for every $\varphi \in C_0^\infty(\Omega)$ there is a $t_0 > 0$ with

$$f + t\varphi \in A \quad \text{for all } t \text{ with } |t| < t_0. \tag{17}$$

Assume that H satisfies for almost all x and all f, p

$$|H_f(x, f, p)| + \sum_{i=1}^{d} |H_{p_i}(x, f, p)| \leq c_4 |p|^2 + c_5 |f|^2 + c_6 \qquad (18)$$

with constants c_4, c_5, c_6; here, the subscripts denote partial derivatives and
$p = (p_1, \ldots p_d)$. *Then for all $\varphi \in C_0^\infty(\Omega)$ we have*

$$\int_\Omega \{H_f(x, f(x), Df(x))\varphi(x) + \sum_{i=1}^{d} H_{p_i}(x, f(x), Df(x))D_i\varphi(x)\}dx = 0$$

$$(19)$$

(ii)   *Under the same assumptions as in (i) assume that even for any $\varphi \in H_0^{1,2}(\Omega)$ there is a t_0 such that (17) holds. Furthermore, assume instead of (18) the inequality*

$$|H_f(x, f, p)| + \sum_{i=1}^{d} |H_{p_i}(x, f, p)| \leq c_7 |p| + c_8 |f| + c_9, \qquad (20)$$

with constants c_7, c_8, c_9. Then the condition (19) holds for all $\varphi \in H_0^{1,2}(\Omega)$.

(iii)   *Under the same assumptions as in (i), let now H be continuously differentiable in all the variables. Then, if f is also twice continuously differentiable, we have*

$$\sum_{i,j=1}^{d} H_{p_i p_j}(x, f(x), Df(x)) \cdot \frac{\partial^2 f(x)}{\partial x^i \partial x^j}$$

$$+ \sum_{i=1}^{d} H_{p_i f}(x, f(x), Df(x))\frac{\partial f(x)}{\partial x^i} + \qquad (21)$$

$$+ \sum_{i=1}^{d} H_{p_i x^i}(x, f(x), Df(x)) - H_f(x, f(x), Df(x)) = 0$$

or, abbreviated,

$$\sum_{i=1}^{d} \frac{d}{dx^i}(H_{p_i}(x, f(x), Df(x)) - H_f(x, f(x), Df(x)) = 0 \qquad (22)$$

(here $\frac{d}{dx^i}$ is to be distinguished from $\frac{\partial}{\partial x^i}$!).

Definition 22.11 The equations (21) are called the Euler-Lagrange equations of the variational problem $I(f) \to \min$.

The equation (21) was first established by Euler for the case $d = 1$ by means of approximation by difference equations and then by Lagrange in the general case by a method essentially similar to the one used here.

Proof of theorem 22.10

(i)   We have

$$I(f) \leq I(f + t\varphi) \quad \text{for } |t| < t_0. \tag{23}$$

Now

$$I(f + t\varphi) = \int_\Omega H(x, f(x) + t\varphi(x), Df(x) + tD\varphi(x))dx.$$

As for $\varphi \in C_0^\infty(\Omega)$, φ and $D\varphi$ are bounded we can apply theorem 16.11 on account of (16) and (18) and conclude that $I(f + t\varphi)$ is differentiable in t for $|t| < t_0$ with derivative

$$\frac{d}{dt}I(f + t\varphi) = \int_\Omega \{H_f(x, f(x) + t\varphi(x), Df(x) + tD\varphi(x))\varphi(x)$$

$$+ \sum_{i=1}^d H_{p_i}(x, f(x) + t\varphi(x), Df(x)$$

$$+ tD\varphi(x)) \cdot D_i\varphi(x)\}dx.$$

From (23) it follows that

$$0 = \frac{d}{dt}I(f + t\varphi)_{|t=0}$$

$$= \int_\Omega \{H_f(x, f(x), Df(x))\varphi(x) + \sum_{i=1}^d H_{p_i}(x, f(x), Df(x))$$

$$\cdot D_i\varphi(x)\}dx.$$

This proves (i).

If (20) holds, we can differentiate under the integral with respect to t in case $\varphi \in H_0^{1,2}(\Omega)$, for then the integrand of the derivative is bounded by

$$(c_7|Df(x) + tD\varphi(x)| + c_8|f(x) + t\varphi(x)| + c_9)(|\varphi(x)| + |D\varphi(x)|)$$

the integral of which, by the Schwarz inequality, is bounded by

$$\text{const. } \|f + t\varphi\|_{H^{1,2}} \cdot \|\varphi\|_{H^{1,2}}.$$

Therefore theorem 16.11 can indeed again be applied to justify differentiation under the integral sign. Thus (ii) follows.

For the proof of (21) we notice that, due to the assumptions of continuous differentiability, there exists for every $x \in \Omega$ a neighborhood $U(x)$ in which $H_{p_ip_j}\frac{\partial^2 f}{\partial x^i \partial x^j}, H_{p_if}\frac{\partial f}{\partial x^i}$ and $H_{p_ix^i}$ are bounded.

For $\varphi \in C_0^\infty(U(x))$ we can then integrate (19) by parts and obtain

$$0 = \int_\Omega \{H_f(x, f(x), Df(x)) - \sum_{i,j=1}^d H_{p_i p_j}(x, f(x), Df(x)) \frac{\partial}{\partial x^i} (\frac{\partial f(x)}{\partial x^j})$$

$$- \sum_{i=1}^d H_{p_i f}(x, f(x), Df(x)) \frac{\partial f}{\partial x^i}(x) - \sum_{i=1}^d H_{p_i x^i}(x, f(x), Df(x))\} \varphi(x) dx.$$

As this holds for all $\varphi \in C_0^\infty(U(x))$ it follows from corollary 19.19 that the expression in the curly brackets vanishes in $U(x)$, and as this holds for every $x \in \Omega$, the validy of (21) in Ω follows. $\qquad\square$

Remark. By the Sobolev embedding theorem, one can substitute the term $c_2|f|^2$ in (16) by $c_2|f|^{\frac{2d}{d-2}}$ for $d > 2$ and by $c_2|f|^q$ with arbitrary $q < \infty$ for $d = 2$, and similarly $c_5|f|^2$ in (18) etc., without harming the validity of the conclusions. (Note, however that the version of the Sobolev embedding theorem proved in the present book is formulated only for $H_0^{1,2}$ and not for $H^{1,2}$, and so is not directly applicable here.)

One can also consider more general variational problems for vector-valued functions: Let

$$H : \Omega \times \mathbb{R}^c \times \mathbb{R}^{dc} \to \mathbb{R}$$

be given, and for $f : \Omega \to \mathbb{R}^c$ consider the problem

$$I(f) := \int_\Omega H(x, f(x), Df(x)) dx \to \min.$$

In this case, the Euler-Lagrange differential equations are

$$\sum_{i=1}^d \frac{d}{dx^i}(H_{p_i^\alpha}(x, f(x), Df(x))) - H_{f^\alpha}(x, f(x), Df(x)) = 0 \quad \text{for } \alpha = 1, \ldots c$$

or, written out,

$$\sum_{i,j=1}^d \sum_{\beta=1}^c H_{p_i^\alpha p_j^\beta}(x, f(x), Df(x)) \frac{\partial^2}{\partial x^i \partial x^j} f^\beta$$

$$+ \sum_{\beta=1}^c \sum_{i=1}^d H_{p_i^\alpha f^\beta}(x, f(x), Df(x)) \frac{\partial f^\beta}{\partial x^i}$$

$$+ \sum_{i=1}^d H_{p_i^\alpha x^i}(x, f(x), Df(x)) - H_{f^\alpha}(x, f(x), Df(x)) = 0 \quad \text{for } \alpha = 1, \ldots, c.$$

So this time, we obtain a *system* of partial differential equations.

For the rest of this paragraph, H will always be of class C^2.

Examples. We shall now consider a series of examples:

1)   For $a, b \in \mathbb{R}$, $f : [a, b] \to \mathbb{R}$, we want to minimize the arc length of the graph of f, thus of the curve $(x, f(x)) \subset \mathbb{R}^2$, hence

$$\int_a^b \sqrt{1 + f'(x)^2}\, dx \to \min.$$

The Euler-Lagrange equations are

$$0 = \frac{d}{dx} \frac{f'(x)}{\sqrt{1 + f'(x)^2}}$$

$$= \frac{f''(x)}{\sqrt{1 + f'(x)^2}} - \frac{f'(x)^2 f''(x)}{[1 + f'(x)^2]^{\frac{3}{2}}}$$

$$= \frac{f''(x)}{(1 + f'(x)^2)^{\frac{3}{2}}},$$

so

$$f''(x) = 0. \tag{24}$$

Of course, the solutions of (24) are precisely the straight lines, and we shall see below that these indeed give the minimum for given boundary conditions $f(a) = \alpha$, $f(b) = \beta$.

2)   The so-called Fermat principle says that a light ray traverses its actual path between two points in less time than any other path joining those two points. Thus the path of light in an inhomogeneous two dimensional medium with speed $\gamma(x, f)$ is determined by the variational problem

$$I(f) = \int_a^b \frac{\sqrt{1 + f'(x)^2}}{\gamma(x, f(x))}\, dx \to \min.$$

The Euler-Lagrange equations are

$$0 = \frac{d}{dx} \frac{f'(x)}{\gamma(x, f(x))\sqrt{1 + f'(x)^2}} + \frac{\gamma_f}{\gamma^2}\sqrt{1 + f'(x)^2}$$

$$= \frac{f''(x)}{\gamma\sqrt{1 + f'(x)^2}} - \frac{(f'(x))^2 f''(x)}{\gamma(1 + f'(x)^2)^{\frac{3}{2}}} - \frac{\gamma_x}{\gamma^2}\frac{f'(x)}{\sqrt{1 + f'(x)^2}}$$

$$- \frac{\gamma_f}{\gamma^2}\frac{f'(x)^2}{\sqrt{1 + f'(x)^2}} + \frac{\gamma_f}{\gamma^2}\sqrt{1 + f'(x)^2},$$

so

$$0 = f''(x) - \frac{\gamma_x}{\gamma} f'(x)(1 + f'(x)^2) + \frac{\gamma_f}{\gamma}(1 + f'(x)^2). \tag{25}$$

Obviously, example 2 is a generalization of example 1.

3)   The brachistochrone problem is formally a special case of the preceding example. Here, two points $(x_0, 0)$ and (x_1, y_1) are joined by a curve on which a particle moves, without friction, under the influence of a gravitational field directed along the y-axis, and it is required that the particle moves from one point to the other in the shortest possible time.

Denoting acceleration due to gravity by g, the particle attains the speed $(2gy)^{\frac{1}{2}}$ after falling the height y and the time required to fall by the amount $y = f(x)$ is therefore

$$I(f) = \int_{x_0}^{x_1} \sqrt{\frac{1 + f'(x)^2}{2gf(x)}}\, dx.$$

We consider this as the problem $I(f) \to$ min. subject to the boundary conditions $f(x_0) = 0$, $f(x_1) = y_1$. Setting $\gamma = \sqrt{2gf(x)}$, equation (25) becomes

$$0 = f''(x) + (1 + f'(x)^2)\frac{1}{2f(x)}. \tag{26}$$

We shall solve (26) explicitly. Consider the integrand

$$H(f(x), f'(x)) = \sqrt{\frac{1 + f'(x)^2}{2gf(x)}}.$$

From the Euler-Lagrange equations

$$\frac{d}{dx}H_p - H_f = 0$$

it follows, as H does not depend explicitly on x, that

$$\frac{d}{dx}(f' \cdot H_p - H) = f'' \cdot H_p + f'\frac{d}{dx}H_p - H_p \cdot f'' - H_f \cdot f'$$

$$= f'(\frac{d}{dx}H_p - H_f) = 0,$$

so $f' \cdot H_p - H \equiv$ const. $\equiv c$.

From this, f' can be expressed as a function of f and c, and in case $f' \neq 0$, the inverse function theorem gives, with $f' = \varphi(f, c)$,

$$x = \int \frac{df}{\varphi(f, c)}.$$

In our case

$$c = f' \cdot H_p - H = -\frac{1}{\sqrt{2gf(1 + f'^2)}},$$

so

$$f' = \pm\sqrt{\frac{1}{2gc^2 f} - 1}.$$

We set $2gc^2 f = \frac{1}{2}(1 - \cos t)$, so that $f' = \sqrt{\frac{1+\cos t}{1-\cos t}} = \frac{\sin t}{1-\cos t}$, and then

$$x = \int \frac{df}{f'} = \int \frac{1 - \cos t}{\sin t} \frac{df}{dt} dt$$

$$= \frac{1}{4gc^2} \int (1 - \cos t)dt = \frac{1}{4gc^2}(t - \sin t) + c_1. \tag{27}$$

Thereby f and x have been determined as functions of t. If one solves (27) for $t = t(x)$ and puts this in the equation for f, then one also obtains $f(x)$.

In the preceding example, we have learnt an important method for solving ordinary differential equations, namely, that of finding an expression which by the differential equation, must be constant as a function of the independent variable. From the constancy of this expression x and $f(x)$ can then be obtained as a function of a parameter. One can proceed similarly in the case where h does not contain the dependent variable f; then the Euler-Lagrange equation is simply

$$\frac{d}{dx} H_p = 0$$

and therefore $H_p = $ const., and from this one can again obtain f' and then x and $f(x)$ by integration.

All the above examples were concerned with the simplest possible situation, namely the case where only one independent and one dependent variable occured. If one considers, for example, in 1) an arbitrary curve $g(x) = (g_1(x), \ldots, g_c(x))$ in $\mathbb{R}^c$, then we have to minimize

$$I(g) = \int_a^b \|g'(x)\|dx = \int_a^b \left(\sum_{i=1}^c (\frac{d}{dx} g_i(x))^2 \right)^{\frac{1}{2}} dx,$$

and we obtain as the Euler-Lagrange equations

$$0 = \frac{d}{dx} \frac{g_i'(x)}{(\sum_{j=1}^c g_j'(x)^2)^{\frac{1}{2}}} = \frac{g_i'' \sum_{j=1}^c (g_j')^2 - g_i' \cdot \sum_{j=1}^c g_j' g_j''}{(\sum_{j=1}^c (g_j')^2)^{\frac{3}{2}}} \quad \text{for } i = 1, \ldots, c. \tag{28}$$

From this, one can at first not see too much, and this is not surprising as we had already seen earlier that the length $I(g)$ of the curve $g(x)$ is invariant under reparametrizations. Thus, if $x \mapsto g(x)$ is a solution of (28) then so is $\gamma(t) := g(x(t))$ for every bijective map $t \mapsto x(t)$. In other words, there are just too many solutions. On the other hand, we know that for a smooth curve g we can always arrange $\|\frac{d}{dt} g(x(t))\| \equiv 1$ by a reparametrization $x = x(t)$. The equations (28) then become

$$\frac{d}{dt}\left(\frac{d}{dt}g_i(x(t))\right) = 0 \quad \text{for } i = 1, \ldots, c,$$

and it follows that $g(x(t))$ is a straight line. Then $g(x)$ is also a straight line, only here $g(x)$ does not necessarily describe the arc length.

In physics, stable equilibria are characterized by the principle of minimal potential energy, whereas dynamical processes are described by Hamilton's principle. In both, it is a question of variational principles. Let a physical system with d degrees of freedom be given; let the parameters be $q^1, \ldots, q^d$. We want to determine the state of the system by expressing the parameters as functions of the time t. The mechanical properties of the system may be described by:

—   the kinetic energy $T = \sum\limits_{i,j=1}^{d} A_{ij}(q^1, \ldots q^d, t)\dot{q}^i \dot{q}^j$

(thus T is a function of the velocities $\dot{q}^1, \ldots, \dot{q}^d$ – a point " · " always denotes derivative with respect to time –, the coordinates $q^1, \ldots, q^d$, and time t; often, T does not depend anymore explicitly on t (see below): Here, T is a quadratic form in the generalized velocities $q^1, \ldots, q^d$)

—   and the potential energy $U = U(q^1, \ldots q^d, t)$.

Both U and T are assumed to be of class C^2.

Hamilton's principle now postulates that motion between two points in time t_0 and t_1 occurs in such a way that the integral

$$I(q) := \int_{t_0}^{t_1} (T - U)dt \tag{29}$$

is stationary in the class of all functions $q(t) = (q^1(t), \ldots, q^d(t))$ with fixed initial and final states $q(t_0)$ and $q(t_1)$ respectively .

Thus one does not necessarily look for a minimum under all motions which carry the system from an initial state to a final state, rather only for a stationary value of the integral. For a stationary value, the Euler-Lagrange equations must hold exactly as for a minimum, thus

$$\frac{d}{dt}\frac{\partial T}{\partial \dot{q}^i} - \frac{\partial}{\partial q^i}(T - U) = 0 \quad \text{for } i = 1, \ldots, d. \tag{30}$$

If U and T do not depend explicitly on time t, then equilibrium states are characterized by all the quantities being moreover constant in time, so in particular $\dot{q}^i = 0$ for $i = 1, \ldots, d$, and thereby $T = 0$, therefore by (30)

$$\frac{\partial U}{\partial q^i} = 0 \quad \text{for } i = 1, \ldots, d. \tag{31}$$

Thus in a state of equilibrium, U must have a critical point and in order for this equilibrium to be stable U must even have a minimum there.

We shall now derive the theorem of conservation of energy in the case where T and U do not depend explicitly on time (though they depend implicitly as they depend on q^i, $\dot{q}^i$ which in turn depend on t).

By observing that

$$\sum_{i,j=1}^{d} A_{ij}\dot{q}^i\dot{q}^j = \frac{1}{2}\sum_{i,j=1}^{d}(A_{ij}+A_{ji})\dot{q}^i\dot{q}^j$$

and, if necessary, replacing A_{ij} by $\frac{1}{2}(A_{ij}+A_{ji})$, we may assume that

$$A_{ij} = A_{ji}.$$

Now

$$T = \sum_{i,j=1}^{d} A_{ij}(q^1,\ldots,q^d)\dot{q}^i\dot{q}^j$$

$$U = U(q^1,\ldots,q^d).$$

Introducing the Lagrangian

$$L = T - U,$$

the Euler-Lagrange equations become

$$0 = \frac{d}{dt}L_{\dot{q}^i} - L_{q^i} \quad (i = 1,\ldots,d).$$

As above, one calculates that

$$\frac{d}{dt}\left(\sum_{i=1}^{d}\dot{q}^i L_{\dot{q}^i} - L\right) = \sum_{i=1}^{d}\left(\ddot{q}^i L_{\dot{q}^i} + \dot{q}^i\frac{d}{dt}L_{\dot{q}^i} - L_{\dot{q}^i}\ddot{q}^i - L_{q^i}\dot{q}^i\right) = 0,$$

so

$$\sum_{i=1}^{d}\dot{q}^i L_{\dot{q}^i} - L = \text{ const. (independent of } t\text{)}.$$

On the other hand

$$\sum_{i=1}^{d}\dot{q}^i L_{\dot{q}^i} = \sum_{i=1}^{d}2\dot{q}^i\sum_{k=1}^{d}A_{ik}\dot{q}^k = 2T,$$

and it follows that

$$2T - L = T + U$$

is constant in t. $T + U$ is called the total energy of the system and we have therefore shown the time conservation of energy, in case T and U do not depend explicitly on t.

A special case is the motion of a point of mass m in three dimensional space; let its path be $q(t) = (q^1(t), q^2(t), q^3(t))$. In this case

$$T = \frac{m}{2} \sum_{i=1}^{3} \dot{q}^i(t)^2$$

and U is determined by Newton's law of gravitation, for example,

$$U = -m \frac{g}{\|q\|}$$

in case an attracting mass is situated at the origin of coordinates ($g = $ const.)

We shall now consider motion in the neighborhood of a stable equilibrium. Here we will again assume that T and U do not depend explicitly on time t. Without loss of generality, assume that the equilibrium point is at $t = 0$ and also that $U(0) = 0$ holds. As motion occurs in a neighborhood of a stationary state, we ignore terms of order higher than two in the $\dot{q}^i$ and q^i; thus, we set

$$T = \sum_{i,j=1}^{d} a_{ij} \dot{q}^i \dot{q}^j$$

$$U = \sum_{i,j=1}^{d} b_{ij} q^i q^j \tag{32}$$

with constant coefficients a_{ij}, b_{ij}. We have therefore substituted U by the second order terms of its Taylor series (the first order terms vanish because of (31)). In particular, we can assume $b_{ij} = b_{ji}$. By writing

$$T = \sum_{i,j=1}^{d} \frac{1}{2}(a_{ij} + a_{ji}) \dot{q}^i \dot{q}^j,$$

we can likewise assume that the coefficients of T are symmetric. As U is to have a minimum at 0, we shall also assume that the matrix

$$B = (b_{ij})_{i,j=1,\ldots,d}$$

is positive definite.

Finally, we also assume that

$$A = (a_{ij})_{i,j=1,\ldots,d}$$

is positive definite.

Equation (30) transforms to

$$\sum_{j=1}^{d} a_{ij} \ddot{q}^j + \sum_{j=1}^{d} b_{ij} q^j = 0 \quad \text{for } i = 1, \ldots, d, \tag{33}$$

so in vector notation to

$$\ddot{q} + Cq = 0 \tag{34}$$

with the positive definite symmetric matrix $C = A^{-1}B$. As C is symmetric, it can be transformed to a diagonal matrix by an orthogonal matrix, hence

$$S^{-1}CS =: D = \begin{pmatrix} \lambda_1 & & 0 \\ & \ddots & \\ 0 & & \lambda_d \end{pmatrix}$$

for an orthogonal matrix S. As C is positive definite, all the eigenvalues $\lambda_1, \ldots \lambda_d$ are positive. We set $y = S^{-1}q$, and (34) then becomes

$$\ddot{y} + Dy = 0,$$

thus

$$\ddot{y}^i + \lambda_i y^i = 0 \quad \text{for } i = 1, 2, \ldots, d. \tag{35}$$

The general solution of (35) is

$$y^i(t) = \alpha_i \cos(\sqrt{\lambda_i}t) + \beta_i \sin(\sqrt{\lambda_i}t)$$

with arbitrary real constants $\alpha_i, \beta_i (i = 1, \ldots, d)$.

We now come to the simplest problems of continuum mechanics. States of equilibrium and motion can be characterized formally as before, however the state of a system can no longer be determined by finitely many coordinates. Instead of $q^1(t), \ldots, q^d(t)$ we now must determine a (real or vector-valued) function $f(x, t)$ or $f(x)$ describing states of motion or rest, respectively.

First we consider the simplest example of a homogeneous vibrating string. The string is under a constant tension μ and executes small vibrations about a stable state of equilibrium. This state corresponds to the segment $0 \leq x \leq \ell$ of the x-axis and the stretching perpendicular to the x-axis is described by the function $f(x, t)$. The string is fixed at the end points and therefore $f(0, t) = 0 = f(\ell, t)$ for all t.

Now the kinetic energy is

$$T = \frac{\rho}{2} \int_0^\ell f_t^2 dx \quad (\rho \text{ means density of the string}), \tag{36}$$

and the potential energy is

$$U = \mu\{ \int_0^\ell \sqrt{1 + f_x^2} dx - \ell \},$$

thus proportional to the increase in length relative to the state of rest. We shall consider a small stretching from the equilibrium position and therefore ignore terms of higher order and set, as before,

$$U = \frac{\mu}{2} \int_0^\ell f_x^2 dx. \tag{37}$$

By Hamilton's principle, the motion is characterized by

$$I(f) = \int_{t_0}^{t_1} (T - U)dt = \frac{1}{2} \int_{t_0}^{t_1} \int_0^\ell (\rho f_t^2 - \mu f_x^2) dx dt \tag{38}$$

being stationary in the class of all functions with $f(0,t) = f(\ell,t) = 0$ for all t.

The Euler-Lagrange equation is now

$$\rho f_{tt} - \mu f_{xx} = 0. \tag{39}$$

This is the so-called *wave equation*. For simplicity we shall take $\rho = \mu = 1$. The weak form of the Euler-Lagrange equation is then

$$\int_{t_0}^{t_1} \int_0^\ell (f_t \varphi_t - f_x \varphi_x) dx dt = 0 \text{ for all } \varphi \in C_0^\infty((0,\ell) \times (t_0, t_1)) \tag{40}$$

(we have not required any boundary conditions for $t = t_0$ and $t = t_1$ and therefore this holds even for functions φ which do not necessarily vanish at $t = t_0$ and $t = t_1$, but this we do not want to investigate here in detail).

Now let $\gamma \in C^1(\mathbb{R})$. Then the function g defined by

$$g(x,t) := \gamma(x - t)$$

is in $C^1([0,\ell] \times [t_0, t_1])$ and satisfies

$$g_x = -g_t.$$

Therefore, for all $\varphi \in C_0^\infty((0,\ell) \times (t_0, t_1))$ we have

$$\int_{t_0}^{t_1} \int_0^\ell (g_t \varphi_t - g_x \varphi_x) dx dt = \int_{t_0}^{t_1} \int_0^\ell (-g_x \varphi_t + g_t \varphi_x) dx dt$$

$$= \int_{t_0}^{t_1} \int_0^\ell g(\varphi_{tx} - \varphi_{xt}) dx dt = 0.$$

Thus g is a solution of (40) although g is not necessarily twice differentiable and therefore not necessarily a classical solution of the Euler-Lagrange equation

$$f_{tt} - f_{xx} = 0. \tag{41}$$

Hence a weak solution of the Euler-Lagrange equation need not necessarily be a classical solution.

In this example, the integrand

$$H(p) = p_2^2 - p_1^2 \quad (p_1 \text{ stands for } \frac{\partial f}{\partial x}, p_2 \text{ for } \frac{\partial f}{\partial t})$$

is analytic indeed, but has an indefinite Hessian $(H_{p_i p_j})$, namely

$$\begin{pmatrix} -2 & 0 \\ 0 & 2 \end{pmatrix}.$$

Moreover, the fact behind this example is that

$$g(x, t) = \gamma(x - t) + \delta(x + t)$$

is the general solution of the wave equation

$$g_{tt} - g_{xx} = 0.$$

If the string is subjected to an additional external force $k(x, t)$ then the potential energy becomes

$$U = \frac{\mu}{2} \int\limits_0^\ell f_x^2 dx + \int\limits_0^\ell k(x, t) f(x, t) dx,$$

and the equation of motion becomes

$$\rho f_{tt} - \mu f_{xx} + k = 0. \tag{42}$$

Correspondingly, an equilibrium state (assuming that k depends no longer on t) is given by

$$\mu f_{xx}(x) - k(x) = 0. \tag{43}$$

The situation looks similar for a plane membrane – i.e. an elastic surface that at rest covers a portion Ω of the xy-plane and can move vertically. The potential energy is proportional to the difference of the surface area to the surface area at rest. We set the factor of proportionality as well as the subsequent physical constants equal to 1. If $f(x, y, t)$ denotes the vertical stretching of the surface then

$$U = \int\limits_\Omega \sqrt{1 + f_x^2 + f_y^2} dx dy - \text{Vol}(\Omega). \tag{44}$$

We shall again restrict ourselves to small pertubations and therefore substitute U as before by

$$U = \frac{1}{2} \int\limits_\Omega (f_x^2 + f_y^2) dx dy. \tag{45}$$

The kinetic energy is

$$T = \frac{1}{2} \int_\Omega f_t^2 \, dxdy. \tag{46}$$

The equation of motion is then

$$f_{tt} - \Delta f = 0 \quad (\Delta f = f_{xx} + f_{yy}) \tag{47}$$

and its state of rest is characterized by

$$\Delta f = 0. \tag{48}$$

We had already derived this earlier. Under the influence of an external force $k(x)$, its state of rest is correspondingly given by

$$\Delta f(x, y) = k(x, y). \tag{49}$$

Thus, if the membrane is fixed at the boundary, we have to solve the Dirichlet problem

$$\Delta f(x, y) = k(x, y) \quad \text{for } (x, y) \in \Omega$$
$$f(x, y) = 0 \quad \text{for } (x, y) \in \partial\Omega.$$

We shall now derive the Euler-Lagrange equations for the area functional

$$I(f) = \int_\Omega \sqrt{1 + f_x^2 + f_y^2} \, dxdy.$$

Setting $H(p_1, p_2) = \sqrt{1 + p_1^2 + p_2^2}$ we have

$$H_{p_i} = \frac{p_i}{\sqrt{1 + p_1^2 + p_2^2}}$$

and

$$H_{p_i p_j} = \frac{\delta_{ij}}{\sqrt{1 + p_1^2 + p_2^2}} - \frac{p_i p_j}{(1 + p_1^2 + p_2^2)^{\frac{3}{2}}} \quad \left(\delta_{ij} = \begin{cases} 1 & \text{for } i = j \\ 0 & \text{for } i \neq j \end{cases} \right).$$

Thereby, the Euler-Lagrange equations become

$$0 = \sum_{i,j=1}^{2} H_{p_i p_j} f_{x^i x^j} = \frac{1}{(1 + f_x^2 + f_y^2)^{\frac{3}{2}}} \{ (1 + f_y^2) f_{xx} \\ - 2 f_x f_y f_{xy} + (1 + f_x^2) f_{yy} \},$$

so

$$(1 + f_y^2) f_{xx} - 2 f_x f_y f_{xy} + (1 + f_x^2) f_{yy} = 0.$$

This is the so-called *minimal surface equation*. It describes surfaces with stationary area that can be represented as graphs over a domain Ω in the (x, y)-plane.

Finally, we consider quadratic integrals of the form

$$Q(f) = \int_\Omega \Big\{ \sum_{i,j=1}^d a_{ij}(x) f_{x^i} f_{x^j} + \sum_{i=1}^d 2b_i(x) f \cdot f_{x^i} + c(x) f(x)^2 \Big\} dx; \quad (50)$$

again, without loss of generality, let $a_{ij} = a_{ji}$. The Euler-Lagrange equations are now

$$-\sum_{i=1}^d \frac{\partial}{\partial x^i} \Big(\sum_{j=1}^d a_{ij}(x) \frac{\partial f}{\partial x^j} + b_i(x) f \Big) + \sum_{i=1}^d b_i(x) \frac{\partial f}{\partial x^i} + c(x) f = 0. \quad (51)$$

The Euler-Lagrange equations for a quadratic variational problem are therefore linear in f and its derivatives.

We shall now study the behaviour of the Euler-Lagrange equations under transformations of the independent variables.

So let $\xi \mapsto x(\xi)$ be a diffeomorphism of Ω' onto Ω; we set $D_x f = (\frac{\partial f}{\partial x^1}, \dots, \frac{\partial f}{\partial x^d})$, $D_\xi x = (\frac{\partial x^i}{\partial \xi^j})_{i,j=1,\dots,d}$ etc.

$$H(x, f, D_x f) = H(x(\xi), f, D_\xi f \cdot (D_\xi x)^{-1}) =: \Phi(\xi, f, D_\xi f).$$

By the change of variables in integrals we have

$$\int_\Omega H(x, f, D_x f) dx = \int_{\Omega'} \Phi(\xi, f, D_\xi f) |\det(D_\xi x)| d\xi. \quad (52)$$

We now write for the sake of abbreviation

$$[H]_f = -\Big(\sum_{i=1}^d \frac{d}{dx^i} H_{p^i} - H_f \Big). \quad (53)$$

We then have for $\varphi \in C_0^\infty(\Omega)$, on account of the derivation of Euler-Lagrange equations,

$$\int_\Omega [H]_f \varphi dx = \frac{d}{dt} \int_\Omega H(x, f + t\varphi, D_x f + t D_x \varphi) dx_{|t=0}$$

$$= \frac{d}{dt} \int_{\Omega'} \Phi(\xi, f + t\varphi, D_\xi f + t D_\xi \varphi) |\det(D_\xi x)| d\xi_{|t=0}$$

$$= \int_{\Omega'} [\Phi | \det(D_\xi x)|]_f \varphi d\xi$$

$$= \int_\Omega [\Phi | \det(D_\xi x)|]_f \varphi |\det(D_x \xi)| dx.$$

As this holds for all $\varphi \in C_0^\infty(\Omega)$, it follows, as usual, from corollary 19.20 that

$$[H]_f = [\Phi | \det(D_\xi x)|]_f |\det(D_x \xi)|. \quad (54)$$

(Under the assumption $H \in C^2$, we consider

$$I(f) = \int_\Omega H(x, f(x), Df(x))dx$$

as a function

$$I : C^2(\Omega) \to \mathbb{R}$$

and $[H]_f$ is then the gradient of I, as the derivative of I is given by

$$\varphi \mapsto DI(\varphi) = \int_\Omega [H]_f \varphi dx.$$

Thus equation (54) expresses that the behaviour under transformations of this gradient is quite analogous to that of a gradient in the finite dimensional case.)

We shall use this to study the transformation of the Laplace operator; the advantage of (54) lies precisely in this that one does not have to transform derivatives of second order. Now the Laplace equation, as we have already seen at the beginning, is precisely the Euler-Lagrange equation for the Dirichlet integral.

So let $\xi \mapsto x(\xi)$ be again a diffeomorphism of Ω' onto Ω; we set

$$g_{ij} := \sum_{k=1}^d \frac{\partial x^k}{\partial \xi^i} \frac{\partial x^k}{\partial \xi^j}$$

and

$$g^{ij} := \sum_{k=1}^d \frac{\partial \xi^i}{\partial x^k} \frac{\partial \xi^j}{\partial x^k}.$$

Thus

$$\sum_{i=1}^d g_{ik}g^{i\ell} = \delta_{k\ell} \left(= \begin{cases} 1 & \text{for } k = \ell \\ 0 & \text{for } k \neq \ell \end{cases} \right).$$

Furthermore, let

$$g := \det(g_{ij}).$$

Now

$$\sum_{i=1}^d \left(\frac{\partial f}{\partial x^i} \right)^2 = \sum_{i=1}^d \sum_{j,k=1}^d \frac{\partial f}{\partial \xi^j} \frac{\partial \xi^j}{\partial x^i} \frac{\partial f}{\partial \xi^k} \frac{\partial \xi^k}{\partial x^i} = \sum_{j,k=1}^d g^{jk} \frac{\partial f}{\partial \xi^j} \frac{\partial f}{\partial \xi^k}.$$

Formula (54) now gives directly, together with (50) and (51),

$$\Delta f(x) = \frac{1}{\sqrt{g}} \sum_{j=1}^d \frac{\partial}{\partial \xi^j} \left(\sqrt{g} \sum_{k=1}^d g^{jk} \frac{\partial f}{\partial \xi^k} \right). \tag{55}$$

This is the desired transformation formula for the Laplace operator.

For plane polar coordinates

$$x = r \cos \varphi, y = r \sin \varphi$$

one calculates from this

$$\Delta f(x,y) = \frac{1}{r} \left(\frac{\partial}{\partial r} \left(r \frac{\partial f}{\partial r} \right) + \frac{\partial}{\partial \varphi} \left(\frac{1}{r} \frac{\partial f}{\partial \varphi} \right) \right), \tag{56}$$

and for spatial polar coordinates

$$x = r \cos \varphi \sin \theta, y = r \sin \varphi \sin \theta, z = r \cos \theta$$

$$\Delta f(x,y,z) = \frac{1}{r^2 \sin \theta} \left(\frac{\partial}{\partial r} \left(r^2 \sin \theta \frac{\partial f}{\partial r} \right) + \frac{\partial}{\partial \varphi} \left(\frac{1}{\sin \theta} \frac{\partial f}{\partial \varphi} \right) + \frac{\partial}{\partial \theta} \left(\sin \theta \frac{\partial f}{\partial \theta} \right) \right) \tag{57}$$

(cf. §18 for the discussion of polar coordinates).

Exercises for § 22

1)   Let $\Omega \subset \mathbb{R}^d$ be open and bounded. For $f \in H^{2,2}(\Omega)$, put

$$E(f) := \int_{\Omega} |D^2 f(x)|^2 dx.$$

(Here, $D^2 f$ is the matrix of weak second derivatives $D_i D_j f$, $i,j = 1, \ldots, d$, and

$$|D^2 f(x)|^2 = \sum_{i,j=1}^{d} |D_i D_j f(x)|^2.)$$

Discuss the following variational problem: For given $g \in H^{2,2}(\Omega)$, minimize $E(f)$ in the class

$$A_g := \{ f \in H^{2,2}(\Omega) : f - g \in H^{1,2}_0(\Omega),$$
$$D_i f - D_i g \in H^{1,2}_0(\Omega), i = 1, \ldots, d \}.$$

2)   Let $H : \Omega \times \mathbb{R} \times \mathbb{R}^d \to \mathbb{R}$ be nonnegative, measurable w.r.t. the first variable, and convex w.r.t. the second and third variables jointly, i.e. for all $f, g \in \mathbb{R}, p, q \in \mathbb{R}^d, 0 \le t \le 1, x \in \Omega$, we have

$$H(x, tf + (1-t)g, tp + (1-t)q) \le tH(x, f, p) + (1-t)H(x, g, q).$$

For $f \in H^{1,2}(\Omega)$, we put

$$I(f) := \int_{\Omega} H(x, f(x), Df(x)) dx.$$

Show that I is lower semicontinuous w.r.t. weak $H^{1,2}$ convergence.

3)

a)   Let A be a $(d \times d)$ matrix with $\det(A) \neq 0$. Consider the coordinate transformation

$$\xi \mapsto x = A\xi.$$

How does the Laplacian $\Delta = \sum_{i=1}^{d} \frac{\partial^2}{(\partial x^i)^2}$ transform under this coordinate transformation?

b)   Discuss the coordinate transformation $(\xi, \eta) \mapsto (x, y)$ with

$$x = \sin \xi \cosh \eta$$
$$y = \cos \xi \sinh \eta$$

(planar elliptic coordinates) and express the Laplacian in these coordinates.

4)   Determine all rotationally symmetric harmonic functions $f : \mathbb{R}^3 \backslash \{0\} \to \mathbb{R}$.

5)   For $m \in \mathbb{N}$, define the Legendre polynomial as

$$P_m(t) := \frac{1}{2^m m!} \left(\frac{d}{dt}\right)^m (t^2 - 1)^m.$$

Show that

$$f(r, \theta) := r^m P_m(\cos \theta)$$

satisfies $\Delta f = 0$ (in spatial polar coordinates).

6)   Let $a, b \in \mathbb{R}, g_1, g_2 > 0$. For functions $f : [a, b] \to \mathbb{R}$ with $f(a) = g_1, f(b) = g_2$, we consider

$$K(f) := 2\pi \int_a^b f(x) \sqrt{1 + f'(x)^2} dx \to \min.$$

($I(f)$ yields the area of the surface obtained by revolving the graph of f about the x-axis. Thus, we are seeking a surface of revolution with smallest area with two circles given as boundary.) Solve the corresponding Euler-Lagrange equations!

7)   We define a plate to be a thin elastic body with a planar rest position. We wish to study small transversal vibrations of such a body, induced by an exterior force K. Let us first consider the equilibrium position. Let $f(x, y)$ be the vertical displacement. The potential energy of a deformation is

$$U = U_1 + U_2,$$

where

$$U_1 = \int\limits_{\Omega} \left(\left(\frac{1}{2} \Delta f(x,y) \right)^2 + \mu \left(f_{xx} f_{yy} - f_{xy}^2 \right) \right) dx dy$$

(here, $\Omega \subset \mathbb{R}^2$ is the rest position, $\mu = $ const.),

$$U_2 = \int\limits_{\Omega} K(x,y) f(x,y) dx dy.$$

Derive the Euler-Lagrange equations

$$\Delta(\Delta f) + K = 0.$$

For the motion, $f(x,y,t)$ is the vertical displacement, and the kinetic energy is

$$T = \frac{1}{2} \int\limits_{\Omega} f_t^2 dx dy.$$

Derive the differential equation that describes the motion of the plate.

23. Regularity of Weak Solutions

It is shown that under appropriate ellipticity assumptions, weak solutions of partial differential equations (PDEs) are smooth. This applies in particular to the Laplace equation for harmonic functions, thereby justifying Dirichlet's principle introduced in the previous paragraph.

In the last sections we had constructed weak solutions of the Laplace and Poisson equations, that is, solutions $f \in H^{1,2}(\Omega)$, $f - g \in H_0^{1,2}(\Omega)$ (g a prescribed boundary value) of the equation

$$\int_\Omega \sum_{i=1}^d D_i f(x) D_i \varphi(x) dx + \int_\Omega k(x) \varphi(x) dx = 0 \tag{1}$$

for all $\varphi \in C_0^\infty(\Omega)$ or even $\varphi \in H_0^{1,2}(\Omega)$, with given $k \in L^2(\Omega)$.

In this section we shall show that a solution of (1) is regular. We shall show, for example that if k is C^∞ on Ω then so is f. In particular, f is then a classical solution of

$$\Delta f(x) = k(x) \quad \text{for } x \in \Omega.$$

Similarly, we shall show that if $\partial\Omega$ is of class C^∞, in a sense yet to be defined, and g is also of class C^∞, then f is even in $C^\infty(\overline{\Omega})$ and for every $x \in \partial\Omega$,

$$f(x) = g(x)$$

holds.

The idea of the proof consists in showing that f has square integrable derivatives of arbitrarily high order and therefore $f \in W^{k,2}(\Omega)$ for every $k \in \mathbb{N}$. With the Sobolev embedding theorem one then concludes easily that $f \in C^\infty(\Omega)$.

In order to bring forth the idea of the proof clearly, let us assume that we could set $\varphi = D_j D_j f$ in (1). Naturally, this does not work as, first of all, this φ does not have zero boundary values and then we do not know whether φ and its first derivatives are square integrable. But we assume, as said, that we could nevertheless employ this φ and obtain

$$\int_\Omega \sum_{i=1}^d D_i f(x) D_i D_j D_j f(x) dx + \int_\Omega k(x) D_j D_j f(x) dx = 0,$$

and from this we obtain, using integration by parts (where we have again made the false assumption that this is possible without further ado) and summing over j,

$$\int_\Omega \sum_{i,j=1}^d D_j D_i f(x) D_j D_i f(x) dx + \int_\Omega k(x) \cdot \sum_{j=1}^d D_j D_j f(x) dx = 0,$$

and then using Hölder's inequality

$$\|D^2 f\|_{L^2}^2 \le \|k\|_{L^2} \|D^2 f\|_{L^2};$$

so

$$\|D^2 f\|_{L^2} \le \|k\|_{L^2}. \tag{2}$$

We thus obtain an estimate for the second derivatives. In the sequel, we shall first substitute the second derivatives of f by difference quotients and then multiply these difference quotients by a suitable $\eta(x) \in C_0^\infty(\Omega)$, and obtain in this way a test function φ, which we may substitute. We shall then obtain an estimate analogous to (2) for the second difference quotient of f, and this estimate would be independent of the difference parameter h. We shall then see that in the limit $h \to 0$ we obtain first the existence and then an estimate for the second weak derivatives of f.

This procedure can then be iterated, provided k fulfils appropriate assumptions, to obtain estimates for higher derivatives.

Definition 23.1 For $f : \Omega \to \mathbb{R}$ we define the difference quotients by

$$\Delta_i^h f(x) := \frac{f(x + he^i) - f(x)}{h} \quad (h \ne 0),$$

where e^i is the ith unit vector of $\mathbb{R}^d$ ($i \in \{1, \ldots, d\}$).

We shall need the following formula for a kind of integration by parts for difference operators.

Lemma 23.2 *Let* $f, \varphi \in L^2(\Omega), \text{supp}\varphi \subset\subset \Omega, |h| < \text{dist}(\text{supp}\varphi, \partial\Omega)$; *we have*

$$\int_\Omega \varphi(x) \Delta_i^h f(x) dx = - \int_\Omega f(x) \Delta_i^{-h} \varphi(x) dx. \tag{3}$$

Proof. For the proof we calculate

$$\int_\Omega \varphi(x)\frac{f(x+he^i)-f(x)}{h}dx$$

$$= \frac{1}{h}\int_\Omega \varphi(x)f(x+he^i)dx - \frac{1}{h}\int_\Omega \varphi(x)f(x)dx$$

$$= \frac{1}{h}\int_\Omega \varphi(y-he^i)f(y)dy - \frac{1}{h}\int_\Omega \varphi(y)f(y)dy$$

(where we have set $y = x + he^i$ in the first integral and $y = x$ in the second (note the assumption on h))

$$= \int_\Omega f(y)\frac{\varphi(y-he^i)-\varphi(y)}{h}dy$$

$$= -\int_\Omega f(y)\frac{\varphi(y+(-h)e^i)-\varphi(y)}{-h}dy,$$

and this proves the required formula.   □

Lemma 23.3 Let $f \in W^{1,2}(\Omega), \Omega' \subset\subset \Omega, |h| < $ dist $(\Omega', \partial\Omega)$. Then $\Delta_i^h f \in L^2(\Omega')$ and

$$\|\Delta_i^h f\|_{L^2(\Omega')} \le \|D_i f\|_{L^2(\Omega)} \quad (i = 1, \ldots, d).$$

Proof. By an approximation argument we can again restrict ourselves to the case $f \in C^1(\Omega) \cap W^{1,2}(\Omega)$. Then

$$\Delta_i^h f(x) = \frac{f(x+he^i)-f(x)}{h}$$

$$= \frac{1}{h}\int_0^h D_i f(x^1, \ldots, x^{i-1}, x^i + \xi, x^{i+1}, \ldots, x^d)d\xi$$

and by the Hölder inequality

$$|\Delta_i^h f(x)|^2 \le \frac{1}{h}\int_0^h |D_i f(x^1, \ldots, x^i + \xi, \ldots, x^d)|^2 d\xi,$$

and further

$$\int_{\Omega'} |\Delta_i^h f(x)|^2 dx \le \frac{1}{h}\int_0^h \int_\Omega |D_i f|^2 dx d\xi$$

$$= \int_\Omega |D_i f|^2 dx.$$

□

Conversely, we have

Lemma 23.4 *Let $f \in L^2(\Omega)$ and assume that there exists a $K < \infty$ with $\Delta_i^h f \in L^2(\Omega')$ and*

$$\|\Delta_i^h f\|_{L^2(\Omega')} \le K$$

for all $h > 0$ and all $\Omega' \subset\subset \Omega$ with $|h| <$ dist $(\Omega', \partial\Omega)$. Then there exists the weak derivative $D_i f$ and

$$\|D_i f\|_{L^2(\Omega)} \le K.$$

Proof. By theorem 21.8, the bounded set $\{\Delta_i^h f\}$ in $L^2(\Omega')$ contains a weakly convergent sequence. As this holds for every Ω', there exists a sequence $h_n \to 0$ and a $v \in L^2(\Omega)$ with $\|v\|_2 \le K$ (see, e.g., corollary 21.9) and

$$\int_\Omega \varphi \Delta_i^{h_n} f \to \int_\Omega \varphi v \quad \text{for all } \varphi \in C_0^1(\Omega).$$

If $h_n <$ dist $(\text{supp}\varphi, \partial\Omega)$ then by (3) (lemma 23.2)

$$\int_\Omega \varphi \Delta_i^{h_n} f = -\int_\Omega f \Delta_i^{-h_n} \varphi \to -\int_\Omega f D_i \varphi \quad \text{as } n \to \infty.$$

It follows that

$$\int_\Omega \varphi v = -\int_\Omega f D_i \varphi,$$

thus $v = D_i f$. □

Theorem 23.5 *Let $f \in W^{1,2}(\Omega)$ be a weak solution of $\Delta f = k$, with $k \in L^2(\Omega)$. Then for every $\Omega' \subset\subset \Omega$, $f \in W^{2,2}(\Omega')$, and we have*

$$\|f\|_{W^{2,2}(\Omega')} \le \text{const.} \, (\|f\|_{L^2(\Omega)} + \|k\|_{L^2(\Omega)}), \tag{4}$$

where the constant depends only on d and on $\delta :=$ dist $(\Omega', \partial\Omega)$. Furthermore, $\Delta f = k$ almost everywhere in Ω.

Proof. Let $\Omega' \subset\subset \Omega'' \subset\subset \Omega$, dist $(\Omega'', \partial\Omega) \ge \delta/4$, dist $(\Omega', \partial\Omega'') \ge \delta/4$. Now

$$\int_\Omega Df \cdot Dw = -\int_\Omega k \cdot w \quad \text{for all } w \in H_0^{1,2}(\Omega). \tag{5}$$

as f is a weak solution.

In the following we consider a w with

supp $w \subset\subset \Omega''$ (i.e. $w \in H_0^{1,2}(\Omega''')$ for some $\Omega''' \subset\subset \Omega''$

and always choose $h > 0$ with

$$|2h| < \text{dist}(\text{supp}w, \partial\Omega'').$$

We can then also substitute $\Delta_\ell^{-h} w (\ell \in \{1, \ldots d\})$ in (5) (as $w \in H^{1,2} \Rightarrow \Delta_\ell^{-h} w \in H^{1,2}$).

It follows that

$$\int_{\Omega''} D\Delta_\ell^h f Dw = \int_{\Omega''} \Delta_\ell^h (Df) \cdot Dw \tag{6}$$

$$= -\int_{\Omega''} Df \Delta_\ell^{-h} Dw = -\int_{\Omega''} Df \cdot D(\Delta_\ell^{-h} w)$$

$$= \int_{\Omega''} k \Delta_\ell^{-h} w$$

$$\leq \|k\|_{L^2(\Omega)} \cdot \|Dw\|_{L^2(\Omega'')}$$

by lemma 23.3 and choice of h. Now let $\eta \in C_0^1(\Omega''), 0 \leq \eta \leq 1, \eta(x) = 1$ for $x \in \Omega', |D\eta| \leq \frac{8}{\delta}$.

Now we set

$$w := \eta^2 \Delta_\ell^h f. \tag{7}$$

By (6) we get

$$\int_{\Omega''} |\eta D\Delta_\ell^h f|^2 = \int_{\Omega''} D\Delta_\ell^h f \cdot Dw - 2 \int_{\Omega''} \eta \Delta_\ell^h f D\Delta_\ell^h f \cdot D\eta$$

$$\leq 2\|k\|_{L^2(\Omega)} (\|\eta D\Delta_\ell^h f\|_{L^2(\Omega'')} + \|\Delta_\ell^h f D\eta\|_{L^2(\Omega'')}) \tag{8}$$

$$+ 2\|\eta D\Delta_\ell^h f\|_{L^2(\Omega'')} \|\Delta_\ell^h f D\eta\|_{L^2(\Omega'')}.$$

In the following estimates, we apply the Schwarz inequality in the form

$$ab \leq \frac{1}{2\varepsilon} a^2 + \frac{\varepsilon}{2} b^2$$

where ε is chosen suitably for our purposes.

Thus, applying lemma 23.3 (note the choice of h) we obtain

$$\|\eta D\Delta_\ell^n f\|_{L^2(\Omega'')}^2 \leq 4\|k\|_{L^2(\Omega)}^2 + \frac{1}{4} \|\eta D\Delta_\ell^h f\|_{L^2(\Omega'')}^2$$

$$+ \|k\|_{L^2(\Omega)}^2 + \sup |D\eta|^2 \|D_\ell f\|_{L^2(\Omega'')}^2 \tag{9}$$

$$+ \frac{1}{4} \|\eta D\Delta_\ell^h f\|_{L^2(\Omega'')}^2 + 4 \sup |D\eta|^2 \|D_\ell f\|_{L^2(\Omega'')}^2.$$

We can get Ω'' here as η has compact support in Ω'', by choosing $h < \text{dist}(\partial\Omega'', \text{supp}\eta)$ and observing that the corresponding norm in (8) must be taken only over suppη.

The important trick here in applying the Schwarz inequality consists in multiplying the expression $\|\eta D \Delta_\ell^h f\|_{L^2(\Omega'')}^2$ on the right side by a smaller factor than that which appears on the left side and therefore the contribution of the right side can be absorbed in the left side. We therefore obtain, as $\eta \equiv 1$ on Ω' and $(a^2 + b^2)^{\frac{1}{2}} \leq a + b$,

$$\|D \Delta_\ell^h f\|_{L^2(\Omega')} \leq \|\eta D \Delta_\ell^h f\|_{L^2(\Omega'')}$$
$$\leq \text{ const } (\|k\|_{L^2(\Omega)} + \sup |D\eta| \cdot \|D_\ell f\|_{L^2(\Omega'')}). \tag{10}$$

Letting $h \to 0$, it follows from lemma 23.4 that

$$\|D^2 f\|_{L^2(\Omega')} \leq \text{ const } (\|k\|_{L^2(\Omega)} + \frac{1}{\delta}\|Df\|_{L^2(\Omega'')}). \tag{11}$$

We now claim that

$$\|Df\|_{L^2(\Omega'')} \leq \text{ const } (\frac{1}{\delta}\|f\|_{L^2(\Omega)} + \delta\|k\|_{L^2(\Omega)}). \tag{12}$$

For this, let $\zeta \in C_0^1(\Omega)$ with $\zeta(x) = 1$ for $x \in \Omega''$, $|D\zeta| \leq \frac{8}{\delta}$. We set $w = \zeta^2 f$ in (5) and obtain with the help of the Schwarz inequality

$$\int_\Omega \zeta^2 |Df|^2 = -2 \int_\Omega \zeta f Df \cdot D\zeta - \int_\Omega k \cdot \zeta^2 f$$
$$\leq \frac{1}{2} \int_\Omega \zeta^2 |Df|^2 + 2 \int_\Omega f^2 |D\zeta|^2 + \delta^2 \int_\Omega k^2 + \frac{1}{\delta^2} \int f^2 \tag{13}$$

and therefore

$$\int_{\Omega''} |Df|^2 \leq \int_\Omega \zeta^2 |Df|^2 \leq \text{ const } (\frac{1}{\delta^2}\|f\|_{L^2(\Omega)}^2 + \delta^2\|k\|_{L^2(\Omega)}^2) \tag{14}$$

and this is just (12). Now (4) follows from (11) and (12).   □

If now k is even in $W^{1,2}(\Omega)$ then one can substitute $D_i w$ for w in (5) and obtain

$$\int_\Omega D(D_i f) Dw = -\int_\Omega D_i k \cdot w.$$

Theorem 23.5 thereby gives $D_i f \in W^{2,2}(\Omega')$, thus $f \in W^{3,2}(\Omega')$. Iteratively, one obtains

Theorem 23.6 Let $f \in W^{1,2}(\Omega)$ be a weak solution of $\Delta f = k$, $k \in W^{m,2}(\Omega)$. Then for every $\Omega' \subset\subset \Omega$ the function $f \in W^{m+2,2}(\Omega')$, and

$$\|f\|_{W^{m+2,2}(\Omega')} \leq \text{ const } (\|f\|_{L^2(\Omega)} + \|k\|_{W^{m,2}(\Omega)}),$$

where the constant depends on d, m and dist $(\Omega', \partial\Omega)$.

Corollary 23.7 If $f \in W^{1,2}(\Omega)$ is a weak solution of $\Delta f = k$ with $k \in C^\infty(\Omega)$, then $f \in C^\infty(\Omega)$.

Proof. This follows from theorem 23.6 and the Sobolev embedding theorem (in particular corollary 20.19), as for every $\eta \in C_0^\infty(\Omega)$ we have $\eta f \in H_0^{m,2}(\Omega)$ for all m, and for any $x_0 \in \Omega$ there is an $r > 0$ with $B(x_0, r) \subset \Omega$ and an $\eta \in C_0^\infty(\Omega)$ with $\eta(x) = 1$ for $x \in B(x_0, r)$, and then for every $x \in B(x_0, r)$ $\eta(x)f(x) = f(x)$. $\qquad\square$

Corollary 23.8 Let $f \in W^{1,2}(\Omega)$ be an eigenfunction of the Laplace operator, thus
$\Delta f + \lambda f = 0$ in the weak sense for some $\lambda \in \mathbb{R}$. Then $f \in C^\infty(\Omega)$.

Proof. We write the equation in the form

$$\Delta f = -\lambda f.$$

By assumption the right side is in $W^{1,2}$ and so $f \in W^{3,2}$ by theorem 23.5. But then, the right side is in $W^{3,2}$ and thereby $f \in W^{5,2}$ by theorem 23.6. Iteratively, it follows that $f \in W^{m,2}$ for all $m \in \mathbb{N}$ and then, as in corollary 23.7, $f \in C^\infty(\Omega)$. $\qquad\square$

For later purposes we must consider more general operators then the Laplace operator, namely linear elliptic operators. We consider thus

$$Lf(x) := \sum_{i=1}^{d} \frac{\partial}{\partial x^i}\left(\sum_{j=1}^{d} a^{ij}(x)\frac{\partial}{\partial x^j}f(x)\right)$$

The functions $a^{ij}(x)$ are assumed to be measurable on Ω. We assume that the operator L is uniformly elliptic in the following sense:
There exist constants $0 < \lambda \leq \mu$ with

$$\lambda\|\xi\|^2 \leq \sum_{i,j=1}^{d} a^{ij}(x)\xi_i\xi_j \leq \mu\|\xi\|^2 \tag{15}$$

for all $x \in \Omega$ and all $\xi = (\xi_1, \ldots, \xi_d) \in \mathbb{R}^d$.

Obviously the Laplace operator satisfies (15) with $\lambda = \mu = 1$. $f \in H^{1,2}(\Omega)$ is called a weak solution of

$$Lf = k \quad (k \in L^2(\Omega) \text{ prescribed}),$$

if for all $v \in H_0^{1,2}(\Omega)$

$$\int_{\Omega} \sum_{i,j=1}^{d} a^{ij}(x) D_j f(x) D_i v(x) dx$$

$$= - \int_{\Omega} k(x) v(x) dx. \tag{16}$$

Theorem 23.5 generalizes to

Theorem 23.9 *Let $f \in W^{1,2}(\Omega)$ be a weak solution of $Lf = k$ in Ω, where L is uniformly elliptic and $k \in L^2(\Omega)$. Furthermore, let the functions $a^{ij}(x)$ be in $C^1(\Omega)$ and assume that*

$$\|a^{ij}\|_{C^1(\Omega)} \leq K \quad \text{for all } i, j, \tag{17}$$

K being a constant.
     Then $f \in W^{2,2}(\Omega')$ for every $\Omega' \subset\subset \Omega$ and

$$\|f\|_{W^{2,2}(\Omega')} \leq c_1(\|f\|_{L^2(\Omega)} + \|k\|_{L^2(\Omega)}), \tag{18}$$

where the constant c_1 again depends on d, $\delta = \text{dist}(\Omega', \partial\Omega)$, λ and K.

(Notice that the requirement $\|a^{ij}\|_{C^1(\Omega)} \leq K$ implies the second inequality in (15) with $\mu = Kd$.)

Proof. We shall modify the proof of theorem 23.5 suitably. For this, we also need a rule for the difference quotients of a product. We have

$$(\Delta_\ell^h(\rho\sigma))(x) = \frac{1}{h}\{\rho(x + he^\ell)\sigma(x + he^\ell) - \rho(x)\sigma(x)\} \tag{19}$$

$$= \{\rho(x + he^\ell)\Delta_\ell^h\sigma(x) + (\Delta_\ell^h\rho(x))\sigma(x)\}.$$

In particular

$$\Delta_\ell^h(\sum_{j=1}^{d} a^{ij}(x) D_j f(x)) = \sum_{j=1}^{d}(a^{ij}(x + he^\ell)\Delta_\ell^h D_j f(x) \tag{20}$$

$$+ (\Delta_\ell^h a^{ij}(x)) D_j f(x)).$$

Now we proceed exactly as before and substitute $\Delta_\ell^{-h}v$ for v in (16), where v satisfies the same conditions as in the proof of theorem 23.5.
     We obtain

$$\int_{\Omega''} \sum_{i,j=1}^{d} \Delta_\ell^h(a^{ij}(x) D_j f(x)) D_i v(x) dx$$

$$= - \int k(x)\Delta_\ell^{-h}v(x) dx, \tag{21}$$

and now when we take (20) into account and use lemma 23.3 we obtain, with

$$\psi^i(x) := \sum_{j=1}^{d}(\Delta_\ell^h a^{ij}(x))D_j f(x)$$

the inequality

$$\int\limits_{\Omega''} \sum_{i,j=1}^{d} a^{ij}(x+he^\ell)D_j\Delta_\ell^h f(x) \cdot D_i v(x)dx \qquad (22)$$

$$\leq (\sum_{i=1}^{d} \|\psi^i\|_{L^2(\Omega'')} + \|k\|_{L^2(\Omega'')})\|Dv\|_{L^2(\Omega'')}$$

$$\leq c_2(\|f\|_{W^{1,2}(\Omega'')} + \|k\|_{L^2(\Omega)}) \cdot \|Dv\|_{L^2(\Omega'')}$$

where c_2 depends on d and K, thus the analogue of (6).

For the rest of the proof, one needs only to observe further that

$$\lambda \int\limits_{\Omega} |\eta D\Delta_\ell^h f(x)|^2 dx \qquad (23)$$

$$\leq \int\limits_{\Omega} \eta^2 \sum_{i,j=1}^{d} a^{ij}(x+he^\ell)\Delta_\ell^h D_i f(x)\Delta_\ell^h D_j f(x)dx$$

and proceed as in the proof of theorem 23.5. One obtains only the extra factor $\frac{1}{\lambda}$ in the estimates.

As an example, we want to derive the analogue of (12):
We substitute as before $\zeta^2 f$, this time in (16), and obtain

$$\lambda^2 \int\limits_{\Omega} \zeta^2(x)|Df(x)|^2 dx \leq \int\limits_{\Omega} \sum_{i,j=1}^{d} a^{ij}(x)D_j f(x)D_i f(x)\zeta^2(x)dx$$

$$= -2 \int\limits_{\Omega} \sum_{i,j=1}^{d} a^{ij}(x)D_j f(x) \cdot \zeta(x)D_i\zeta(x)f(x)dx \qquad (24)$$

$$- \int\limits_{\Omega} k(x)\zeta^2(x)f(x)dx$$

$$\leq \frac{\lambda^2}{2} \int\limits_{\Omega} \zeta^2(x)|Df(x)|^2 dx + \frac{2}{\lambda^2}K^2 \int\limits_{\Omega} f^2(x)|D\zeta(x)|^2 dx$$

$$+ \delta^2 \int\limits_{\Omega} \zeta^2(x)k(x)^2 dx + \frac{1}{\delta^2} \int\limits_{\Omega} f^2(x)dx$$

,

as usual by applying the Schwarz inequality

$$2ab \leq \varepsilon a^2 + \frac{1}{\varepsilon}b^2.$$

Therefore we can again absorb the term $\frac{\lambda^2}{2}\int \zeta^2(x)|Df(x)|^2 dx$ in the left hand side and obtain

$$\int_{\Omega''} |Df(x)|^2 dx \le c_3\left(\frac{1}{\delta^2}\|f\|^2_{L^2(\Omega)} + \delta^2\|k\|^2_{L^2(\Omega)}\right), \tag{25}$$

which is the analogue of (12), whereby c_3 depends on λ and K. $\square$

Iteratively one proves, as in theorem 23.6 and corollary 23.7, the following theorem

Theorem 23.10 *Let $f \in W^{1,2}(\Omega)$ be a weak solution of $Lf = k$, with $k \in W^{m,2}(\Omega)$. Furthermore, let the coefficients a^{ij} of L be in $C^{m+1}(\Omega)$ with*

$$\|a^{ij}\|_{C^{m+1}(\Omega)} \le K_m \quad \text{for all } i,j.$$

Then for every $\Omega' \subset\subset \Omega$, $f \in W^{m+2,2}(\Omega')$ with

$$\|f\|_{W^{m+2,2}(\Omega')} \le c(\|f\|_{L^2(\Omega)} + \|k\|_{W^{m,2}(\Omega)}),$$

where c depends on d, λ, m, K_m and dist $(\Omega', \partial\Omega)$. If k and the coefficients a^{ij} are in $C^\infty(\Omega)$ then f is also in $C^\infty(\Omega)$. $\square$

From this, the regularity of solutions of an arbitrary elliptic operator in divergence form can easily be recovered:

Corollary 23.11 *Let $f \in W^{1,2}(\Omega)$ be a weak solution of*

$$Mf(x) := \sum_{i=1}^{d}\left(\frac{\partial}{\partial x^j}\left(\sum_{j=1}^{d} a^{ij}(x)\frac{\partial}{\partial x^j}f(x) + b^i(x)f(x)\right)\right)$$

$$+ \sum_{i=1}^{d} c^i(x)\frac{\partial}{\partial x^i}f(x) + d(x)f(x) = k(x).$$

If $k(x), d(x)$ and all coefficients $a^{ij}(x)$ are of class C^∞ then $f \in C^\infty(\Omega)$.

Here, weak solution means that for all $v \in H^{1,2}_0(\Omega)$

$$\int_\Omega \{\sum_{i=1}^{d}\sum_{j=1}^{d}(a^{ij}(x)D_j f(x) + b^i(x)f(x))D_i v(x)$$

$$- (\sum_{i=1}^{d} c^i(x)D_i f(x) + d(x)f(x))v(x)\}dx \tag{26}$$

$$= -\int_\Omega k(x)v(x)dx.$$

Proof. We write (26) as

$$\int_\Omega \sum_{i,j=1}^d a^{ij}(x)D_j f(x)D_i v(x)dx = \int_\Omega \psi(x)v(x)dx \qquad (27)$$

with

$$\psi = \sum_{i=1}^d (b^i + c^i)D_i f + \left(\sum_{i=1}^d D_i b^i + d\right)f - k.$$

By assumption, $\psi \in L^2(\Omega)$ for every $\Omega_0 \subset\subset \Omega$, so f is a weak solution of

$$Lf = \psi \quad \text{in } \Omega_0$$

with right side in L^2. Theorem 23.9 gives $f \in W^{2,2}$ locally. But then (still locally) $\psi \in W^{1,2}$, therefore $f \in W^{3,2}$ by theorem 23.10 and so $\psi \in W^{2,2}$; again by theorem 23.10 $f \in W^{4,2}$ and so on. □

In order to show that f is also regular on $\partial\Omega$ we must first formulate suitable assumptions on $\partial\Omega$.

Definition 23.12 A bounded open set $\Omega \subset \mathbb{R}^d$ is said to be of class $C^k(k = 0,1,2,\ldots\infty)$ if for every $x_0 \in \partial\Omega$ there is an $r > 0$ and a mapping $\Phi : U(x_0, r) \to \mathbb{R}^d$ with the following properties

(i) Φ and Φ^{-1} are of class C^k

(ii) $\Phi(\Omega \cap U(x_0, r)) \subset \{x = (x^1, \ldots, x^d) \in \mathbb{R}^d : x^d > 0\}$

(iii) $\Phi(\partial\Omega \cap U(x_0, r)) \subset \{x = (x^1, \ldots, x^d) \in \mathbb{R}^d : x^d = 0\}$

In particular, $\partial\Omega$ is thus of dimension $d-1$, in the sense that locally there exists a C^k-diffeomorphism to an open subset of $\mathbb{R}^{d-1}$. A set of class C^k can therefore, in particular, have no isolated boundary points if $d > 1$.

Definition 23.13 Let $\Omega \subset \mathbb{R}^d$ be of class C^k. A mapping $g : \overline{\Omega} \to \mathbb{R}$ is said to be of class $C^\ell(\overline{\Omega})$, for $\ell \le k$, if $g \in C^\ell(\Omega)$ and for every $x_0 \in \partial\Omega$ $g \circ \Phi^{-1} : \{x = (x^1, \ldots, x^d) = x^d \ge 0\} \to \mathbb{R}$ is of class C^ℓ, where Φ is chosen as in the definition 23.12.

We can now prove the following global statement ("global" here means that the statement refers to $\overline{\Omega}$ and not just to relative compact subsets of Ω; such a statement would only be "local").

Theorem 23.14 *Let $g \in W^{2,2}(\Omega)$, $k \in L^2(\Omega)$ and let the open, bounded set Ω be of class C^2. Let f be a weak solution of $\Delta f = k$, with $f - g \in H_0^{1,2}(\Omega)$. Then $f \in W^{2,2}(\Omega)$ and*

$$\|f\|_{W^{2,2}(\Omega)} \le c(\|k\|_{L^2(\Omega)} + \|g\|_{W^{2,2}(\Omega)}), \qquad (28)$$

where c depends on d and Ω.

Proof. First, we reduce the proof to the case $f \in H_0^{1,2}(\Omega)$. Otherwise, we consider $\tilde{f} = f - g \in H_0^{1,2}(\Omega)$. Then we have

$$\Delta \tilde{f} = \Delta f - \Delta g = k - \Delta g = \tilde{k} \in L^2(\Omega)$$

and $\|\tilde{k}\|_{L^2} \leq \|k\|_{L^2} + \|g\|_{W^{2,2}(\Omega)}$ and $\tilde{f}$ therefore satisfies the same assumptions as f, and obviously the $W^{2,2}$-norm of f can be approximated by those of $\tilde{f}$ and g.

Thus we assume that $f \in H_0^{1,2}(\Omega)$.

Now let $x_0 \in \partial\Omega$. As Ω, by assumption, is of class C^2, there exists a C^2-diffeomorphism Φ with $\Phi(x_0) = 0$,

$$\Phi : U(x_0, r) \rightarrow \Phi(U(x_0, r)) \subset \mathbb{R}^d \quad (r > 0)$$

with the properties described in definition 23.12. We now transform the Laplace operator Δ by the coordinate transformation Φ^{-1}, as described in §22. The mapping $\overline{f} := f \circ \Phi^{-1}$ then fulfils, by the considerations at the end of §22, a weak equation of the form

$$\sum_{i,j=1}^{d} \frac{1}{\sqrt{g}} \frac{\partial}{\partial x^i} (\sqrt{g} g^{ij} \frac{\partial}{\partial x^j} \overline{f}) = k,$$

so with $a^{ij} := \sqrt{g} g^{ij}, \overline{k} := \sqrt{g} k$ (in particular $a^{ij} \in C^1$) we obtain

$$\sum_{i,j=1}^{d} \frac{\partial}{\partial x^i} (a^{ij} \frac{\partial}{\partial x^j} \overline{f}) = \overline{k}. \tag{29}$$

As $\Phi(U(x_0, r))$ is an open set containing 0, equation (29) holds in $U^+(0, R) := \{(x^1, \ldots, x^d) \in U(0, R) : x^d > 0\}$ for some $R > 0$. Moreover, there exist constants $\lambda > 0$, $K < \infty$, such that (15) and (17) hold in $U^+(0, R)$. The constants λ and K here depend of course on the special geometry of $\partial\Omega$ (the matrix (g^{ij}) is positive definite as Φ is a diffeomorphism; for the same reason, $g = \det(g_{ij})$ is positive).

We now make the following observation:

If $\eta \in C_0^1(U(0, R))$, then $\eta \overline{f} \in H_0^{1,2}(U^+(0, R))$.

Namely, $\overline{f}$ vanishes for $x^d = 0$, so we do not need to demand this of η. If now $1 \leq i \leq d - 1$ and $|h| <$ dist $(\text{supp}\,\eta, \partial U(0, R))$, then also

$$\eta^2 \Delta_i^h \overline{f} \in H_0^{1,2}(U^+(0, R)).$$

But this means that we can argue exactly as in the proof of theorem 23.9 to obtain a corresponding estimate for $D_{ij}\overline{f} \in L^2(U(0, \frac{R}{2}))$, as long as i and j are not both equal to d. Finally, in order to control $D_{dd}\overline{f}$ we rewrite (29) as

$$\frac{\partial^2}{(\partial x^d)^2}\overline{f} = \frac{1}{a^{dd}}\{\overline{k} - \sum_{\substack{i,j=1 \\ (i,j)\neq(d,d)}}^{d} \frac{\partial}{\partial x^i}(a^{ij}\frac{\partial}{\partial x^j}\overline{f}) - (\frac{\partial}{\partial x^d}a^{dd})\frac{\partial\overline{f}}{\partial x^d}\}. \tag{30}$$

As we have just controlled the L^2-norm of the right side of (30) the L^2-norm of $D_{dd}\overline{f}$ has also thereby been controlled.

It follows that

$$\|\overline{f}\|_{W^{2,2}(U^+(0,\frac{R}{2}))} \leq c_4(\|\overline{k}\|_{L^2(U^+(0,R))} + \|\overline{f}\|_{W^{1,2}(U^+(0,R))}).$$

However, now

$$\|f\|_{W^{2,2}(\Phi^{-1}(U^+(0,\frac{R}{2})))} \leq c_5\|\overline{f}\|_{W^{2,2}(U^+(0,\frac{R}{2}))},$$

where c_5 depends on Φ and its first and second derivatives. We have thus found for every $x_0 \in \partial\Omega$ a neighborhood on which we can control the $W^{2,2}$-norm of f by the L^2-norm of k and the $W^{1,2}$-norm of f. Since Ω is bounded, $\partial\Omega$ is compact and can therefore be covered by finitely many such neighborhoods.

It follows, together with the interior estimate from theorem 23.5, that

$$\|f\|_{W^{2,2}(\Omega)} \leq c_6(\|k\|_{L^2(\Omega)} + \|f\|_{W^{1,2}(\Omega)}). \tag{31}$$

But now, by the Poincaré inequality (corollary 20.16), we have

$$\|f\|_{L^2(\Omega)} \leq c_7\|Df\|_{L^2(\Omega)} \tag{32}$$

as $f \in H_0^{1,2}(\Omega)$, where c_7 depends on the size of Ω.

Finally, on account of $f \in H_0^{1,2}(\Omega)$,

$$\|Df\|_{L^2(\Omega)}^2 = \int_\Omega Df \cdot Df = \int_\Omega k \cdot f$$

$$\leq \frac{1}{\varepsilon}\int_\Omega k^2 + \varepsilon\int_\Omega f^2, \tag{33}$$

for every $\varepsilon > 0$. Choosing ε sufficiently small, it follows from (32) and (33) that

$$\|f\|_{W^{1,2}(\Omega)} \leq c_8\|k\|_{L^2(\Omega)}, \tag{34}$$

and therefore from (31)

$$\|f\|_{W^{2,2}(\Omega)} \leq c_9(\|k\|_{L^2(\Omega)}). \tag{35}$$

This is (28) for the case $f \in H_0^{1,2}(\Omega)$, to which we had reduced the proof at the beginning.   □

Theorem 23.15 *Let Ω be an open, bounded subset of $\mathbb{R}^d$ of class C^∞. Let $k, g \in C^\infty(\overline{\Omega})$. Then there exists a unique solution of the Dirichlet problem*

$$\Delta f(x) = k(x) \ \text{for} \ x \in \Omega$$
$$f(x) = g(x) \ \text{for} \ x \in \partial\Omega,$$

and this solution is of class $C^\infty(\overline{\Omega})$.

Proof. We had already shown in the last paragraph that there is a unique weak solution of the problem

$$\Delta f = k \ \text{on} \ \Omega,$$
$$f - g \in H_0^{1,2}(\Omega).$$

So we must only show the regularity. This can be done as before by iterating the argument of theorem 23.14 together with the Sobolev embedding theorem. In order that we can iterate the proof of theorem 23.14, we must first prove, in case we make the same reduction as at the beginning of theorem 23.14, that f is even in $H_0^{2,2}(\Omega)$. For this it again suffices to show, using the notations of theorem 23.14, that for $\eta \in C_0^1(U(0,R))$ and $1 \le i \le d-1$,

$$\eta D_i \overline{f} \in H_0^{1,2}(U^+(0,R)).$$

Now by lemma 23.3

$$\eta \Delta_i^h \overline{f} \in H_0^{1,2}(U^+(0,R)) \ \text{for sufficiently small} \ |h|,$$

and

$$\|\eta \Delta_i^h \overline{f}\|_{W^{1,2}(U^+(0,R))} \le \|\eta\|_{C^1(U^+(0,R))} \cdot \|\overline{f}\|_{W^{2,2}(U^+(0,R))}.$$

Therefore, by theorem 21.8, there exists a sequence $(h_n)_{n\in\mathbb{N}}$, $\lim_{n\to\infty} h_n = 0$, for which $\eta \Delta_i^{h_n} \overline{f}$ converges weakly in $H_0^{1,2}(U^+(0,R))$.

However, the limit of such a sequence is $\eta D_i \overline{f}$ and as $H_0^{1,2}$ is closed under weak convergence, it follows that indeed $\eta D_i \overline{f} \in H_0^{1,2}(U^+(0,R))$. Thus $f \in H_0^{2,2}(\Omega)$ and we can thereby apply the argument of the proof of theorem 23.14 to $D_i f$, $i = 1, \ldots, d-1$. $D_{dd} f$ is then controlled as before by the differential equation.

In this way one can iterate the proof of theorem 23.14 to show that $f \in W^{m,2}(\Omega)$, $f - g \in H_0^{m,2}(\Omega)$ for every $m \in \mathbb{N}$. Corollary 20.18 as usual then yields the assertion. $\qquad\square$

Entirely similar statements can be proven for the operators

$$Lf = \sum_{i,j=1}^{d} \frac{\partial}{\partial x^i}\left(a^{ij}\frac{\partial}{\partial x^j}f\right) + \sum_{i=1}^{d} \frac{\partial}{\partial x^i}(b^i f) + \sum_{i=1}^{d} c^i \frac{\partial}{\partial x^i}f + ef \qquad (36)$$

with corresponding regular coefficients a^{ij}, b^i, c^i, e. Certainly one obtains an extra term $\|f\|_{L^2(\Omega)}$ on the right side of (28).

That such a term in general is necessary, one already sees from the differential equation

$$f''(t) + \lambda^2 f(t) = 0, \quad 0 < t < \pi$$
$$f(0) = 0, f(\pi) = 0$$

with $\lambda > 0$. The solutions for $\lambda \in \mathbb{Z}$ are $f(t) = \alpha \sin(\lambda t)$, with arbitrary α.

The details of the regularity proof for such an operator L are recommended to the reader as an exercise.

Exercises for § 23

1)   Carry out the regularity proof for an operator L as given in (36).

2)   Define a notion of weak solution for the equation

$$\Delta f = \sum_{i=1}^{d} D_i k^i, \quad \text{where } k^i \in L^2(\Omega) \text{ for } i = 1, \dots, d.$$

Show regularity results for weak solutions under appropriate assumptions on k^i.

24. The Maximum Principle

The strong maximum principle of E. Hopf says that a solution of an elliptic PDE cannot assume an interior maximum. This leads to further results about solutions of such PDEs, like removability of singularities, gradient bounds, or Liouville's theorem saying that every bounded harmonic functions defined on all of Euclidean space is constant.

Let Ω be, as usual, an open subset of $\mathbb{R}^d$. In this paragraph, we consider linear elliptic differential operators of the form

$$Lf(x) = \sum_{i,j=1}^{d} a^{ij}(x)\frac{\partial^2 f}{\partial x^i \partial x^j} + \sum_{i=1}^{d} b^i(x)\frac{\partial f}{\partial x^i}$$

which fulfil the following conditions:

(i) $a^{ij}(x) = a^{ji}(x)$ for all i, j, x (this is not an essential restriction, as already remarked occasionally)

(ii) uniform ellipticity: There are constants $0 < \lambda \le \mu < \infty$ with

$$\lambda|\xi|^2 \le \sum_{i,j=1}^{d} a^{ij}(x)\xi^i\xi^j \le \mu|\xi|^2$$

for all $x \in \Omega, \xi \in \mathbb{R}^d$

(iii) there is a constant K such that

$$|b^i(x)| \le K \quad \text{for all } x \in \Omega, i \in \{1, \ldots, d\}$$

We point out explicitly that the type of the operators considered here is different from that of the operators considered in the previous paragraph, as the operators there were of the form

$$\sum_{i=1}^{d} \frac{\partial}{\partial x^i}\left(\sum_{j=1}^{d} a^{ij}(x)\frac{\partial}{\partial x^j}\right) + \ldots$$

We now come to the so-called *weak maximum principle*.

Theorem 24.1 *Let Ω be bounded and let $f \in C^2(\Omega) \cap C^0(\overline{\Omega})$ satisfy*

$$Lf \geq 0 \text{ in } \Omega.$$

Then f assumes its maximum on $\partial\Omega$, i.e.

$$\sup_{x \in \Omega} f(x) = \max_{x \in \partial\Omega} f(x). \tag{1}$$

If $Lf \leq 0$ then the corresponding statement holds for the minimum.

Proof. We first consider the case that $Lf > 0$ in Ω. We claim that in this case, f cannot have a maximum in the interior of Ω. Namely, at an interior maximum x_0,

$$Df(x_0) = 0,$$

and

$$D^2 f(x_0) = (\frac{\partial^2 f}{\partial x^i \partial x^j}(x_0))_{i,j=1,\dots,d}$$

is negative semi-definite.

As the matrix $A = (a^{ij}(x_0))$ is, by assumption, positive definite,

$$Lf(x_0) = \sum_{i,j=1}^{d} a^{ij}(x_0)\frac{\partial^2 f}{\partial x^i \partial x^j}(x_0)$$
$$= \text{Tr} \, (A \cdot D^2 f(x_0)) \leq 0$$

(Tr   denotes the trace of a matrix), in contradiction to the assumption $Lf(x_0) > 0$. Thus, in fact, f can in this case have no maximum in the interior of Ω. We now consider for $\alpha = $ const.

$$Le^{\alpha x^1} = (\alpha^2 a^{11}(x) + \alpha b^1(x))e^{\alpha x^1} \geq (\lambda\alpha^2 - K\alpha)e^{\alpha x^1},$$

by (ii) and (iii).

So, for sufficiently large α

$$Le^{\alpha x^1} > 0. \tag{2}$$

We now fix an α which satisfies (2). Then, for every $\varepsilon > 0$,

$$L(f(x) + \varepsilon e^{\alpha x^1}) > 0.$$

Therefore by what has already been shown

$$\sup_{x \in \Omega}(f(x) + \varepsilon e^{\alpha x^1}) = \max_{x \in \partial\Omega}(f(x) + \varepsilon e^{\alpha x^1}).$$

Now (1) follows by letting $\varepsilon \to 0$. $\qquad\square$

Corollary 24.2 *Let $f, g \in C^2(\Omega) \cap C^0(\overline{\Omega})$, Ω bounded. If*

$$Lf \geq Lg \text{ in } \Omega \tag{3}$$

and $f(y) \le g(y)$ for $y \in \partial\Omega$ (4)

then

$f(x) \le g(x)$ for $x \in \Omega$. (5)

If, instead of (3), even $Lf = Lg$ holds in Ω and instead of (4) $f = g$ on $\partial\Omega$, then $f = g$ in Ω.

In particular, the Dirichlet problem

$$Lf(x) = k(x) \quad \text{for} \quad x \in \Omega \quad (k \in C^0(\Omega) \quad \text{given})$$
$$f(x) = \varphi(x) \quad \text{for} \quad x \in \partial\Omega \quad (\varphi \in C^0(\partial\Omega) \text{ given})$$

is uniquely solvable.

Proof. We have $L(f - g) \ge 0$. Thus, by theorem 24.1,

$$\sup_{x \in \Omega}(f(x) - g(x)) = \max_{y \in \partial\Omega}(f(y) - g(y)) \le 0,$$

on account of (4). This is (5). The remaining assertions follow easily.   □

Definition 24.3 $f \in C^2(\Omega)$ is called a subsolution of $Lf = 0$ if $Lf \ge 0$, and a supersolution of $Lf = 0$, if $Lf \le 0$ in Ω. A subsolution, resp. supersolution, of $\Delta f = 0$ is called subharmonic and superharmonic, respectively.

This terminology is motivated as follows:

If $\Delta h = 0, \Delta f \ge 0$ in Ω and $f = h$ on $\partial\Omega$ then $f \le h$, by corollary 24.2.

A subharmonic function therefore always lies below a harmonic function with the same boundary values.

Examples.

1)   For $\beta \in \mathbb{R}$ and $x \in \mathbb{R}^d$, $\Delta|x|^\beta = (d\beta + \beta(\beta - 2))|x|^{\beta-2}$; for $\beta \ge 2$ this is also defined for $x = 0$ and $\Delta|x|^\beta \ge 0$. Thus $|x|^\beta$ is subharmonic for $\beta \ge 2$.

2)   Let $f : \Omega \to \mathbb{R}$ be harmonic and positive and $\alpha \ge 1$. We compute

$$\Delta f^\alpha = \sum_{i=1}^d \frac{\partial}{\partial x^i}(\alpha f^{\alpha-1} f_{x^i})$$

$$= \sum_{i=1}^d (\alpha(\alpha - 1)f^{\alpha-2} f_{x^i} f_{x^i} + \alpha f^{\alpha-1} f_{x^i x^i})$$

$$= \sum_{i=1}^d \alpha(\alpha - 1)f^{\alpha-2} f_{x^i} f_{x^i}, \quad \text{as } f \text{ is harmonic}$$

$$\ge 0.$$

Therefore for $\alpha \geq 1$ and a positive harmonic f, f^α is subharmonic.

3)   We compute for a positive $f : \Omega \to \mathbb{R}$

$$\Delta \log f = \sum_{i=1}^{d} \frac{\partial}{\partial x^i} \left(\frac{f_{x^i}}{f} \right) = \sum_{i=1}^{d} \left(\frac{f_{x^i x^i}}{f} - \frac{f_{x^i} f_{x^i}}{f^2} \right).$$

Therefore if f is harmonic then $\log f$ is superharmonic.

As an application of the maximum principle, we can prove a result on the removability of isolated singularities of the Laplace operator.

Theorem 24.4 Let $x_0 \in \mathbb{R}^d, d \geq 2, R > 0$ and

$$f : U(x_0, R) \backslash \{x_0\} \to \mathbb{R}$$

be harmonic and bounded. Then f has a harmonic extension through the point x_0, i.e. there is a harmonic function

$$h : U(x_0, R) \to \mathbb{R}$$

such that $h = f$ in $U(x_0, R) \backslash \{x_0\}$.

Proof. Since the Laplace equation is invariant under translations and homotheties, we can assume that $x_0 = 0$ and $R = 2$.
    Let

$$\overline{f} : B(0, 1) \to \mathbb{R}$$

be a solution of the Dirichlet problem

$$\Delta \overline{f} = 0 \text{ on } U(0, 1) \tag{6}$$
$$\overline{f}(x) = f(x) \text{ for } x \in \partial B(0, 1).$$

The existence of $\overline{f}$ has been proven in §23 (theorem 23.15: notice that by corollary 23.7, f is C^∞ in some neighborhood of $\partial B(0, 1)$).
    We recall, moreover, that for $x \neq 0$

$$g(x) := \begin{cases} \log \frac{1}{\|x\|} & \text{for } d = 2 \\ (\|x\|^{2-d} - 1) & \text{for } d \geq 3 \end{cases}$$

is harmonic (see §9).
    We now consider

$$f_\varepsilon(x) := \overline{f}(x) + \varepsilon g(x). \tag{7}$$

For all ε we have, as $g(x) = 0$ for $\|x\| = 1$

$$f_\varepsilon(x) = \overline{f}(x) = f(x) \quad \text{for } \|x\| = 1 \tag{8}$$

and

$$\lim_{x \to 0} f_\varepsilon(x) = \infty \quad \text{for all } \varepsilon > 0. \tag{9}$$

For every $\varepsilon > 0$ there thus exists an $r = r(\varepsilon) > 0$ with

$$f_\varepsilon(x) > f(x) \quad \text{for } \|x\| \le r \tag{10}$$

since, by assumption, $f(x)$ is bounded.

Finally, for sufficiently large ε we also have

$$f_\varepsilon(x) > f(x) \quad \text{for } 0 < \|x\| < 1, \tag{11}$$

as $\frac{\partial}{\partial r} g(x) < 0$ for $\|x\| = 1$ ($r = \|x\|$) and f, as a C^∞-function, has a bounded derivative (corollary 23.7), in particular for $\|x\| = 1$.

We now choose the smallest $\varepsilon_0 \ge 0$ such that

$$f_{\varepsilon_0}(x) \ge f(x) \quad \text{for } \|x\| \le 1. \tag{12}$$

From the above considerations, it follows that such an ε_0 exists.

Now if ε_0 were positive, there would be $y_0 \in U(0,1)\backslash\{0\}$ with

$$f_{\varepsilon_0/2}(y_0) < f(y_0), \tag{13}$$

by (8). But then by (10) we would have

$$\inf_{x \in U(0,1)\backslash B(0, r(\varepsilon_0/2))} (f_{\varepsilon_0/2}(x) - f(x)) < 0$$

$$= \min_{y \in \partial B(0,1) \cup \partial B(0, r(\varepsilon_0/2))} (f_{\varepsilon_0/2}(y) - f(y)).$$

As $f_{\varepsilon_0/2} - f$ is harmonic in $U(0,1)\backslash B(0, r(\varepsilon_0/2))$, this would contradict corollary 24.2.

It follows that $\varepsilon = 0$, so

$$\overline{f} = f_0 \ge f \quad \text{in } U(0,1)\backslash\{0\}.$$

Similarly one proves that

$$\overline{f} \le f \quad \text{in } U(0,1)\backslash\{0\}.$$

Thus $\overline{f} = f$ in $U(0,1)\backslash\{0\}$ and

$$h := \begin{cases} f & \text{in } U(0,2)\backslash\{0\} \\ \overline{f} & \text{in } U(0,1) \end{cases}$$

has the required properties. $\square$

Example. It follows in particular that the following Dirichlet problem has no solution: $f : B(0,1) \to \mathbb{R}$ with

$$\Delta f(x) = 0 \quad \text{in } U(0,1)\backslash\{0\}$$
$$f(y) = 0 \quad \text{for } \|y\| = 1$$
$$f(0) = 1$$

By theorem 24.4, such a solution would admit a harmonic extension $h :$ $B(0,1) \to \mathbb{R}$ with $\Delta h(x) = 0$ in $U(0,1)$, $h(y) = 0$ for $\|y\| = 1$. But then by corollary 24.2 $h \equiv 0$ holds in $B(0,1)$, in particular $h(0) = 0 \neq 1$. (For this example, recall also the discussion before definition 22.8.)

A further consequence of the maximum principle is a gradient estimate for solutions of the Poisson equation.

Theorem 24.5 *Let* $f \in C^2(U(x_0,r)) \cap C^0(B(x_0,r))$ *satisfy*

$$\Delta f(x) = k(x)$$

in $U(x_0,r)$, *with* k *a bounded function. Then for* $i = 1,\dots,d$ *we have*

$$|\frac{\partial}{\partial x^i} f(x_0)| \leq \frac{d}{r} \sup_{\partial B(x_0,r)} |f| + \frac{r}{2} \sup_{B(x_0,r)} |k|. \tag{14}$$

Proof. We may assume $x_0 = 0$ and $i = d$. We set

$$U^+(0,r) := \{x = (x^1,\dots,x^d) : \|x\| < r, x^d > 0\},$$

and for $x = (x^1,\dots,x^d)$ let $x' := (x^1,\dots,x^{d-1})$. Let $M := \sup_{\partial B(0,r)} |f|, K := \sup_{B(0,r)} |k|$. Set

$$g(x) := \frac{1}{2}(f(x',x^d) - f(x',-x^d))$$

and

$$\varphi(x) := \frac{M}{r^2}\|x\|^2 + x^d(r - x^d)(\frac{dM}{r^2} + \frac{K}{2}).$$

We then have

$$\Delta\varphi(x) = -K \quad \text{in } U^+(0,r)$$
$$\varphi(x',0) \geq 0 \quad \text{for all } x'$$
$$\varphi(x) \geq M \quad \text{for } \|x\| = r, x^d \geq 0.$$

Also

$$|\Delta g(x)| \leq K \quad \text{in } U^+(0,r)$$
$$g(x',0) = 0 \quad \text{for all } x'$$
$$|g(x)| \leq M \quad \text{for } \|x\| = r, x^d \geq 0.$$

Altogether this gives

$$\Delta(\varphi \pm g) \leq 0 \quad \text{in } U^+(0,r)$$
$$\varphi \pm g \geq 0 \quad \text{on } \partial U^+(0,r).$$

The maximum principle therefore implies

$$|g(x)| \leq \varphi(x) \quad \text{in } U^+(0,r).$$

It follows that

$$\left|\frac{\partial}{\partial x^d}f(0)\right| = \lim_{\substack{x^d \to 0 \\ x^d > 0}} \left|\frac{g(0, x^d)}{x^d}\right| \le \lim_{\substack{x^d \to 0 \\ x^d > 0}} \frac{\varphi(0, x^d)}{x^d} = \frac{dM}{r} + \frac{r}{2}K.$$

□

As a direct consequence, we obtain Liouville's theorem:

Corollary 24.6 *Every bounded harmonic function defined on all of $\mathbb{R}^d$ is constant.*

Proof. Let $f : \mathbb{R}^d \to \mathbb{R}$ be harmonic with $\sup_{x \in \mathbb{R}^d} |f(x)| \le M$.

By (14),

$$\left|\frac{\partial}{\partial x^i}f(x_0)\right| \le \frac{dM}{r} \tag{15}$$

holds for every $i \in \{1, \ldots, d\}$, every $x_0 \in \mathbb{R}^d$ and every $r > 0$.

Letting $r \to \infty$ gives $\frac{\partial}{\partial x^i}f \equiv 0$. Thus f is constant.   □

We can also easily derive estimates for higher order derivatives, as all derivatives of a harmonic function are again harmonic.

Corollary 24.7 *Let $f : \Omega \to \mathbb{R}$ be harmonic, $\Omega' \subset\subset \Omega$, $\delta := \text{dist}(\Omega', \partial\Omega)$. For every multiindex $\alpha \in \mathbb{N}^d$ we have*

$$\sup_{\Omega'} |D^\alpha f| \le \left(\frac{d|\alpha|}{\delta}\right)^{|\alpha|} \sup_{\Omega} |f|.$$

Proof. We consider the case $\alpha = (1, 1, 0, \ldots 0)$. The general case follows iteratively according to the same pattern. Let $x_0 \in \Omega'$, $r = \frac{\delta}{2}$. Then by (14) applied to the harmonic function $\frac{\partial}{\partial x^1}f$ we obtain

$$\left|\frac{\partial}{\partial x^2}\left(\frac{\partial}{\partial x^1}f(x_0)\right)\right| \le \frac{2d}{\delta} \sup_{y \in \partial B(x_0, \frac{\delta}{2})} \left|\frac{\partial f}{\partial x^1}(y)\right|. \tag{16}$$

But as $B(y, \frac{\delta}{2}) \subset \Omega$ for every $y \in \partial B(x_0, \frac{\delta}{2})$, we can apply (14) again to f itself and obtain

$$\sup_{y \in \partial B(x_0, \frac{\delta}{2})} \left|\frac{\partial f}{\partial x^1}(y)\right| \le \frac{2d}{\delta} \sup_{z \in \Omega} |f(z)|. \tag{17}$$

Combining (16) and (17) gives the result for the case considered.   □

This implies in its turn

Corollary 24.8 *Let $f_n : \Omega \to \mathbb{R}$ be a bounded sequence of harmonic functions (so $\sup\limits_{x \in \Omega} |f_n(x)| \leq K$ for all n). Then a subsequence converges uniformly on any compact subset to a harmonic function. In particular, a uniform limit of harmonic functions is again harmonic.*

Proof. By corollary 24.7, all the partial derivatives of (f_n) are equicontinuous on any $\Omega' \subset\subset \Omega$. By the theorem of Arzela-Ascoli, we obtain in particular a subsequence which is C^2 convergent. This implies that the limit is again harmonic. In order to conclude that the same subsequence has the required property for any $\Omega' \subset\subset \Omega$, we apply the usual diagonal process.

Let $\Omega_1 \subset\subset \Omega_2 \subset\subset \ldots \Omega_n \subset\subset \ldots \Omega$ with

$$\Omega = \bigcup_{n=1}^{\infty} \Omega_n. \tag{18}$$

We then find a subsequence $(f_{1,n})$ of (f_n) which converges on Ω_1, then a subsequence $(f_{2,n})$ of $(f_{1,n})$ which converges on Ω_2, and so on. The subsequence $(f_{n,n})$ then converges on every Ω_m and thus, on account of (18), on any $\Omega' \subset\subset \Omega$. $\qquad\square$

Remark. Corollary 24.8 can also be obtained by means of the Sobolev embedding theorem, using the integral estimates of § 23.

For what follows we also need the following generalization of theorem 24.1.

Corollary 24.9 *Let $c : \Omega \to \mathbb{R}$ be nonnegative. Let L satisfy the assumptions formulated before theorem 24.1. Let $f \in C^2(\Omega) \cap C^0(\overline{\Omega})$ be such that*

$$Lf(x) - c(x)f(x) \geq 0.$$

Then for $f^+ := \max(f, 0)$ we have

$$\sup_{\Omega} f^+ \leq \max_{\partial\Omega} f^+.$$

Proof. Let $\tilde{\Omega} := \{x \in \Omega : f(x) > 0\}$. In $\tilde{\Omega}$ we have, because c is nonnegative,

$$Lf \geq 0,$$

so by theorem 24.1

$$\sup_{\tilde{\Omega}} f \leq \max_{\partial\tilde{\Omega}} f.$$

The boundary of $\tilde{\Omega}$ consists of two parts:

$$\partial_1 \tilde{\Omega} := \partial \tilde{\Omega} \cap \Omega$$

$$\partial_2 \tilde{\Omega} := \partial \tilde{\Omega} \cap \partial \Omega \subset \partial \Omega.$$

We have

$$f_{|\partial_1 \tilde{\Omega}} = 0$$

and

$$\max_{\partial_2 \tilde{\Omega}} f \leq \max_{\partial \Omega} f.$$

Therefore

$$\max_{\partial \tilde{\Omega}} f \leq \max_{\partial \Omega} f^+,$$

and as also

$$\sup_{\Omega} f^+ = \sup_{\tilde{\Omega}} f,$$

the result follows.   □

We now prove the *strong maximum principle*. In order to give the proof in as transparrent a manner as possible, we consider only the case of the Laplace operator. The corresponding assertion for an operator L considered at the beginning of this article can then be proved entirely analogously.

Theorem 24.10 *Let $\Omega \subset \mathbb{R}^d$ be open and let $f : \Omega \to \mathbb{R}$ satisfy*

$$\Delta f \geq 0 \text{ in } \Omega. \tag{19}$$

If f assumes its maximum in the interior of Ω then f is constant. More generally, if $c : \Omega \to \mathbb{R}$ is a nonnegative function and if

$$\Delta f(x) - c(x) f(x) \geq 0 \text{ for } x \in \Omega, \tag{20}$$

and f assumes a nonnegative maximum in the interior of Ω, then f is constant.

We first prove the *boundary point lemma of E. Hopf*.

Lemma 24.11 *Let $\Delta f - cf \geq 0$ in $\tilde{\Omega} \subset \mathbb{R}^d$ for a nonnegative function $c : \Omega \to \mathbb{R}$. Let $x_0 \in \partial \tilde{\Omega}$.*
   Assume further that

(i)   *f is continuous at x_0*

(ii)   *$f(x_0) \geq 0$, if $c(x) \not\equiv 0$.*

(iii)   *$f(x_0) > f(x)$ for all $x \in \tilde{\Omega}$*

(iv)   *There exists a ball $U(y, R) \subset \tilde{\Omega}$ with $x_0 \in \partial U(y, R)$.*

   Then for $r = \|x - y\|$

$$\frac{\partial f}{\partial r}(x_0) > 0, \tag{21}$$

if this derivative exists (in the direction of the outer normal to $\tilde{\Omega}$).

Proof. We may assume that y and R are so chosen in (iv) that $\partial U(y, R) \cap \partial \tilde{\Omega} = \{x_0\}$. For $x \in U(y, R) \backslash B(y, \rho)$ $(0 < \rho < R)$ we consider the auxiliary function

$$g(x) := e^{-\gamma \|x-y\|^2} - e^{-\gamma R^2}.$$

We have

$$\Delta g - cg = (4\gamma^2 \|x - y\|^2 - 2d\gamma - c)e^{-\gamma \|x-y\|^2} + ce^{-\gamma R^2}.$$

Therefore for sufficiently large γ we have

$$\Delta g - cg \geq 0 \text{ in } U(y, R) \backslash B(y, \rho). \tag{22}$$

By (iii) and (iv),

$$f(x) - f(x_0) < 0 \text{ for } x \in \partial B(y, \rho).$$

Therefore there exists an $\varepsilon > 0$ with

$$f(x) - f(x_0) + \varepsilon g(x) \leq 0 \text{ for } x \in \partial B(y, \rho). \tag{23}$$

As $g = 0$ on $\partial B(y, R)$, (23) holds in turn because of (iii) and (iv) also for $x \in \partial B(y, R)$.

On the other hand,

$$\Delta(f(x) - f(x_0) + \varepsilon g(x)) - c(x)(f(x) - f(x_0) + \varepsilon g(x))$$
$$\geq c(x)f(x_0) \geq 0$$

for $x \in U(y, R) \backslash B(y, \rho)$, because of (ii). Since $f(x) - f(x_0) + \varepsilon g(x) \leq 0$ on $\partial(B(y, R) \backslash B(y, \rho))$, it follows from corollary 24.9 that

$$f(x) - f(x_0) + \varepsilon g(x) \leq 0 \text{ for } x \in B(y, R) \backslash B(y, \rho).$$

It follows (in case this derivative exists) that

$$\frac{\partial}{\partial r}(f(x) - f(x_0) + \varepsilon g(x)) \geq 0 \text{ at the point } x = x_0 \in \partial B(y, R),$$

thus

$$\frac{\partial}{\partial r}f(x_0) \geq -\varepsilon \frac{\partial g(x_0)}{\partial r} = \varepsilon(2\gamma Re^{-\gamma R^2}) > 0.$$

Proof of theorem 24.10 To prove the theorem, we assume that f is not constant and yet it assumes a maximum $m(\geq 0$ if $c \neq 0)$ in the interior of Ω. Then

$$\tilde{\Omega} := \{x \in \Omega : f(x) < m\} \neq \emptyset$$

and
$$\partial\tilde{\Omega}\cap\Omega\neq\emptyset.$$

Let $y\in\tilde{\Omega}$ be closer to $\partial\tilde{\Omega}$ than to $\partial\Omega$ and $U(y,R)$ the largest ball contained in $\tilde{\Omega}$ with center y. Then

$$f(x_0)=m \text{ for some } x_0\in\partial B(y,R),$$

and $f(x)<f(x_0)$ for $x\in\tilde{\Omega}$.

Now (21) is applicable and gives

$$df(x_0)\neq 0,$$

which, however, is not possible at an interior maximum. $\qquad\qquad\square$

The assumption $c\geq 0$ in theorem 24.10 is clearly required, as the existence of non-trivial eigenfunctions shows. For example, $f(x)=\sin x$ satisfies on $(0,\pi)$

$$f''(x)+f(x)=0,$$

but f assumes a positive maximum at $x=\frac{\pi}{2}$.

Exercises for § 24

1) Show the strong maximum principle for an operator of the form

$$L=\sum_{i,j=1}^{d}a^{ij}(x)\frac{\partial^2}{\partial x^i\partial x^j}+\sum_{i=1}^{d}b^i(x)\frac{\partial}{\partial x^i}$$

under the assumptions (i) – (iii) made at the beginning of this paragraph.

2) Let $f:\mathbb{R}^d\to\mathbb{R}$ be harmonic with bounded gradient. Then f is affine linear, i.e. $f(x)=\sum_{i=1}^{d}a_ix^i+b$, with constants $a_1,\ldots,a_d,b$.

25. The Eigenvalue Problem for the Laplace Operator

We use Rellich's embedding theorem to show that every L^2 function on an open $\Omega \subset \mathbb{R}^d$ can be expanded in terms of eigenfunctions of the Laplace operator on Ω.

We first recall the following result from linear algebra:

Let $\langle \cdot, \cdot \rangle$ denote the Euclidean scalar product in $\mathbb{R}^d$ and let A be a symmetric $d \times d$-matrix, so

$$\langle Ax, y \rangle = \langle x, Ay \rangle \quad \text{for all } x, y \in \mathbb{R}^d. \tag{1}$$

Then $\mathbb{R}^d$ has an orthonormal basis $v_1, \ldots, v_d$ consisting of eigenvectors of A. Thus

$$Av_i + \lambda_i v_i = 0 \quad \text{with } \lambda_i \in \mathbb{R} \tag{2}$$

$$\langle v_i, v_j \rangle = \delta_{ij} \quad \text{for all } i, j. \tag{3}$$

We shall now study an analogous situation in an infinite dimensional Hilbert space. Even though the following considerations can be made in far greater generality, we shall restrict ourselves to a concrete case, namely to the eigenvalue problem for the Laplace operator.

Let Ω be an open bounded subset of $\mathbb{R}^d$. We want to study the eigenfunctions of Δ in $H_0^{1,2}(\Omega)$, thus $f \in H_0^{1,2}(\Omega)$ with

$$\Delta f(x) + \lambda f(x) = 0 \quad \text{for all } x \in \Omega,$$

λ being real.

In the following we set

$$H := H_0^{1,2}(\Omega), \quad \langle f, g \rangle := \int_\Omega f(x)g(x)dx$$

for $f, g \in L^2(\Omega)$,

$$\|f\| := \|f\|_{L^2(\Omega)} = \langle f, f \rangle^{\frac{1}{2}}.$$

We note that the Laplace operator is symmetric in a certain sense. For example for φ, ψ in $C_0^\infty(\Omega)$ we have

$$\langle \Delta \varphi, \psi \rangle = -\langle D\varphi, D\psi \rangle = \langle \varphi, \Delta \psi \rangle.$$

We now define
$$\lambda_1 := \inf_{f \in H \setminus \{0\}} \frac{\langle Df, Df \rangle}{\langle f, f \rangle}.$$

From the Poincaré inequality (corollary 20.16) it follows that

$$\lambda_1 > 0. \tag{4}$$

Now let $(f_n)_{n \in \mathbb{N}}$ be a minimizing sequence, so that

$$\lim_{n \to \infty} \frac{\langle Df_n, Df_n \rangle}{\langle f_n, f_n \rangle} = \lambda_1.$$

Here, we may assume that

$$\|f_n\| = 1 \text{ for all } n \tag{5}$$

and then also

$$\|Df_n\| \leq K \text{ for all } n. \tag{6}$$

By theorem 21.8, after a choice of a subsequence, the sequence $(f_n)_{n \in \mathbb{N}}$ converges weakly in the Hilbert space H to some $v_1 \in H$, and by the Rellich compactness theorem (theorem 20.20) $(f_n)_{n \in \mathbb{N}}$ then also converges strongly in $L^2(\Omega)$ to v_1; by (5) it follows that

$$\|v_1\| = 1.$$

Furthermore, it follows, because of lower semicontinuity of $\|Df\|_{L^2(\Omega)}$ for weak convergence in H (corollary 21.9; notice that by the Poincaré inequality $\|Df\|_{L^2(\Omega)}$ defines a norm in H), and the definition of λ_1 that

$$\lambda_1 \leq \langle Dv_1, Dv_1 \rangle \leq \lim_{n \to \infty} \langle Df_n, Df_n \rangle = \lambda_1,$$

so

$$\frac{\langle Dv_1, Dv_1 \rangle}{\langle v_1, v_1 \rangle} = \lambda_1.$$

Now assume that $(\lambda_1, v_1), \ldots, (\lambda_{m-1}, v_{m-1})$ have already been determined iteratively, with $\lambda_1 \leq \lambda_2 \leq \ldots \leq \lambda_{m-1}$,

$$\Delta v_i(x) + \lambda_i v_i(x) = 0 \text{ in } \Omega,$$

and

$$\langle v_i, v_j \rangle = \delta_{ij} \text{ for } i, j = 1, \ldots, m - 1. \tag{8}$$

We set

$$H_m := \{f \in H : \langle f, v_i \rangle = 0 \text{ for } i = 1, \ldots, m - 1\}$$

and

$$\lambda_m := \inf_{f \in H_m \setminus \{0\}} \frac{\langle Df, Df \rangle}{\langle f, f \rangle}.$$

We make two simple remarks

(i)

$$\lambda_m \geq \lambda_{m-1}, \text{ as } H_m \subset H_{m-1} \tag{9}$$

(ii) H_m, being the orthogonal complement of a finite dimensional sub-space, is closed (if $(f_n)_{n\in\mathbb{N}} \subset H_m$ converges to f then, as $\langle f_n, v_i \rangle = 0$ for all $n \in \mathbb{N}$, $\langle f, v_i \rangle = 0$ for $i = 1, \ldots, m$, so $f \in H_m$) and therefore it is also a Hilbert space.

With the same argument as before we now find a $v_m \in H_m$ with $\|v_m\| = 1$ and

$$\lambda_m = \langle Dv_m, Dv_m \rangle = \frac{\langle Dv_m, Dv_m \rangle}{\langle v_m, v_m \rangle}. \tag{10}$$

Now we claim that

$$\Delta v_m + \lambda_m v_m = 0 \quad \text{in } \Omega. \tag{11}$$

For a proof we observe that for all $\varphi \in H_m, t \in \mathbb{R}$

$$\frac{\langle D(v_m + t\varphi), D(v_m + t\varphi) \rangle}{\langle v_m + t\varphi, v_m + t\varphi \rangle} \geq \lambda_m$$

and this expression is differentiable in t (this is seen as in the derivation of the Euler-Lagrange equations) and has a minimum at $t = 0$; so

$$\begin{aligned}
0 &= \frac{d}{dt} \frac{\langle D(v_m + t\varphi), D(v_m + t\varphi) \rangle}{\langle v_m + t\varphi, v_m + t\varphi \rangle}|t = 0 \\
&= 2(\frac{\langle Dv_m, D\varphi \rangle}{\langle v_m, v_m \rangle} - \frac{\langle Dv_m, Dv_m \rangle}{\langle v_m, v_m \rangle} \frac{\langle v_m, \varphi \rangle}{\langle v_m, v_m \rangle}) \\
&= 2(\langle Dv_m, D\varphi \rangle - \lambda_m \langle v_m, \varphi \rangle)
\end{aligned}$$

for all $\varphi \in H_m$.

However, for $i = 1, \ldots, m - 1$

$$\langle v_m, v_i \rangle = 0$$

and

$$\langle Dv_m, Dv_i \rangle = \langle Dv_i, Dv_m \rangle = \lambda_i \langle v_i, v_m \rangle = 0.$$

It follows that

$$\langle Dv_m, D\varphi \rangle - \lambda_m \langle v_m, \varphi \rangle = 0 \tag{12}$$

even for all $\varphi \in H$.

This means that v_m is a solution of

$$\int_\Omega Dv_m(x) D\varphi(x) dx - \lambda_m \int_\Omega v_m(x) \varphi(x) dx = 0 \text{ for all } \varphi \in H_0^{1,2}(\Omega).$$

By corollary 23.8, $v_m \in C^\infty(\Omega)$ and

$$\Delta v_m(x) + \lambda_m v_m(x) = 0 \text{ for all } x \in \Omega. \tag{13}$$

Lemma 25.1 $\lim\limits_{m \to \infty} \lambda_m = \infty$.

Proof. Otherwise, by (10), we would have

$$\|Dv_m\| \le K \text{ for all } m \in \mathbb{N}.$$

By the Rellich compactness theorem (theorem 20.20) the sequence $(v_m)_{m \in \mathbb{N}}$, after choosing a subsequence, would converge in $L^2(\Omega)$, say to the limit v.
Thus

$$\lim_{m \to \infty} \|v_m - v\| = 0.$$

However, this is not compatible with the fact that, using (8), i.e. $\langle v_\ell, v_m \rangle = \delta_{m\ell}$

$$\|v_\ell - v_m\|^2 = \langle v_\ell, v_\ell \rangle - 2\langle v_\ell, v_m \rangle + \langle v_m, v_m \rangle = 2 \text{ for } \ell \ne m$$

which violates the Cauchy property. This contradiction proves the lemma.
$\qquad\square$

Theorem 25.2 *Let $\Omega \subset \mathbb{R}^d$ be open and bounded. Then the eigenvalue problem*

$$\Delta f + \lambda f = 0, f \in H_0^{1,2}(\Omega)$$

has countably many eigenvalues with pairwise orthonormal vectors v_m, also

$$\langle v_m, v_\ell \rangle = \delta_{m\ell}, \tag{14}$$
$$\Delta v_m + \lambda_m v_m = 0 \text{ in } \Omega$$
$$\langle Dv_m, Dv_\ell \rangle = \lambda_m \delta_{m\ell}. \tag{15}$$

The eigenvalues are all positive and

$$\lim_{m \to \infty} \lambda_m = \infty.$$

For $f \in H_0^{1,2}(\Omega)$ we have

$$f = \sum_{i=1}^{\infty} \langle f, v_i \rangle v_i, \tag{16}$$

where this series converges in $L^2(\Omega)$ and

$$\langle Df, Df \rangle = \sum_{i=1}^{\infty} \lambda_i \langle f, v_i \rangle^2. \tag{17}$$

Remark. Equation (16) means that the eigenvectors form a complete orthonormal basis in $L^2(\Omega)$. This generalizes the fact referred to at the beginning that in the finite dimensional case, a symmetric operator has an orthonormal basis of eigenvectors.

Proof. First we notice that (15) follows from (12) and (14). It remains to show (16) and (17). We set for $f \in H$ as abbreviation

$$\alpha_i := \langle f, v_i \rangle \quad (i \in \mathbb{N})$$

and

$$f_m := \sum_{i=1}^{m} \alpha_i v_i, \quad \varphi_m := f - f_m.$$

φ_m is thus the orthogonal projection of f onto H_{m+1}, the subspace of H orthogonal to $v_1, \ldots, v_m$. Hence

$$\langle \varphi_m, v_i \rangle = 0 \text{ for } i = 1, \ldots, m \tag{18}$$

and by definition of λ_{m+1}

$$\langle D\varphi_m, D\varphi_m \rangle \geq \lambda_{m+1} \langle \varphi_m, \varphi_m \rangle. \tag{19}$$

By (12) and (18) we also have

$$\langle D\varphi_m, Dv_i \rangle = 0 \text{ for } i = 1, \ldots, m. \tag{20}$$

From (18), we obtain

$$\langle \varphi_m, \varphi_m \rangle = \langle f, f \rangle - \langle f_m, f_m \rangle, \tag{21}$$

and from (20)

$$\langle D\varphi_m, D\varphi_m \rangle = \langle Df, Df \rangle - \langle Df_m, Df_m \rangle. \tag{22}$$

Now (19) and (20) give

$$\langle \varphi_m, \varphi_m \rangle \leq \frac{1}{\lambda_{m+1}} \langle Df, Df \rangle$$

and on account of lemma 25.1, the sequence φ_m therefore converges to 0 in $L^2(\Omega)$. This means that

$$f = \lim_{m \to \infty} f_m = \sum_{i=1}^{\infty} \langle f, v_i \rangle v_i \text{ in } L^2(\Omega),$$

hence (16). Furthermore

$$Df_m = \sum_{i=1}^{m} \alpha_i Dv_i,$$

so by (15)

$$\langle Df_m, Df_m \rangle = \sum_{i=1}^{m} \alpha_i^2 \langle Dv_i, Dv_i \rangle \tag{23}$$

$$= \sum_{i=1}^{m} \lambda_i \alpha_i^2.$$

Now, as by (22), $\langle Df_m, Df_m \rangle \le \langle Df, Df \rangle$ and all the λ_i are positive, the series

$$\sum_{i=1}^{\infty} \lambda_i \alpha_i^2$$

converges.

Now for $m \le n$

$$\langle D\varphi_m - D\varphi_n, D\varphi_m - D\varphi_n \rangle = \langle Df_n - Df_m, Df_n - Df_m \rangle$$
$$= \sum_{i=m+1}^{n} \lambda_i \alpha_i^2.$$

Therefore, not only (φ_m) but also $(D\varphi_m)$ is a Cauchy sequence in $L^2(\Omega)$ and φ_m therefore converges in H to 0, with respect to the $H^{1,2}$ norm.

Hence by (22)

$$\langle Df, Df \rangle = \lim_{m \to \infty} \langle Df_m, Df_m \rangle = \sum_{i=1}^{\infty} \lambda_i \alpha_i^2 \text{ (compare (23)).}$$

We finally want to verify still that we have found all the eigenvalues and that all the eigenvectors are linear combinations of the v_i.

First, the eigenvectors corresponding to different eigenvalues are L^2-orthogonal: Namely, if for $v, w \ne 0$

$$\Delta v + \lambda v = 0 \text{ and } \Delta w + \mu w = 0,$$

then for all $\varphi \in H$

$$\langle Dv, D\varphi \rangle = \lambda \langle v, \varphi \rangle, \quad \langle Dw, D\varphi \rangle = \mu \langle w, \varphi \rangle$$

and therefore

$$\lambda \langle v, w \rangle = \langle Dv, Dw \rangle = \langle Dw, Dv \rangle = \mu \langle w, v \rangle,$$

and so, if $\lambda \ne \mu$, we must have

$$\langle v, w \rangle = 0.$$

Now if there were an eigenvalue λ not contained in $\{\lambda_m\}$, say with an eigenvector $v \ne 0$ that is linearly independent of all the v_i, then $\langle v, v_i \rangle$ would be 0 for all i and therefore by (16), $v = 0$, a contradiction. $\square$

The decisive result in the proof above was the Rellich compactness theorem:

If for a sequence $(f_n)_{n \in \mathbb{N}} \subset H_0^{1,2}$, $\langle Df_n, Df_n \rangle$ is uniformly bounded, then a subsequence converges in L^2. One can then easily generalize the considerations above, by considering instead of $\langle Df, Df \rangle$ and $\langle f, f \rangle$ two bilinear forms $K(f, f)$ and $B(f, f)$ on a Hilbert space, of which the first has a compactness property analogous to $\langle Df, Df \rangle$.

Index

Universitext